Advances in Nemertean Biology

Proceedings of the Third International Meeting on Nemertean Biology,
Y Coleg Normal, Bangor, North Wales, August 10–15, 1991

Edited by

R. Gibson, J. Moore & P. Sundberg

Reprinted from Hydrobiologia, vol. 266 (1993)

SPRINGER-SCIENCE+BUSINESS MEDIA, B.V.

Library of Congress Cataloging-in-Publication Data

A C.I.P. Catalogue record for this book is available from the Library of Congress.

ISBN 978-94-010-4909-2 ISBN 978-94-011-2052-4 (eBook)
DOI 10.1007/978-94-011-2052-4

Printed on acid-free paper

Softcover reprint of the hardcover 1st edition 1993
Originally published by Kluwer Academic Publishers in 1993

Contents

PART THREE: GENERAL BIOLOGY

Preface

In any scientific discipline, meetings with presented papers and discussions are the most effective stimulus to the advancement of knowledge. Nemerteans have for long largely been neglected because their taxonomy is difficult: the need for histological study of serial sections has proved inhibiting to most zoologists. During the last twenty years, however, this intriguing phylum has attracted the attention not only of increasing numbers of taxonomists but also of workers interested in many aspects of, for example, their ecology, evolution, physiology and fine structure. The enthusiasm stemming from the First International Meeting held in Philadelphia during December 1983 made it abundantly clear that regular meetings of this type should be continued. The Second Meeting, at the Tjärnö Marine Biological Laboratory in Sweden (August 1986), emphasised the importance of such international gatherings. The Third Meeting, originally scheduled for August 1990, had to be postponed owing to the extreme difficulty of raising funds, but was held in Britain, at Y Coleg Normal, Bangor, North Wales, from August 10th to 15th 1991, with 24 participants from 6 countries. The present volume records most of the papers given at this conference, although other presentations by N. Anadon, J. Bierne, W. R. Kem, R. K. Okazaki and G. Vernet are, for a variety of reasons, not included.

The last day of the conference was organized as a Practical Workshop on Systematic Theory and Nemertean Taxonomy. Different approaches to taxonomic studies of nemerteans, including electrophoretic and cladistic analyses as well as traditional methods, were discussed in a lively manner, and the use of computer programs in the analysis of taxonomic data was led by P. Sundberg.

Support for the Meeting came from various sources; the editors gratefully acknowledge American Airlines for generously providing free transatlantic flights for three of the North American participants, Land Rover Ltd. for the use of two Range Rovers during the Meeting, and the AppleCenter, Warrington, for the loan of computer equipment. We most warmly thank Y Coleg Normal, Bangor, and in particular Mr Geoffrey Lincoln, for very great help, providing a setting and an atmosphere such that both the scientific and social activities were thoroughly enjoyed by all participants. We are also most grateful to Sophie Knott for her willing and efficient work as secretary to the conference. Finally we thank all those who participated for their professional and social contributions to the success of the Meeting.

Bangor, August 1991

RAY GIBSON
JANET MOORE
PER SUNDBERG

Hydrobiologia **266**, 1993.
R. Gibson, J. Moore & P. Sundberg (eds), Advances in Nemertean Biology.

List of participants and contributors

ANADÓN Nuria, Departamento de Biología de Organismos y Sistemas, Laboratorio de Zoología, Facultad de Biología, Universidad de Oviedo, 33005 Oviedo, Spain

BIERNE Jacques, Laboratoire de Biologie Cellulaire, UR4R, Faculté des Sciences, Université de Reims Champagne Ardenne, BP 347, 51062 Reims Cedex, France

CRANDALL Frank B., Turkey Run Research Institute, 900 Turkey Run Road, McLean, Virginia 22101-1705, U.S.A.

ENVALL Mats, Department of Zoology, University of Gothenburg, Medicinaregat. 18, 41390 Göteborg, Sweden

FERRARIS Joan D., National Institutes of Health, National Heart Lung Blood Institute, Laboratory of Kidney and Electrolyte Metabolism, Bldg 10, Rm 6N307, Bethesda, Maryland 20892, U.S.A.

GIBSON Ray, School of Natural Sciences, Liverpool Polytechnic, Byrom Street, Liverpool L3 3AF, U.K.

HYLBOM Richard, Nygatan 41 C, S-46231 Vänersborg, Sweden

IWATA Fumio, Laboratory of Biology, Division of Liberal Arts, Kushiro Public University, Kushiro 085, Japan

JENSEN Donald D., Department of Psychology, University of Nebraska-Lincoln, Lincoln, Nebraska 68588-0308, U.S.A.

JUNOY Juan, Departamento de Biología Animal, Facultad de Ciencias, Universidad de Alcalá de Henares, 28871 Alcalá de Henares, Madrid, Spain

KEM William R., Department of Pharmacology, J267 J. H. Miller Health Center, College of Medicine, University of Florida, Gainesville, Florida 32610, U.S.A.

KNOTT Sophie, 4 Salem Place, Llanllechid, Bangor, Gwynedd LL57 3ES, U.K.

KURIS Armand M., Department of Biological Sciences, University of California, Santa Barbara, California 93106, U.S.A.

MCDERMOTT John J., Department of Biology, Franklin & Marshall College, Lancaster, Pennsylvania 17604-3003, U.S.A.

MCEVOY Eric G., School of Natural Sciences, Liverpool Polytechnic, Byrom Street, Liverpool L3 3AF, England

MOORE Janet, Department of Zoology, University of Cambridge, Downing Street, Cambridge CB2 3EJ, U.K.

NORENBURG Jon L., Smithsonian Oceanographic Sorting Center, Smithsonian Institution, Washington, D.C. 20560, U.S.A.

OKAZAKI Robert K., Department of Biological Sciences, Southeastern Louisiana University, P.O. Box 814, Hammond, Louisiana 70402, U.S.A.

RISER Nathan W., Marine Science Institute, Northeastern University, East Point, Nahant, Massachusetts 01908, U.S.A.

ROE Pam, Department of Biological Sciences, California State University, Stanislaus, Turlock, California 95380, U.S.A.

ROGERS Alex D., Department of Evolutionary & Environmental Biology, University of Liverpool, Port Erin Marine Laboratory, Port Erin, Isle of Man, U.K.

SUNDBERG Per, Department of Zoology, University of Gothenburg, Medicinaregat. 18, 41390 Göteborg, Sweden

SVENSSON Mikael, Department of Zoology, University of Gothenburg, Medicinaregat. 18, 41390 Göteborg, Sweden

VERNET Guy, Laboratoire de Zoologie et des Sciences de l'Environnement, UR4R, Faculté des Sciences, Université de Reims Champagne Ardenne, BP 347, 51062 Reims Cedex, France

WERN Judy, Department of Biological Sciences, Texas A&M University at Galveston, P.O. Box 1675, Galveston, Texas 77553-1675, U.S.A.

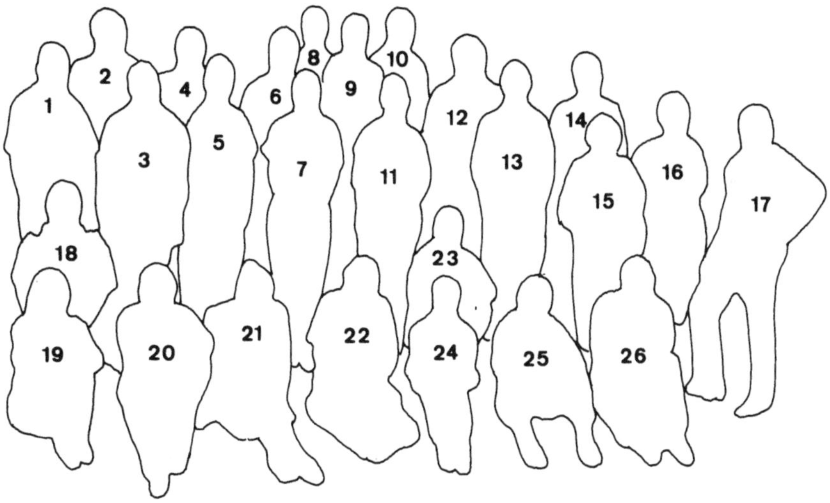

1. Jacques Bierne
2. John McDermott
3. Fumio Iwata
4. William Kem
5. Mikael Svensson
6. Armand Kuris
7. Joan Ferraris
8. Geoffrey Lincoln
9. Jon Norenburg
10. Mats Envall
11. Nathan Riser
12. Don Jensen
13. Ray Gibson
14. Per Sundberg
15. Nuria Anadón
16. Robert Okazaki
17. Frank Crandall
18. Guy Vernet
19. Judy Wern
20. Alex Rogers
21. Sophie Knott
22. Janet Moore
23. Juan Junoy
24. Pam Roe
25. Richard Hylbom
26. Eric McEvoy

Hydrobiologia **266**: 1–14, 1993.
R. Gibson, J. Moore & P. Sundberg (eds), Advances in Nemertean Biology

Life cycles of nemerteans that are symbiotic egg predators of decapod Crustacea: adaptations to host life histories

Armand M. Kuris
Department of Biological Sciences and Marine Science Institute, University of California, Santa Barbara, CA 93106, USA

Key words: Carcinonemertes, Pseudocarcinonemertes, Nemertea, life cycle, egg predator, adaptation, Crustacea

Abstract

Life cycles of five species of nemerteans that are symbiotic egg predators of decapod crustaceans are reviewed and compared. These life cycles range from the relatively simple life cycle of *Carcinonemertes regicides* from the red king crab, *Paralithodes camtschatica*, to the complex life cycle of *C. errans* on the Dungeness crab, *Cancer magister.* Increased complexity of these life cycles is evaluated as a series of adaptations to cope with life history features of their respective hosts. These host attributes include seasonal brooding, rapid embryogenesis and brooding in alternate years. Lengthy host brood periods permit autoinfection of the respective nemertean symbionts with resultant loss of many host eggs. Adaptations associated with short or intermittent brooding periods include migrations of regressing adult worms to the gills and their return to the egg mass at the next oviposition, transfer of worms from old to new cuticles at ecdysis, use of male and juvenile female crabs as hosts, and sexual transmission at host copulation. Future research needs are briefly discussed.

Introduction

Over the past twenty years it has become evident that the egg mass of decapod crustaceans has provided a rich habitat supporting a major adaptive radiation of the phylum Nemertea. Six species in 3 genera have been described in this period but the rate of reports of new species (Wickham & Kuris, 1985, 1988; Kuris *et al.*, 1991) considerably exceeds the rate of species description. I venture the prediction that by the time this symbiotic fauna is well described it will encompass about half of the species in the phylum. Studies on the biology of these worms have enabled us to drop the albatross that economic importance is 'nil' which has often accompanied textbook summaries of the phylum (e.g., Laverack & Dando 1987). Nemertean egg predators cause substantial losses to the reproductive output of several commercially fished crustacean species. (Wickham, 1979, 1986; Shields *et al.*, 1990; Kuris *et al.*, 1991).

As nemertean symbionts on decapod crustaceans that have been carefully investigated are egg predators (but see McDermott & Gibson, 1993), it seems probable that the crustacean egg attached to the ovigerous host has provided the resource base for this impressive radiation. Hosts now include several families of Brachyuran crabs, Anomurans and both spiny lobsters (Panuliridae) and true lobsters (Nephropidae). While most of the described species appear to be related at the family level (Carcinonemertidae) another type of monostiliferoidean hoplonemertean (*Pseudocarcinonemertes*) has also assumed this life style. The stylet enables these worms to pierce the egg membranes of their host. The attached egg is such an easy prey item that most described species of nemertean egg predators lack a reserve stylet sac. This suggests that the stylets are rarely detached during penetration of the egg. However, undescribed species have been reported that have typical monostiliferoidean reserve stylet sacs (Wickham & Kuris, 1988, Kuris *et al.*, 1991, Shields *et al.*, in prep.; Shields, in prep.) as does *P. homari* (Fleming & Gibson, 1981).

The symbiotic egg predators in the family Carcinonemertidae share many features associated with small size among the Nemertea. Other families that consist predominantly of species of small body size (*i.e.*, Tetrastemmidae, Ototyphlonemertidae) cluster near the Carcinonemertidae in the cladistic analysis of Sundberg (1990). Indeed, Uhazy *et al.* (1985) have argued for the placement of *Pseudocarcinonemertes* in the Tetrastemmidae.

Species included in the Carcinonemertidae share a striking morphological feature that substantially extends the morphological definition of the phylum. Instead of simple paired gonads, each with a lateral pore, the male carcinonemertid reproductive system resembles that of a parasitic flatworm, particularly a cestode. Testes are small and numerous, vasa efferentia leading from the testes collect in larger sperm ducts. Near the terminal part of this system the vas deferens is enlarged as a seminal vesicle to store sperm which exit through a posterior gonopore.

Most recently we have collected the first heteronemertean symbiotic egg predator (Kuris & Lafferty, in prep.). This species has cephalic slits and engulfs entire eggs when feeding. When one considers what a small proportion of potential decapod host species have been examined, even cursorily, for brood inhabitants it is apparent that many novel forms await discovery for those willing to explore crustacean egg masses.

Here I will summarize and compare the known life cycles of symbiotic egg predators. Distinctive adaptations to the host life cycle will be discussed and interesting avenues for future investigation will be noted. As transmission has not been obtained experimentally for any of the carcinonemertid species, some of the conclusions are based on indirect evidence.

Life cycles

Carcinonemertes regicides *on the red king crab,* Paralithodes camtschatica

The large lithodid anomurous crab, *Paralithodes camtschatica,* supports major fisheries in the North Pacific and the Bering Sea off the coasts of Alaska and Russia. In these cold waters the crabs grow slowly, entering the fishery 8–10 years after postlarval settlement (Blau, 1986). The female crabs brood a clutch of eggs for about 11 months (March or April to February or March) thus some ovigerous hosts are sometimes available throughout the year (Powell & Nickerson, 1965). In any case the time between the end of the last brood and the start of the next brood is brief, less than one month.

The life cycle of *C. regicides* is simple and appears to involve only ovigerous female hosts (Fig. 1). As for all carcinonemertids, male and female worms eat crab eggs, grow, mature, mate and reproduce in the egg mass of their hosts. While female worms are sometimes found in a poorly defined mucus sheath, the males do not appear to have a fixed site in the egg mass. Circumstantial evidence suggests that larval transmission is quite brief, perhaps resulting in autoinfection of the same ovigerous crab (Kuris *et al.*, 1991). Early in the red king crab brooding season infestation intensity is low. In the summer worm intensity increases to very high levels. Mean intensities of over 300000 worms per crab were estimated from Terror Bay, Kodiak I., Alaska in June, 1984. High worm intensities are sustained throughout the remainder of the brooding season but worm populations crash to negligible numbers by the start of the next brooding period. It is presumed that larval transmission accounts for the spread of the infestation between hosts. Males

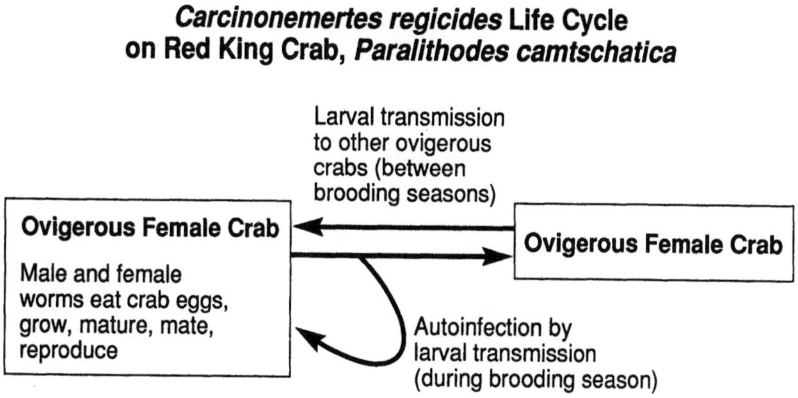

Fig. 1. Life cycle of *Carcinonemertes regicides* on the red king crab, *Paralithodes camtschatica*. Arrows represent movement of worms between crabs.

and juvenile female crabs are not infested. Host brood mortality is often very high (up to 100%) and high worm intensities have persisted in some areas for several years (Kuris *et al.*, 1991; Shields *et al.*, 1990b).

Carcinonemertes regicides has also been recovered at low intensities from the egg masses of the tanner crab, *Chionoecetes bairdi* by Wickham & Kuris (1990). This crab also has a very long brooding period and may serve as an important alternative host during the brief season when ovigerous adult female *P. camtschatica* are not available.

The life cycle of an undescribed species of nemertean, form #4 of Wickham & Kuris (1988), appears similar (Kuris *et al.*, 1991). It is abundant on ovigerous red king crabs at some locations but does not appear to use juvenile or male hosts. Undescribed worms from other hosts with long brooding seasons or repeated ovipositions, such as recently observed on the majid crab, *Loxorhynchus grandis*, in California (Kuris & Kushner, unpubl. obs.), also seem to infest only female hosts.

Carcinonemertes carcinophila *on the blue crab,* Callinectes sapidus

The portunid crab, *Callinectes sapidus,* supports an important fishery along the Atlantic and Gulf of Mexico coasts of the United States. Young crabs migrate to waters of low salinity where they grow, mature and mate. Males stay in low salinity habitats for the remainder of their lives. Female crabs mature in their second year, cease to molt after maturity and mate at the end of the summer (Williams, 1984). Oviposition is delayed to the following spring and the brooding period is only two weeks long (Davis, 1965). Females may produce two or more broods during each summer, and may live for 2 more years. In the tropics brooding may be year round (Williams, 1984). *Carcinonemertes carcinophila* grow, mate and mature in the egg mass (Humes, 1942). The life cycle of this worm involves only adult female hosts and has only been studied in warm temperate waters. Although young female and all male worms are errant in the egg mass, large female worms secrete a sheath along the endopods of the host. This sheath becomes partly filled with eggs after the female worms oviposit. After eclosion of the host egg mass, the worms migrate to the gill chambers where they secrete a hard capsule on the gill lamellae, decrease in size and dedifferentiate their reproductive organs. When the host oviposits again, these worms emerge from their capsules and migrate to the egg masses to renew feeding, growth and reproduction (Fig. 2). On the gills they are joined by recently settled worms from the plankton. The regressed adults may be distinguished from newly settled worms by their color, the former being red, the latter colorless (Hopkins, 1947). As very small worms are more numerous in advanced egg masses (Humes, 1942),

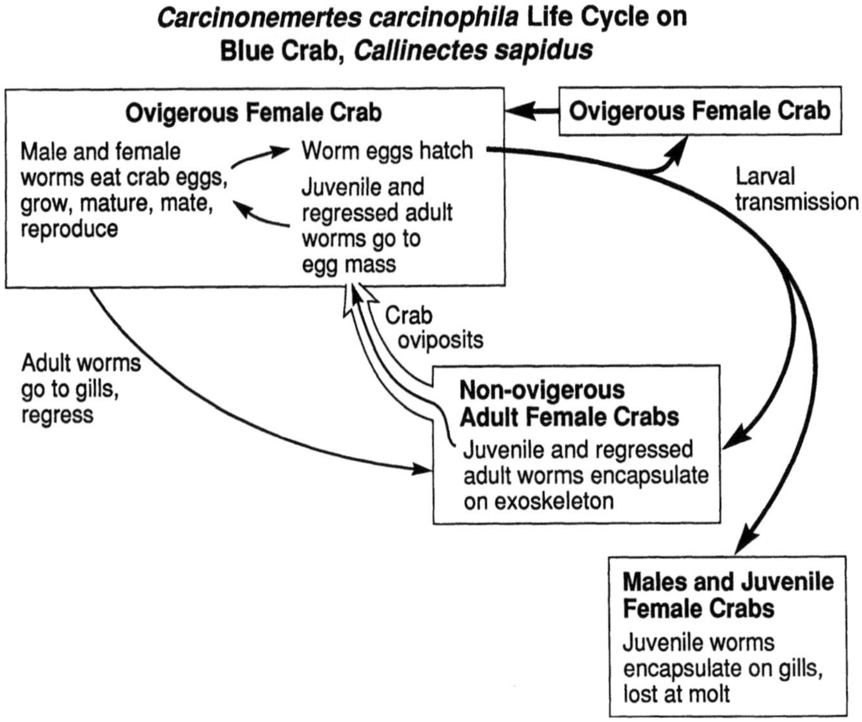

Fig. 2. Life cycle of *Carcinonemertes carcinophila* on the blue crab, *Callinectes sapidus*. Heavy arrows represent movement of worms between crabs; light arrows movement on an individual crab; the broad open arrow denotes that a female crab has oviposited her eggs without an intervening molt.

this suggests that larval development is quite brief, perhaps including autoinfection. Abbreviated larval development is supported by the early observations of Van Beneden (1861) on another species of crab.

Worms are less abundant, in terms of both prevalence and intensity, on the gills of males and juvenile females. Humes (1942) stated that they are shed at molt of juvenile crabs. Thus, such hosts are a dead end in terms of transmission (Fig. 2).

The patterns of infestation among sexes and maturation states of the blue crab suggest that these worms are able to detect and recruit more strongly to adult female hosts. As worms are retained on females between broods, Humes (1942) suggested that intensity builds up on crabs over the two or more years of adult life. The increase in size of the worms on the gills of adult non-ovigerous females from the time of the pubertal molt in the late summer to oviposition the next spring (Humes, 1942) indicates that, while encapsulated, these worms are able to obtain sufficient nutrition to achieve growth. This is remarkable; as the only source of nutrition for juvenile nemerteans appears to be uptake of dissolved organic substances leaked through the gills and arthrodial membranes of the hosts, as demonstrated for *Car. errans* on *Cancer magister* (Roe *et al.*, 1981).

Although Humes observed that egg fragments were common in the guts of the worms, brood mortality appears to be slight. Comely & Ansell (1989) found no evidence of host egg predation for *Car. carcinophila* on *Liocarcinus depurator* in Scotland, Lipcius (pers. commun.) noted no differential fecundity of infested blue crabs compared to uninfested blue crabs and Overstreet (1978) also felt that egg losses were minimal. However, as Kuris *et al.*, (1991) showed for *Car. regicides*, brood mortality, even when very substantial, may only be evident late in the host embryogenic period and was intensity dependent.

Prevalence (presence/absence) of worms was a very weak predictor for brood losses. Brood losses were also best estimated by direct count of the empty egg membranes retained on the pleopods (Wickham, 1979) although indirect methods are also available (Kuris & Wickham 1987). Estimates of egg mortality are not yet available for *Car. carcinophila* on any of its hosts. Shields (in prep.), using direct estimates, showed that the related species, *Car. mitsukurii* on *Portunus pelagicus* is a minor source of brood mortality in Queensland, Australia.

Host specificity of *Car. carcinophila* is a matter requiring further study. Originally described from a 'small crab', probably *Carcinus maenas*, by Kölliker (1845), it has been reported from a wide variety of portunid crabs (over 20 species) on both sides of the Atlantic Ocean (Wickham & Kuris, 1985, Campbell *et al.*, 1989, Comely & Ansell, 1989). Consistent small differences in body size and stylet morphology led Humes (1942) to describe the North American form as a separate variety, *Carcinonemertes carcinophila imminuta*. Based on the hypothesis that these worms exhibited little host specificity, worms from the many portunid species have been grouped with worms recovered from several species of xanthid, grapsid, dromiid, leucosiid, calappid and majid crabs as well as galatheid anomurans. Wickham & Kuris (1988) have raised doubts about inclusion of some of these worms in *Car. carcinophila* as substantial morphological differences have been observed for worms from some of these hosts (e.g., on *Galathea strigosa* by Dieck 1874, and on *Cal. danae* by Humes 1942). Worms on some of these hosts may exhibit significant differences with respect to life cycle. For instance, encapsulated worms are numerous on the gills of male *Liocarcinus depurator* (Comely & Ansell, 1989) but not on male *Cal. sapidus* (Humes, 1942).

In general, carcinonemertid worms recovered from portunid hosts exhibit some consistent similarities. Compared to other nemertean egg predators these worms are long (20 + mm) and juveniles are found encapsulated on gill lamellae. They are generally infrequently reported from male and juvenile female hosts. This pattern has also been noted for *Car. mitsukurii* on *Portunus pelagicus* by Shields (in prep.). Shields also observed that *Car. mitsukurii* returned to the gills after host eggs hatched and that sacculinized females were infested with worms that were retained on the gills. However, worms of an undescribed species of *Carcinonemertes* recovered from a portunid, *Ovalipes trimaculatus*, in Chile by Kuris (pers. obs.), were only found encapsulated on the thoracic and abdominal sterna as well as the pleopods of adult female crabs that were probably post-ovigerous (based on the accumulated anoxic detritus under the abdominal flap). No worms were present on the gill lamellae.

Carcinonemertes epialti *on the shore crab,* Hemigrapsus oregonensis

Hemigrapsus oregonensis is a small intertidal grapsid crab of no commercial importance. It is abundant in bays, harbors and marshes along the Pacific coast of the United States, Canada and Baja California, Mexico. Growth to maturity takes about one year and crabs reach an age of about 4 years (Kuris, 1971). Crabs brood their eggs for about 6 weeks along the Central California coast. Broods in the summer are followed by molts. In the winter two or three successive broods occur in the same instar separated by periods of a few days to two weeks (Kuris, 1971).

The nemertean, *Carcinonemertes epialti*, is abundant at only a few locations (e.g., Bodega Harbor, Elkhorn Slough, in California, Yaquina Bay in Oregon). Adults are only found in egg masses. Juvenile worms are enclosed in a soft mucus sheath along arthrodial membranes; principally those of the limb axillae and under the abdomen of non-ovigerous host crabs of either sex, (Kuris, 1978). The lengthy stay of these juvenile worms on the exoskeleton may be sustained by dissolved organics leaked from the arthrodial membranes of the crabs (Roe *et al.*, 1981). These worms are able to transfer at molt to the new exoskeleton (Wickham *et al.*, 1984), thus they are not lost at host ecdysis (Fig. 3). When an adult female crab oviposits, the juvenile worms exsheath, migrate to the egg mass within two days and begin to feed on eggs. Adult worms secrete

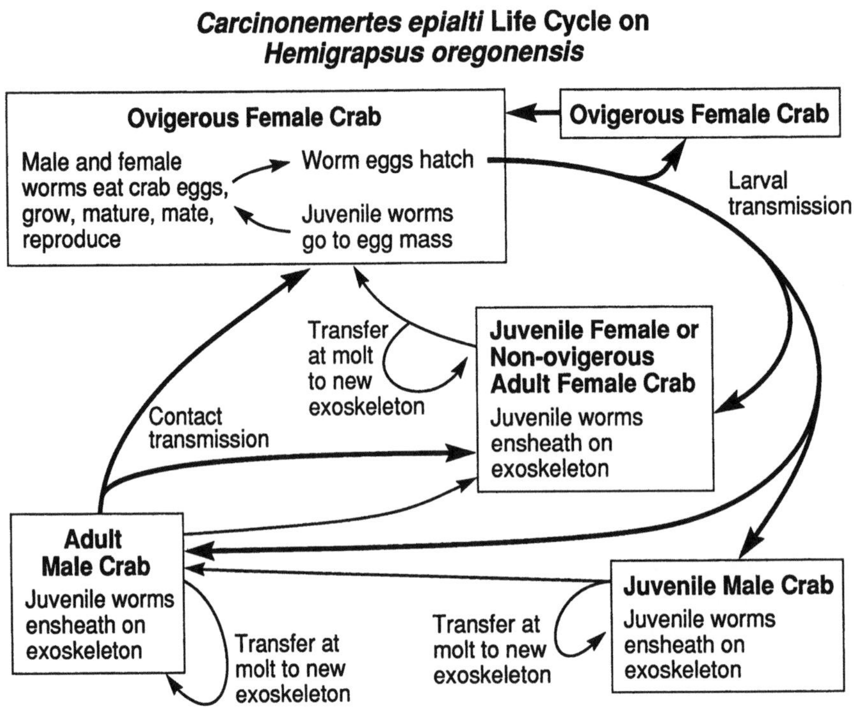

Fig. 3. Life cycle of *Carcinonemertes epialti* on the shore crab, *Hemigrapsus oregonensis*. Arrows as in Fig. 2.

a mucus tube and anchor these to the abdomen or pleopods of the ovigerous crab. Worm eggs are laid in separate mucus sacs and most hatch at about the time of host eclosion (Kuris, 1978). After the host eggs have hatched, an increased frequency of worms in the gills has been noted (Kuris, 1978). Worms are otherwise uncommon in the gill chambers of nonovigerous crabs. On these post-ovigerous female crabs (recognized by the presence of empty egg membranes retained on the pleopods) worms often seem moribund (Kuris, 1978). Whether they will die or are merely regressing and will survive to emerge on to the egg mass again is not known.

The duration of larval development is unknown. Attempts to transmit infections to new crabs in the laboratory have been unsuccessful although larva have been maintained in sea water for almost two months (Roe, 1979). The short period of host embryogenesis does not appear to permit a substantial increase in intensity over the brooding period; so autoinfection does not appear to be epidemiologically important and may not occur if a long interval is required for larvae to attain competency. Transmission of juvenile worms by contact between infested male crabs and uninfested male and female crabs has been demonstrated in the laboratory (Kuris, 1978). This may be an important mechanism as these crabs are sedentary and spend long periods of time under the same rock (Kuris, 1971). Otherwise, the fate of the many worms on male hosts is uncertain. Venereal transmission has not been observed but seems unlikely. Female crabs are only briefly receptive and copulation lasts but a few minutes (Kuris, 1971). Mating behavior of these crabs includes neither a pre- nor post-copulatory embrace.

Roe (1986) has documented parthenogenesis for females of *Car. epialti* reared in isolation *in vitro*. The significance of parthenogenesis in nature is uncertain but as Roe (1986) states, it may be of significance where intensities are low, as is often the case (Kuris, 1978).

For both sexes prevalence and intensity increase with size of the crab. This appears due to transfer at ecdysis and perhaps to the larger surface area available for settlement of planktonic larvae on larger crabs (Kuris, 1978; Roe, 1979). Seasonal changes in worm abundance have been noted (Kuris, 1978; Roe, 1979; Shields & Kuris, 1988), and are marked by a late fall and winter peak followed by a nadir in the spring. An outbreak has been reported with intensities of up to 1000 worms per ovigerous crab (Shields & Kuris, 1988). Nearly all eggs were lost on most crabs. The outbreak coincided with the seasonal peak in abundance. The seasonal peak in abundance also coincided with the highest season of settlement of worms on sentinel hosts (Bauman, 1983).

Host specificity of *Car. epialti*, as with *Car. carcinophila*, is clouded by taxonomic uncertainties. Originally described from the majid crab, *Pugettia producta* by Coe (1902), *Car. epialti* has been reported from several species of grapsid, cancrid and majid crabs probably induding the worm recently observed on *Loxorhynchus grandis*. (Kuris, 1978; Roe, 1979; Wickham & Kuris, 1985; Kuris & Kushner, unpubl. obs.). However, its biology remains uninvestigated on its type host. Although the host is common, infestations are exceedingly rare at Bodega Head in northern California (Kuris, 1971) and uncommon in the Santa Barbara Channel in southern California. It has been suggested (Wickham & Kuris, 1985, 1988) that *Car. epialti* is a species complex. Thus, the life cycle described here is most appropriate for the three grapsid hosts. Observations of worms referred to *Car. epialti* on *Cancer anthonyi* support a similar life cycle. Regressed adult worms appear to survive until the next oviposition on this host (Shields *et al.*, 1990a) and sexual transmission is possible since copulation of these crabs is prolonged and followed by a post-copulatory embrace (Shields, 1991).

Pseudocarcinonemertes homari *on the American lobster,* Homarus americanus

The American lobster is a very important commercial species off the northeast coast of the United States and Atlantic Canada. These lobsters are a long-lived species entering the fishery after 7–12 years in inshore waters (Aiken & Waddy, 1986). The brooding period is extended, lasting 10–12 months. Females generally oviposit every other year and may or may not molt between clutches (Aiken & Waddy, 1980). Oviposition takes place in the spring and this period may overlap with hatching of the eggs from the broods of the preceding year.

Pseudocarcinonemertes homari is something of an enigma. The American lobster is probably the most intensively investigated crustacean with careful studies of egg development and fecundity being conducted repeatedly over the past 100 years (Brooks & Herrick, 1892; Herrick, 1896; Templeman, 1940; Perkins, 1971; Aiken & Waddy, 1980). Yet, until an outbreak of *P. homari* around the Maritime Provinces in the early 1980s that caused great concern for the fishery and mariculture (Aiken *et al.*, 1985), this worm had never been reported. Even when it was abundant in the Maritimes it was never observed in the vicinity of Newfoundland (G. Ennis, pers. commun.) and only once on a single lobster off Massachussetts (B. Estrella, pers. commun.). Just as mysteriously as it appeared it has disappeared from natural populations by the late 1980's (S. Waddy, pers. commun.). It still persists in the lobster culture facility at the Biological Station in St. Andrews, New Brunswick, despite considerable efforts at eradication and decontamination of the facility (S. Waddy, pers. commun.).

Pseudocarcinonemertes homari is not closely related to the other described nemertean symbiotic egg predators. It lacks Takakura's duct, the trematode-like system of sperm ducts that collect the testicular products, as well as several other important features shared by the carcinonemertids (Uhazy *et al.*, 1985; Wickham & Kuris, 1988). It may well represent an independent evolution of the distinctive life style of a symbiotic egg predator. Yet, the key life history feature of the carcinonemertids is shared by *P. homari*; the worms feed on crab eggs in order to grow, mature and reproduce (with one apparent exception, see below).

Juvenile worms ensheathed on the gills and in protected locations of the exoskeleton, exsheath, and migrate to the egg mass shortly after oviposition of adult female lobsters. Brattey *et al.* (1985) estimated that three generations of worms (with overlap) occur in the egg mass. Worm eggs hatch two months after lobster brooding is initiated. Worms tend to avoid the periphery of the egg mass (Campbell & Brattey, 1986). Development is direct; there is no free-swimming larval stage. Autoinfection has been documented and is associated with marked population increases on certain lobsters (Aiken *et al.*, 1985).

After eclosion of the lobster eggs, the worms return to the gills and the exoskeleton and ensheath. They often occur in aggregates (Aiken *et al.*, 1985). In the laboratory, worms may be found on the gills and exoskeletons of males, non-ovigerous females and juvenile lobsters (Fleming & Gibson, 1981; Aiken *et al.*, 1985), but in field samples males were rarely infested (Brattey *et al.*, 1985). Fleming & Gibson (1981) reported that the gills of infested lobsters were often damaged. This association was doubted by Aiken *et al.*, (1985) who have observed the described damage on uninfested lobsters and saw no evidence of damage on many infested lobsters. They did, however, report a single instance of apparent parasitism in the gills of an isolated male lobster that had been maintained in their culture facility for two years without contact with female lobsters. On this specimen the gills were damaged and, most curiously, adult *P. homari* were present and had reproduced, ovipositing unusually small egg sacs.

In addition to recently hatched worms swimming short distances, other lobsters may be infested by direct contact. Continued problems with decontamination of the St. Andrews lobster culture facility strongly suggest that free-living movement between lobsters probably occurs. It is also possible that they have been transferred on the hands of culture personnel (Aiken *et al.*, 1985). Lobsters may also become infested by mature nemerteans displaced from egg masses of certain lobsters that vigorously groom their egg masses (Aiken *et al.*, 1985).

Study of *P. homari* population dynamics is complicated by the idiosyncratic extent of grooming by individual lobsters. Some lobsters will strip off all their eggs in the presence of low intensity infestations while others will groom little if at all and permit the buildup of high intensity infestations. Thus, unlike *Carcinonemertes regicides* on *Paralithodes camtschatica* or *C. errans* on *Cancer magister*, for which most ovigerous individual hosts in a population have comparable worm burdens, *Pseudocarcinonemertes homari* worms are very strongly aggregated among ovigerous lobsters (Brattey *et al.*, 1985). Prevalence of *P. homari* never reached 100% (unlike infested populations of the *Carcinonemertes* spp.) and only a few hosts were so heavily infested that egg losses were substantial (Campbell & Brattey, 1986).

Impact of *P. homari* on individual lobsters has been difficult to directly estimate as lobster grooming removes empty host egg membranes (Campbell & Brattey, 1986). Aside from the relatively few heavily infested lobsters recovered during the outbreak of the early 1980s, hyperactive variability in lobster grooming means that the unknown loss of eggs due to grooming behavior may have been quite considerable.

Homarus americanus is the only reported host for *P. homari* and no closely related worms have been described. Direct development and some potential for parasitism of males and non-ovigerous females are its singular life history features (Fig. 4).

Carcinonemertes errans on the Dungeness crab, *Cancer magister*

Cancer magister is a cancrid crab supporting major fisheries in northern California, Oregon, Washington, British Columbia and Alaska. Crabs enter the fishery after 1.5–2 years of postlarval growth (Butler, 1961; Smith & Jamieson, 1989). Crabs mate in the spring but oviposition is delayed until the autumn. Brooding is completed in about three months and crabs usually molt between broods (Hankin *et al.*, 1989).

The life cycle of *Car. errans* on *Can. magister* is the most complex reported for a symbiotic nemertean egg predator (Fig. 5). The worms grow, mate and reproduce in the host egg mass. Both sexes are errant. They are more abundant at the

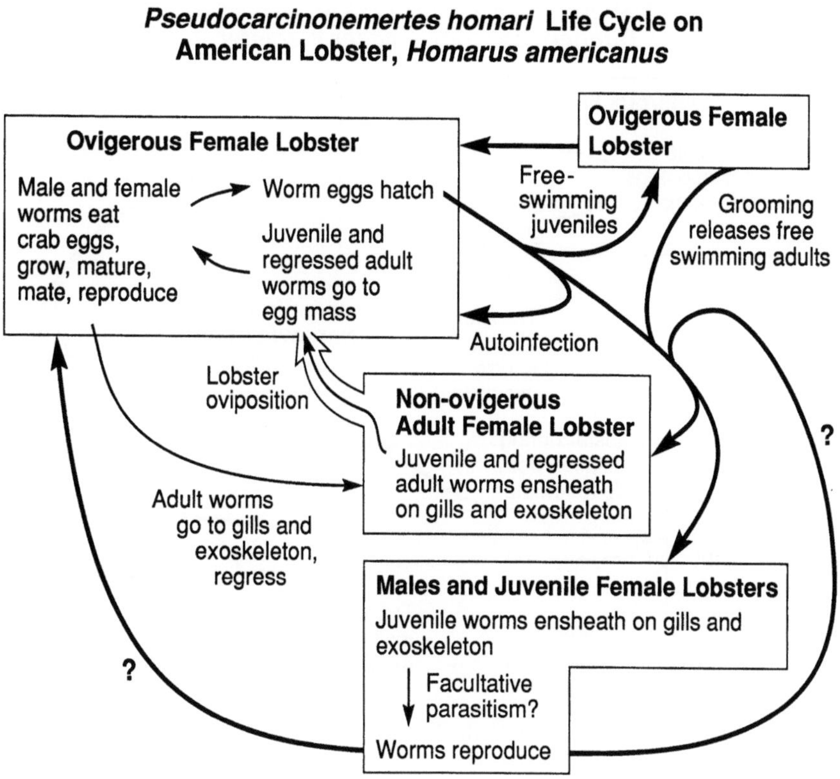

Fig. 4. Life cycle of *Pseudocarcinonemertes homari* on the American lobster, *Homarus americanus*. Arrows as in Fig. 2; ?, implied by laboratory observations, no direct evidence in the field.

periphery of the egg mass early in the brooding period (Wickham, 1979) but tend to move to the interior later. Egg sacs are laid during the last half of the embryogenic period (Kuris & Wickham, 1987). The duration of larval development is unknown. Autoinfection seems unlikely as the host brooding period is relatively short and very seasonal. For nine months there are no ovigerous hosts available for settlement by *Car. errans*.

Juvenile worms appear to be long-lived. They are able to take up dissolved organic matter from seawater. The host crab is the source for this dissolved organic matter. Crabs leak an appreciable quantity of primary amines through their arthrodial membranes (Roe *et al.*, 1981, Crowe *et al.*, 1982). This appears to be the sole source of nutrition for these worms and sustains their quiescent metabolism when ensheathed on the crab exoskeleton. Experimental studies have demonstrated that the juvenile worms are able to detect an impending molt, become active and move to the epimeral suture where they transfer to the new, postmolt exoskeleton (Wickham *et al.*, 1984). Thus, the worms are retained on the crabs through the sequence of molts between the juvenile instars. This study also showed that worms are able to transfer from males to females at copulation. These life cycle attributes appear to contribute to the persistence of the high intensity populations of *Car. errans* on the low density population of *Can. magister* in Central California (Wickham *et al.*, 1984; Kuris & Wickham, 1987). As *Car. errans* intensity increases markedly with juvenile crab size (Wickham, 1980), it is also apparent that larval transmission is important.

The pattern of intensity of *Car. errans* on *Can. magister* and the impact of these worms on the fishery has received considerable attention (reviewed by Wickham, 1986; Kuris & Wickham, 1987; Kuris, 1991) and will only briefly be noted here. High intensity worm populations in Central California kill 50–60% of the crab embryos and may prevent the recovery of this fishery. In northern California, Oregon and Wash-

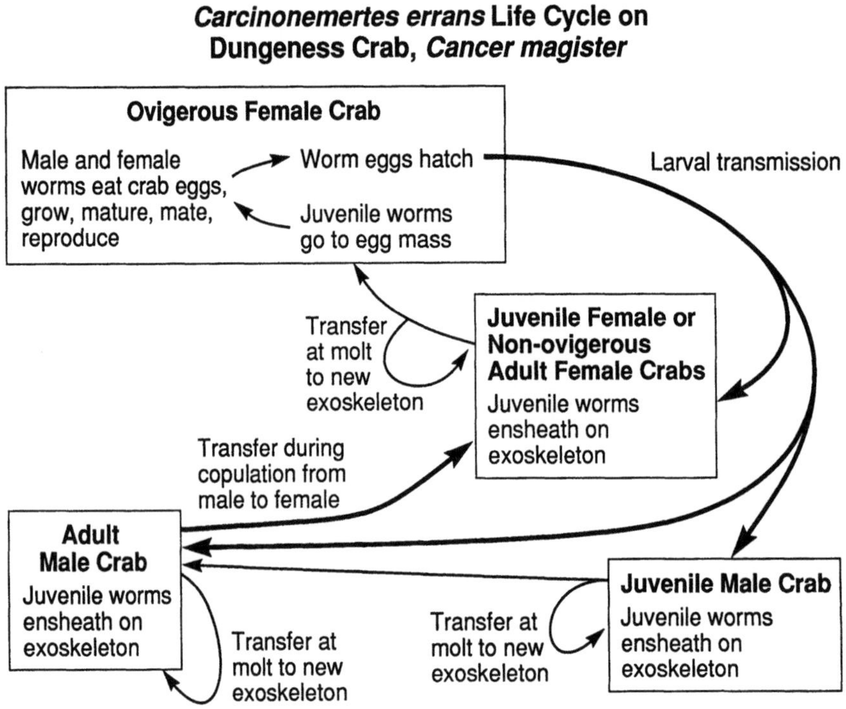

Fig. 5. Life cycle of *Carcinonemertes errans* on the Dungeness crab, *Cancer magister*. Arrows as in Fig. 2.

ington worm intensity appears to track the cyclical peaks in the catch that are a notable feature of these fisheries. Low intensity populations characterize canadian and alaskan populations but even there prevalence is 100%. Current models of this fishery indicate that while the impact of *Car. errans* is insufficient to drive the host population cycles, the worms may contribute to its cyclical behavior (Hobbs & Botsford, 1989).

Carcinonemertes errans was initially reported as *Car. epialti* (Kuris, 1971) until Wickham (1978) noted its unique characteristics. It has only been collected from the Dungeness crab. Other species of *Cancer* in California are infested by worms that are more similar to *Car. epialti* (Wickham & Kuris, 1985).

Discussion

The nemerteans that are symbiotic egg predators of decapod crustacean hosts present a range of life cycle complexity adapted to the brooding patterns of their hosts. The simplest life cycle known is that of *Carcinonemertes regicides* on the red king crab. Ovigerous crabs are often available throughout the year. The nemertean life cycle appears to only involve ovigerous female hosts, larval development appears to be brief and autoinfection is epidemiologically important. This simple and brief life cycle has been modeled in the context of this fishery (Kuris & Lafferty, 1992). The model suggests that the threshold of transmission could be broken, and yield of red king crabs increased, by fishing out local populations of infested female hosts.

Relatively few chemically mediated cues are required to synchronize the life cycle of *C. regicides* with that of its host. The larvae must, of course, be able to detect the ovigerous host in order to settle and begin feeding.

Hosts that exhibit similar features, such as majid crabs which repeatedly oviposit eggs without an intervening molt, or *Homarus* with a very long brood time, are infested with symbiotic egg predators that have a relatively simple life cycle. The skip-brood pattern of the American lobster and its grooming defensive behavior are countered by *Pseudocarcinonemertes homari* which can persist on male lobsters and is found on non-brooding females as an ensheathed juvenile worm. The epidemiological importance of autoinfection is a life cycle similarity shared by *C. regicides* and *P. homari*. Indeed, *P. homari* lacks the typical hoplonemertean ciliated larva and develops directly into swimming juvenile worms competent to settle on hosts soon after hatching.

Some hosts have long periods of time between broods. Ensheathed juvenile worms provide the key counter-adaptation on the part of the worms. Thus, these worms must respond to an additional chemically mediated cue. They must be able to detect oviposition of the host, exsheath, and migrate to the egg mass from the gills or other exoskeletal sites. *Pseudocarcinonemertes homari*, *C. epialti*, *C. errans* and *C. carcinophila* all have ensheathed juveniles that are able to detect and promptly respond to oviposition.

The presence of juvenile and non-ovigerous female hosts poses a distinctive element of environmental complexity for settling worms. Such hosts may molt and shed ensheathed juvenile worms. For *C. errans* and *C. epialti*, but apparently not *C. carcinophila*, the ensheathed juveniles are able to detect the impending molt, migrate to the epimeral ecdysial suture and transfer to the new postmolt cuticular surface at ecdysis (Wickham *et al.*, 1984). *Carcinonemertes pinnotheridophila* from *Pinnixa chaetopterana* is also able to transfer to new cuticule at molt (McDermott, 1966; McDermott & Gibson, 1993). High titers of molting hormones or their metabolites prior to the molt (Chang, 1989) or ecdysial lubricants could provide possible cues to which ensheathed juvenile nemerteans might respond. It is interesting to note that *C. errans* and *C. epialti* juveniles are enveloped in a soft mucus sheath whereas *C. carcinophila* and the related worms on portunid hosts (*C. mitsukurii*, *C.* sp on *Ovalipes trimaculatus* from Chile) are encased in firm capsules which may impede or preclude detection of host ecdysis.

The juveniles of symbiotic egg predators that are ensheathed or encapsulated must also be able to maintain themselves for long periods. These little worms are long-lived, persisting for a year or more without access to the only food source (decapod eggs) known to permit growth, maturation and reproduction. They are able to maintain themselves through the uptake of dissolved organic matter leaked from their hosts (Roe *et al.*, 1981, Crowe *et al.*, 1982). The variation in size shown by juvenile *C. carcinophila* (Humes, 1942) strongly suggests that that species may even be able to appreciably grow using this food source.

The very brief duration of embryogenesis of *Callinectes sapidus* (10–14 days) makes the growth in length of *Car. carcinophila* to 35 mm (Humes, 1942) quite impressive. The short feeding time coupled with a lengthy interbrood period is associated with the return to the gills and other exoskeletal areas. These worms appear to decrease in size and dedifferentiate their reproductive system, returning to a juvenile-like anatomical condition. A short host brooding period implies a short breeding period for *Car. carcinophila*. Thus, regression and subsequent recrudescence of adult worms may be an important adaptation to increase the size and reproductive output of worms using such an ephemeral food source. Negative growth and gonadal regression may also occur in *Car. epialti* and *Car. errans* but the fate of these adult worms after the eggs of the host hatch has not yet been carefully studied.

Lengthy periods between broods may also select for ability to use male hosts as substrates for juvenile worms. This has been best demonstrated for *C. errans* on the Dungeness crab.

The ability to detect mating of their hosts is a fourth event to which some of these symbiotic nemerteans may respond, apparently using a chemically mediated cue. Wickham *et al.*, (1984) have shown that *C. errans* are sexually transmitted with considerable efficiency from males to females at copulation. This may also occur for *C. epialti* on the yellow rock crab. Pheromones produced by female crabs to attract males (Dunham, 1978) may provide the cue. However, sexual transmission seems improbable for *C. epialti* on the shore crab as mating is a brief encounter in the latter species. Hence, the abundance of worms on male shore crabs is enigmatic since a means to complete their life cycle is not readily apparent.

Of the known life cycles, *C. errans* is the most complex, involving adult and juvenile crabs of both sexes. Specific, presumably chemically mediated adaptations include host location, transfer to the new cuticle at molt, transfer from male to female host at copulation and migration to the egg mass shortly after host oviposition. These adaptations are consistent with the strongly seasonal and relatively brief brooding period of *Cancer magister*. The complex life cycle appears to buffer this worm from a crash in host abundance. Thus, *C. errans* has persisted at high intensities for over 20 years in the collapsed Central California Dungeness crab fishery (Wickham, 1986). This phenomenon led Botsford & Wickham (1978) to suggest that an alternative stable state of high worm intensity and low crab abundance had been attained.

Review of the life cycles of these worms brings some future research topics into focus. The most important, from the standpoint of increasing both our understanding of basic nemertean biology, and of symbiont population ecology and infestation dynamics, is the experimental determination of the cues for larval settlement. Knowledge gained from further sentinel host experiments in the field (Bauman, 1983; Wickham & Roe, 1987) might provide clues for the period of larval competency. Observations on accumulation of small worms (*C. errans*) at the base of the eyestalks and antennae (Wickham, pers. commun.) suggest that secretions from organs such as the antennal glands may be possible cues for host detection. Recent molecular approaches to the problem of marine invertebrate larval settlement may be useful here (e.g., Morse, 1986). Also of interest would be the experimental establishment of the nature of the signal for movement to the egg mass following host oviposition, movement to the new cuticles following ecdysis (*C. errans*, *C. epialti*) and movement from males to females at copulation (*C. errans*). Experimental confirmation as to whether contact transmission, sexual transmission or transfer to new cuticles at molt occur in the life cycles where they have not been detected would contribute to population dynamics models. These life history attributes may also be useful clues for the determination of species differences in this group of morphologically similar worms (Gibson & Jones, 1990). More work is needed on the mode of juvenile sustenance. Growth of juvenile worms on the exoskeleton needs to be more critically documented, particularly for *C. carcinophila* and similar worms on portunids. Ultrastructural investigations of the surface of juvenile and adult worms would complement studies of juvenile worm nutrition.

Each worm species secretes several different products during its life cycle: tubes or sheaths for adult worms, capsules or sheaths for juvenile worms, egg sacs, and mucus trails associated with movement. Identification of the responsible glandular structures with histochemical techniques would help reveal the mechanisms for the production of these various structures. Further comparative studies would aid our understanding of the systematics of these worms if specific glandular structures could be associated with their secreted products. Stricker (1986) provided a starting point for this sort of investigation. He noted PAS + material, probably representing incipient egg sacs, being secreted by the submuscular glands at the posterior end of ovipositing female *C. epialti*. Perhaps a comparison of adult male and female *C. epialti* might aid the recognition of these specific secretory cells.

Finally, the formation of the stylet in those carcinonemertids that lack reserve stylet sacs needs to be examined. Stricker & Reed (1981) have shown that larval *C. epialti* lack a stylet. Juveniles always exhibit stylets (Kuris, pers. obs.) but stylet sacs have not been reported. Thus, stylet formation, presumably in an ephemeral stylet sac, must occur soon after larval settlement and metamorphosis to the juvenile. Careful examination of very small juveniles, perhaps in conjunction with transmission studies would help resolve this issue.

References

Aiken, D. E. & S. L. Waddy, 1980. Maturity and reproduction in the American lobster. Can Tech. Rep. Fish. aquat. Sci. 932: 59–71.

Aiken, D. E. & S. L. Waddy, 1986. Environmental influence on recruitment of the American lobster, *Homarus americanus*: a perspective. Can. J. Fish. aquat. Sci. 43: 2258–2270.

Aiken, D. E., S. L. Waddy & L. S. Uhazy, 1985. Aspects of the biology of *Pseudocarcinonemertes homari* and its association with American lobster, *Homarus americanus*. Can. J. Fish. aquat. Sci. 42: 351–356.

Bauman, L, 1983. The larval biology of *Carcinonemertes*. Master's Thesis, University of California, Santa Cruz, California.

Blau, S. F., 1986. Recent declines of red king crab (*Paralithodes camtschatica*) populations and reproductive conditions around the Kodiak Archipelago, Alaska. Can. Spec. Publ. Fish. aquat. Sci. 92: 360–369.

Botsford, L. W. & D. E. Wickham, 1978. Behavior of age-specific, density-dependent models and the northern California Dungeness crab (*Cancer magister*) fishery. J. Fish. Res. Bd Can. 35: 833–843.

Brattey, J., A. Campbell, A. E. Bagnall & L. S. Uhazy, 1985. Geographic distribution and seasonal occurrence of the nemertean *Pseudocarcinonemertes homari* on the American lobster, *Homarus americanus*. Can. J. Fish. aquat. Sci. 42: 360–367.

Brooks, W. K. & F. H. Herrick, 1892. The embryology and metamorphosis of the Macroura Mem. Nat. Acad. Sci. 5: 323–576.

Butler, T. H., 1961. Growth and age determination of the Pacific edible crab *Cancer magister* Dana. J. Fish. Res. Bd Can. 18: 873–889.

Campbell, A. & J. Brattey, 1986. Egg loss from the American lobster, *Homarus americanus*, in relation to nemertean, *Pseudocarcinonemertes homari*, infestation. Can. J. Fish. aquat. Sci. 43: 772–780.

Campbell, A., R. Gibson & L. H. Evans, 1989. A new species of *Carcinonemertes* (Nemertea: Carcinonemertidae) ectohabitat on *Panulirus cygnus* (Crustacea: Palinuridae) from Western Australia. Zool. 95: 257–268.

Chang, E. S., 1989. Endocrine regulation of molting in Crustacea. CRC Critical Rev. aquat. Sci. 1: 131–157.

Coe, W. R., 1902. Nemertean parasites of crabs. Am. Nat. 36: 431–450.

Comely, C. A. & A. D. Ansell, 1989. The incidence of *Carcinonemertes carcinophila* (Kölliker) on some decapod crustaceans from the scottish west coast. *Ophelia* 30: 225–233.

Crowe, J., L. Crowe, P. Roe & D. Wickham, 1982. Uptake of DOM by nemertean worms: Association of worms with arthrodial membranes. Am. Zool. 22: 671–682.

Davis, C. C., 1965. A study of the hatching process in aquatic invertebrates: XX. The blue crab, *Callinectes sapidus*, Rathbun, XXI. The nemertean, *Carcinonemertes carcinophila* (Kölliker). Chesapeake Sci. 6: 201–208.

Dieck, G., 1874. Beitrage zur Entwicklunsgeschichte der Nemertinen. Jena Z. Naturw. 8: 500–520.

Dunham, P. J., 1978. Sex pheromones in Crustacea. Biol. Rev. 53: 555–583.

Fleming, L. C. & R. Gibson, 1981. A new genus and species of monostiliferous hoplonemerteans, ectohabitant on lobsters. J. exp. mar. Biol. Ecol. 52: 79–93.

Gibson, R. & D. S. Jones, 1990. A new species of *Carcinonemertes* (Nemertea: Enopla: Carcinonemertidae) from the egg masses of *Naxia aurita* (Latreille) (Decapoda: Brachyura: Majidae) collected in the Albany region of Western Australia. In F. E. Wells, D. I. Walker, H. Kirkman & R. Lethbridge (eds), Proc. Third Int. Mar. Biol. Workshop. The Marine Flora & Fauna of Albany, Western Australia, Western Australia Mus., Perth. Vol. 1: 195–202.

Hankin, D. G., N. Diamond, M. S. Mohr & J. Ianelli, 1989. Growth and reproductive dynamics of adult female Dungeness crabs (*Cancer magister*) in northern California. J. Cons. int. Explor. Mer. 46: 94–108.

Herrick, F. H., 1896. The American lobster. A study of its habits and development. Bull. U.S. Fish. Comm. 15: 1–252.

Hobbs, R. C. & L. W. Botsford, 1989. Dynamics of an age-structured prey with density- and predator-dependent recruitment: the Dungeness crab and a nemertean egg predator worm. Theor. Pop. Biol. 36: 1–22.

Hopkins, S. H., 1947. The nemertean *Carcinonemertes* as an indicator of the spawning history of the host, *Callinectes sapidus*. J. Parasit. 33: 146–150.

Humes, A. G., 1942. The morphology, taxonomy, and bionomics of the nemertean genus *Carcinonemertes*. Illinois Biol. Monogr. 18: 1–105.

Kölliker, A., 1845. Ueber drei neue Gattungen von Wurmern, Lineola, Chloraima, Polycystia, neue Wurmgattungen, und neue Arten von Nemertes. Verh. Schweiz. Naturf. Ges., Chur, 29 (1844): 86–98.

Kuris, A. M., 1971. Population interactions between a shore crab and two symbionts. Ph.D. Thesis, Univ. California, Berkeley, 431 pp.

Kuris, A. M., 1978. Life cycle, distribution and abundance of *Carcinonemertes epialti*, a nemertean egg predator of the shore crab, *Hemigrapsus oregonensis*, in relation to host size, reproduction and molt cycle. Biol. Bull. 154: 121–137.

Kuris, A. M., 1991. A review of patterns and causes of crustacean brood mortality. In: A. M. Wenner & A. M. Kuris [ed.] Crustacean Egg Production, Crustacean Issues 6, Balkema, Rotterdam, pp. 117–141.

Kuris, A. M., S. F. Blau, A. J. Paul, J. D. Shields & D. E. Wickham, 1991. Infestation by brood symbionts and their impact on egg mortality in the red king crab, *Paralithodes camtschatica*, in Alaska,: geographic and temporal variation. Can. J. Fish aquat. Sci. 48: 559–568.

Kuris, A. M. & K. D. Lafferty, 1992. Modeling crustacean fisheries: effects of parasites on management strategies. Can. J. Fish. aquat. Sci. 327–336.

Kuris, A. M. & D. E. Wickham, 1987. Effect of nemertean egg predators on crustaceans. Bull. mar. Sci. 41: 151–164.

Laverack, M. S. & J. Dando, 1987. Lecture Notes on Invertebrate Zoology, 3rd edn. Blackwell Scientific Publications, Oxford, 203 pp.

McDermott, J. J., 1966. The biology of a nemertean parasite of pinnotherid crabs. Am. Zool. 6: 331.

McDermott, J. J. & R. Gibson, 1993. *Carcinonemertes pinnotheridophila* sp. nov. (Nemertea, Enopla, Carcinonemertidae) from the branchial chambers of *Pinnixa chaetopterana* (Crustacea, Decapoda, Pinnotheridae): description, incidence and biological relationships with the host. Hydrobiologia 266: 57–80.

Morse, D. E., 1986. External molecular signals controlling reproduction, settlement and metamorphosis of benthic marine invertebrates. In M.-F. Thompson, R. Sarojini & R. Nagabhushanam (eds), Biology of Benthic Marine Organisms. Techniques and Methods as Applied to the Indian Ocean, Oxford & IBH Publ. Co., New Delhi: 379–385.

Overstreet, R. M., 1978. Marine Maladies? Worms, Germs, and other Symbionts from the Northern Gulf of Mexico. Mississippi-Alabama Sea Grant Consortium. MASGAP-78-021: 1–140.

Perkins, H. C., 1971. Egg loss during incubation from offshore northern lobsters (Decapoda: Homaridae). Fish. Bull. 60: 451–453.

Powell, G. C. & R. B. Nickerson, 1965. Reproduction of king crabs, *Paralithodes camtschatica* (Tilesius). J. Fish. Res. Bd Can. 22: 101–111.

Roe, P., 1979. Aspects of development and occurrence of *Carcinonemertes epialti* (Nemertea) from shore crabs in Monterey Bay, California. Biol. Bull. 156: 130–140.

Roe, P., 1986. Parthenogenesis in *Carcinonemertes* spp. (Nemertea: Hoplonemertea). Biol. Bull. 171: 700–706.

Roe, P., J. Crowe, L. Crowe & D. Wickham, 1981. Uptake of amino acids by juveniles of *Carcinonemertes errans* (Nemertea). Comp. Biochem. Physiol. 69A: 423–427.

Shields, J. D., 1991. The reproductive ecology and fecundity of *Cancer* crabs. In Wenner, A. M. & A. M. Kuris (eds), Crustacean egg production. Crustacean issues 7. Balkema, Rotterdam, pp. 193–213.

Shields, J. D. & A. M. Kuris, 1988. Temporal variation in abundance of the egg predator *Carcinonemertes epialti* (Nemertea) and its effect on egg mortality of its host, the shore crab, *Hemigrapsus oregonensis*. Hydrobiologia 156: 31–38.

Shields, J. D., R. K. Okazaki & A. M. Kuris, 1990a. Brood mortality and egg predation by the nemertean, *Carcinonemertes epialti*, on the yellow rock crab, *Cancer anthonyi*, in southern California. Can. J. Fish. aquat. Sci. 47: 1269–1275.

Shields, J. D., D. E. Wickham, S. F. Blau & A. M. Kuris, 1990b. Some implications of egg mortality caused by symbiotic nemerteans for data acquisition and management strategies of red king crabs *Paralithodes camtschatica*. Proc. Int. King Tanner Crab Symp. Lowell Wakefield Symp. Ser. Alaska Sea Grant Rep. 90-04: 383–395.

Smith, B. D. & G. S.Jamieson, 1989. Growth of male and female Dungeness crabs near Tofino, British Columbia. Trans. am. Fish. Soc. 118: 556–563.

Stricker, S. A., 1986. An ultrastructural study of oogenesis, fertilization, and egg laying in a nemertean ectosymbiont of crabs, *Carcinonemertes epialti* (Nemertea: Hoplonemertea). Can. J. Zool. 64: 1256–1269.

Stricker, S. & C. G. Reed, 1981. Larval morphology of the nemertean *Carcinonemertes epialti* (Nemertea: Hoplonemertea). J. Morph. 169: 61–70.

Sundberg, P., 1990. Gibson's reclassification of the enoplan Nemerteans (Enopla, Nemertea): a critique and cladistic analysis. Zool. Scripta 19: 133–140.

Templeman, W., 1940. Embryonic developmental rates and egg-laying of Canadian lobsters. J. Fish. Res. Bd Can. 5: 71–83.

Uhazy, L. S., D. E. Aiken & A. Campbell, 1985. Morphology and systematics of the nemertean *Pseudocarcinonemertes homari* (Hoplonemertea: Monostilifera) from the American lobster, *Homarus americanus*. Can. J. Fish. aquat. Sci. 42: 342–350.

Van Beneden, P. J., 1861. Recherches sur la faune littorale de Belgique. Turbellariés. Mem. Acad. Roy. Soc. Belg. 32: 1–56.

Wickham, D. E., 1979. Predation by the nemertean *carcinonemertes errans* on eggs of the Dungeness crab, *Cancer magister*. Mar. Biol. 55: 45–53.

Wickham, D. E., 1978. A new species of *Carcinonemertes* (Nemertea: Carcinonemertidae) with notes on the genus from the pacific coast. Proc. Biol. Soc. Wash. 91: 197–202.

Wickham, D. E., 1980. Aspects of the life history of *Carcinonemertes errans* (Nemertea: Carcinonemertidae), an egg predator of the crab *Cancer magister*. Biol. Bull. 159: 247–257.

Wickham, D. E., 1986. Epizootic infestations by nemertean brood parasites on commercially important crustaceans. Can J. Fish aquat. Sci. 43: 2295–2302.

Wickham, D. E. & A. M. Kuris, 1985. The comparative ecology of nemertean egg predators. Am. Zool. 25: 127–134.

Wickham, D. E. & A. M. Kuris, 1988. Diversity among nemertean egg predators of decapod crustaceans. Hydrobiologia 156: 23–30.

Wickham, D. E. & A. Kuris, 1990. Brood mortality estimation and role of symbionts in egg mortality of the tanner crab *Chionoecetes bairdi* in Alaskan waters. Proc. Int. King Tanner Crab Symp. Lowell Wakefield Symp. Ser. Alaska Sea Grant Rep. 9004: 397–402.

Wickham, D. E. & P. Roe, 1987. Selectivity in transmission to crab hosts by the symbiotic nemertean, *Carcinonemertes epialti*. J. Parasit. 73: 697–701.

Wickham, D. E., P. Roe & A. M. Kuris, 1984. Transfer of nemertean egg predators during host molting and copulation. Biol. Bull. 167: 331–338.

Williams, A. B., 1984. Shrimps, Lobsters, and Crabs of the Atlantic Coast of the Eastern United States, Maine to Florida. Smithsonian Institution Press, Washington, 550 pp.

Hydrobiologia **266**: 15–28, 1993.
R. Gibson, J. Moore & P. Sundberg (eds), *Advances in Nemertean Biology*

Nemertea inhabiting the *Haploops* (Amphipoda) community of the northern Øresund with special reference to the biology of *Nipponnemertes pulcher* (Hoplonemertea)

John J. McDermott
Department of Biology, Franklin and Marshall College, Lancaster, Pennsylvania 17604, USA

Key words: ecology, feeding experiments

Abstract

The densities of nemerteans and associated fauna on a soft-bottom sampling station (27–30 m deep) in the Øresund were determined from 47 cores (each 135 cm^2 in cross-section; 20 + cm deep) collected from September to December 1989; these data were compared with 14 cores taken from the same location in December 1982. Nine species of nemerteans were identified from cores and dredge samples: Palaeonemertea – *Callinera*-like sp.; Heteronemertea – *Cerebratulus fuscus*, *C. marginatus*, *Lineus bilineatus*, *Micrura fasciolata*, *M. purpurea*; Hoplonemertea – *Amphiporus bioculatus*, *A. dissimulans*, *Nipponnemertes pulcher*. Mean numbers of heteronemerteans were 32 and 10 m^{-2} in 1982 and 1989, respectively, and hoplonemerteans were 90 and 71 m^{-2}, respectively. Only one palaeonemertean was collected during both years. Mean densities of the dominant species, *N. pulcher*, were similar for the two years, 74 and 68 m^{-2}. The dominant groups of macrofauna (n m^{-2}) in 1989 were ostracods (1028), amphipods (618), polychaetes (514), and ophiuroids (449). Amphipods (> 90% *Haploops* spp.) and polychaetes (at least 30 spp.) are the major potential prey for hoplonemerteans and heteronemerteans, respectively. Laboratory feeding experiments with *N. pulcher* revealed that it consumed amphipods (*Haploops tenuis* and *H. tubicola*) at a rate of 2.6 $worm^{-1}$ d^{-1} during the first 12 hours, but after 36 hours and beyond the rate was maintained at approximately 0.2 $worm^{-1}$ d^{-1}. Beyond 12 hours this nemertean showed a tendancy to only partially evacuate its prey. It was demonstrated experimentally that *N. pulcher* has a supply of toxin capable of killing six amphipods in approximately one hour. Limited tests showed that *N. pulcher* fed on the cumacean *Diastylis tumida*, but not on the amphipod *Maera loveni* or the ostracod *Philomedes globosus*, and that *Amphiporus dissimulans* readily attacked *Haploops* spp., but not *Maera* or *Philomedes*. Although the results of laboratory experiments are tentative, they do suggest that suctorial hoplonemerteans can exert a *potentially* significant effect on benthic communities. Employing seven species of polychaetes as prey for *Cerebratulus fuscus* and *Micrura fasciolata*, only the latter responded positively to one of them, *Glycera alba*. The hermit crab *Pagurus bernhardus* violently rejected *N. pulcher* in all feeding trials.

Introduction

Although nemerteans are usually observed in most marine benthic communities, they are seldom identified and are often lumped as 'Nemertea', especially in preserved material. Under these conditions worms lose their normal shape, color or color patterns, and unless there is recourse to histological sectioning, most hope for specific identifications is lost. Their trophic role as predators is usually specified, but components of the particular community affected by the worms are usually not known or may be stated incorrectly. Overall, community studies in which the roles of nemertean fauna are considered or defined are rare (McDermott & Roe, 1985, McDermott, 1988; Wilson, 1990). Notable exceptions, however, are the well-defined and extensive studies of Roe (1970, 1976, 1979) dealing with the macrophagous hoplonemertean *Paranemertes peregrina* and its influence on intertidal communities in the northwestern United States. This predator has an enormous effect on the structure and development of mud flat communities; e.g. she estimated that *Paranemertes* consumed 14–35% of the standing crop of *Platynereis canaliculata* (its preferred polychaete prey) per year. Studies by Bartsch (1973, 1975, 1977) in the Wadden Sea stressed the possible importance of the suctorial hoplonemertean *Tetrastemma melanocephalum* as one of the regulators of *Corophium volutator* populations.

Other less specifically oriented studies on nemertean interactions are those of Reise (1985b) and Nordhausen (1988) who employed exclusion cages on intertidal flats in the northern Wadden Sea to evaluate predation by species of the heteronemertean *Lineus*. Polychaetes were their primary prey.

Data presented by McDermott (1984) dealing with the soft-bottom *Haploops* community in the northern Øresund suggested that the suctorial hoplonemertean *Nipponnemertes pulcher* may be instrumental in the regulation of at least the dominant *Haploops* species (*tenuis* and *tubicola*), and perhaps some of the other less abundant species of amphipods. The roles played by other less common nemerteans in the same community were poorly defined.

The purpose of the present study is to further evaluate the predation potential of *N. pulcher* and some of the other species in laboratory experiments, and to relate such information to quantitative data on the abundance of nemerteans and their potential prey in the soft-bottom community of the Øresund.

Material and methods

Benthic sampling was restricted to an area of the Øresund (between Denmark and Sweden) north of the Island of Ven and ±8 km SE of Helsingør (between the latitudes 55° 58.9′ N and 55° 57.5′ N; longitudes 12° 41.1′ E and 12° 42.5′ E). This encompasses the same area that I sampled in 1982 (McDermott, 1984) and also the area being studied for its primary production and the fate of this production (Christensen & Kanneworff, 1985, 1986; Kanneworff & Christensen, 1986). The region has been characterized as a *Haploops* community because of the somewhat dominant or conspicuous tubicolous amphipods, *Haploops tenuis* and *H. tubicola*. The sampling depth ranged from 27–30 m; salinity of the bottom water was ±32‰ from August through November 1989, but dropped to an unusually low 25‰ in the beginning of December. The water temperature ranged from 8 °C in early August to an uncommonly low 6 °C in early December. This abnormal salinity and temperature was the result of wind-driven vertical mixing of the surface brackish water from the Baltic Sea with the deep North Sea water. Primary production in the bottom water is negligible because the maximun compensation depth is only 18–20 m (Christensen & Kanneworff, 1985). Thus many of the macrobenthic animals depend upon phytoplankton sedimenting from the upper layers (Christensen & Kanneworff, 1985, 1986).

Collections of nemerteans and other fauna were made from R/V 'Ophelia' employing a benthic dredge with a rectangular mouth and the 'Haps' bottom corer specifically designed for quantitative sampling of soft-sediment (Kanneworff & Nicolaisen, 1973). Dredged material (sandy mud, only the top 5 cm of which was aerobic) after being partially washed in a 2 mm screen aboard the vessel, was brought back to the laboratory in large tubs covered with seawater. Macrofauna were removed for identification and for use in feeding experiments. As the water stagnated in the tubs, the nemerteans, in particular, crawled up the sides to the water surface from which they were removed and maintained in dishes of seawater.

Forty-seven 'Haps' cores (135 cm^2 in cross-sectional area; 20 + cm deep) were taken at random on five separate dates (7-9-89, 10 cores; 29-9-89, 8 cores; 2-11-89, 9 cores; 16-11-89, 10 cores; 14-12-89, 10 cores). Cores, within their plastic tubes (each covered by ± 5 cm of seawater), were transported to the laboratory where they were maintained at room temperature (± 18–20 °C), and animals were removed periodically over a 24 h period as they moved out of the sediments. Particular attention was given to the nemerteans which were separated from other animals for identification and measurement. Finally the top 5 cm of each core was removed and passed through a 1.3 mm sieve from which all animals were removed and preserved in 10% seawater formalin. The lower anaerobic sediments were checked for signs of additional animals, but were always negative. Material passing through the sieve was also checked periodically by straining it through a 0.5 mm sieve. Only occasional minute juvenile polychaetes and small specimens of the ostracod *Philomedes globosus* were found, but neither was added to the totals. The contents of each core were analyzed separately so that the variation for each sampling date could be appreciated. All polychaete and amphipod tubes in the cores were opened and examined for the normal inhabitants or for secondary occupants such as *Nipponnemertes pulcher*, which is often found in uninhabited *Haploops* tubes.

Following identification and tabulation, animals from each collection in each broad faunal group (e.g. polychaetes, bivalves, etc.) were pooled, drained, blotted and weighed. Nemertean species were measured (length of relaxed moving worms) and weighed individually on an analytical balance. The two species of *Haploops* (Kanneworf, 1966) were measured to the nearest mm with an ocular micrometer (length – head to end of telson) and their pooled wet weight was determined separately from other amphipods in the collection.

Laboratory observations on the predation of *Haploops* spp. by *N. pulcher* were expanded beyond those made several years ago (McDermott, 1984). My primary concern was to verify previous data on predation rates which were based on individual tests of paired prey and predator in intimate association in small dishes; after each amphipod was killed and consumed another was added until termination of the test. It was thought that a more realistic approach to the problem would be to place many amphipods together with several worms in a large container and to monitor predation over time. The numbers of prey and predators in each aquarium would be somewhat in proportion to that of natural densities calculated from quantitative analyses of 14 'Haps' cores taken in 1982 (McDermott, 1984). In each trial, five nemerteans were placed with 100 *Haploops* spp. and the number of the latter killed was determined at 1, 4 and 12 h and at 12 h intervals thereafter for a total of two or more days. Four experiments, each with two replications, were run on different dates with forty different worms. Control aquaria were not necessary because amphipods killed by worms are readily distinguished from those that die naturally (the latter were few in number). Nemerteans were measured and weighed (wet weight) after the experiment, and predated amphipods were identified to species and measured. The difference in the two species of *Haploops* is so subtle that attempts to identify each specimen prior to starting a trial would damage the animals; furthermore, experience in 1982 indicated that worms showed no predilection for one or the other.

Each aquarium was rectangular, 22.5 cm long × 21.5 cm wide × 6.5 cm depth of water. Thus, animals occupied an area of ± 484 cm^2 and a volume of ± 3144 cm^3. Each aquarium was provided with running (± 400 ml min^{-1}) seawater (± 32‰) from the Laboratory's recirculating system; water temperature was 12–13 °C.

A variety of additional laboratory experiments, which attempted to induce feeding in hoplonemerteans and heteronemerteans, were conducted in small vessels containing non-circulating seawater.

It was shown previously that the teleost fish *Myxocephalus scorpius* rejected *N. pulcher* presented to it as food (McDermott, 1984). A series of tests were run with *N. pulcher* and the hermit crab *Pagurus bernhardus* L. to deterrnine if it reacts similarly to the fish.

Regression equations and correlation coefficients (r) for weight – weight and length – weight relationships in *N. pulcher* and *Haploops* spp., to be referred to later in the text, are as follows (length, mm; weight, mg):

Nipponnemertes pulcher		
dry wt = − 0.295 + 0.261 (wet wt)	$r = 0.96$	$N = 93$
proboscis wet wt = 0.588 + 0.051 (total wet wt)	$r = 0.80$	$N = 73$
log wet wt = − 1.397 + 2.188 (log length)	$r = 0.87$	$N = 92$
log dry wt = − 2.114 + 2.272 (log length)	$r = 0.85$	$N = 88$
Haploops spp.		
dry wt = 0.179 + 0.160 (wet wt)	$r = 0.95$	$N = 52$
log wet wt = − 1.233 + 2.548 (log length)	$r = 0.95$	$N = 52$
log dry wt = − 1.364 + 1.821 (log length)	$r = 0.82$	$N = 52$

Results

Benthic studies

Quantitative information from the 47 benthic cores taken in the Øresund in 1989 and a list of species identified from the region are given in Tables 1 and 2, respectively.

Among the five species of heteronemerteans (Table 2), the only relatively common species were *Cerebratulus fuscus* ($\bar{x}$ length 29.3 ± 9.3 mm, $n = 3$) and *Micrura fasciolata* (length 28 mm), both of which were found frequently in dredged material, but infrequently in cores (two cores and one core, respectively) (Table 3). The others were recovered only in dredged samples. *Cerebratulus marginatus* was found in 1989 but not in 1982, whereas the reverse was true for *Lineus bilineatus* and *Micrura purpurea*. Heteronemerteans averaged 10 m^{-2} in 1989 as compared to 32 m^{-2} in the 14 cores analyzed in 1982.

Of the three species of hoplonemerteans that occur at the sampling site (Table 3), only *N. pulcher* was identified from the 1989 cores, and was readily available in dredged samples. The two species of *Amphiporus* were seen only infrequently in dredged material, but were found in a few cores in 1982. Two of the 45 hoplonemerteans recovered from the cores could not be identified with certainty, but one appeared to be *N. pulcher* and the other *Amphiporus*. Thus there were at least 43 *N. pulcher* in 47 cores, or 0.92 $core^{-1}$, giving 68 m^{-2}; $\bar{x}$ length = 13.2 ± 5.3 mm, range = 5–25 mm, $n = 38$. These data are surprisingly similar to those acquired in the more limited sampling of 1982, 1.00 $core^{-1}$ and 74 m^{-2}. The total numbers of *N. pulcher* in each of the five series of cores ranged from 5 to 13. Therefore, I suggest that ± 70 *N. pulcher* m^{-2} may be a reasonable figure for later discussion of the potential impact by this dominant nemertean on prey within the *Haploops* community. Functionally, however, the three species of hoplonemerteans might be considered as a unit because they are all suctorial predators of crustaceans.

Table 1. Analysis of major taxa in 47 'Haps' corer samples taken from the *Haploops* community of the Øresund, September to December 1989. Molluscan shells included in weights, but *Pectinaria* tubes not included.

Taxon	Number		Wet weight (g)	
	In cores	Per m^{-2}	In cores	Per m^{-2}
Anthozoa	4	6	0.02	0.03
Polycladida	3	5	0.05	0.08
Palaeonemertea	1	2	<0.01	<0.01
Heteronemertea	6	10	0.17	0.27
Hoplonemertea	45	71	0.48	0.75
Polychaeta	325	512	9.30	14.65
Hirudinea	2	3	0.02	0.03
Gastropoda	14	22	1.26	1.99
Aplacophora	1	2	0.02	0.04
Bivalvia	109	172	13.40*	21.12
Pycnogonida	1	2	<0.01	<0.01
Ostracoda	652	1028	1.17	1.85
Cumacea	19	30	0.03	0.05
Mysidacea	1	2	<0.01	<0.01
Amphipoda	392	618	2.33	3.68
Ophiuroidea	285	449	14.75	23.25
Asteroidea	2	3	13.84**	21.81
Echinoidea	1	2	2.53	3.98
Enteropneusta	15	23	1.62	2.55
Tunicata	1	2	<0.01	<0.01

* Includes a 7.6 g scallop (*Pseudamussinum*).
** Includes a 13.5 g seastar (*Asterias*).

Table 2. Check-list of marine fauna found in the Øresund in 'Haps' cores and dredge samples (27–30 m) taken in 1982 (McDermott, 1984) and 1989. Species not found in the cores are indicated with an asterisk.

Coelenterata
- cerianthid anemone
- **Metridium senile*

Platyhelminthes
- Turbellaria
  - polyclad

Nemertea
- Palaeonemertea
  - *Callinera*-like sp.
- Heteronemertea
  - *Cerebratulus fuscus* (McIntosh)
  - **Cerebratulus marginatus* Renier
  - **Lineus bilineatus* (Renier)
  - *Micrura fasciolata* Ehrenberg
  - **Micrura purpurea* (Dalyell)
- Hoplonemertea
  - *Amphiporus bioculatus* McIntosh
  - *Amphiporus dissimulans* Riches
  - *Nipponnemertes pulcher* (Johnston)

Priapulida
- **Priapulus caudatus* Lamarck

Sipuncula
- *Phascolion strombi* (Montagu)

Annelida
- Polychaeta
  - *Amphicteis gunneri* (Sars)
  - *Amphitrite cirrata* Muller
  - *Anobothrus gracilis* Malmgren
  - *Aphrodita aculeata* L.
  - *Artacama proboscidea* Malmgren
  - **Eumida sanguinea* (Oersted)
  - *Glycera alba* (Muller)
  - *Goniada maculata* Oersted
  - *Harmothoe* sp.
  - **Hydroides norvegica* Gunnerus
  - *Lumbrinereis fragilis* (Muller)
  - *Maldane sarsi* Malmgren
  - *Mysta barbata* Malmgren
  - *Myxicola infundibulum* (Renier)
  - *Nephtys* sp.
  - *Nicomache lumbricalis* (Fabricius)
  - *Ophelina acuminata* Oersted
  - *Ophiodromus flexuosus* (Delle Chiaje)
  - *Pectinaria auricoma* (Muller)
  - **Pectinaria belgica* (Pallas)
  - *Pherusa plumosa* (Muller)
  - Pista cristata (Muller)
  - *Polyphysia crassa* (Oersted)
  - *Sabella penicillus* L.
  - *Schistomeringos* sp.

Table 2. (Continued)

  - *Sige fusigera* Malmgren
  - *Sphaerodorum gracilis* (Rathke)
  - *Spiophanes kroyeri* Grube
  - *Terebellides stroemi* Sars
  - *Trochochaeta multisetosa* (Oersted)
- Hirudinea
  - piscicolid leech

Mollusca
- Gastropoda
  - **Aporrhais pespelecani* (L.)
  - *Buccinum undatum* L.
  - **Lunatia pallida* (Broderip & G. B. Sowerby I)
  - *Neptunea antiqua* (L.)
  - *Odostomia turrita* Hanley
  - *Oenopota travelliana* (Turton)
  - **Philine aperta* (L.)
  - *Vitreolina philippi* (Rayneval & Ponzi)
- Aplacophora
  - *Chaetoderma nitidulum* Loven
- Bivalvia
  - *Abra nitida* (Muller)
  - **Aequipecten opercularis* (L.)
  - *Arctica islandica* (L.)
  - *Ceratoderma ovale* (G. B. Sowerby II)
  - **Hiatella* sp.
  - *Modiolus modiolus* (L.)
  - *Musculus niger* (J. E. Gray)
  - *Mya truncata* L.
  - *Nuculana minuta* (Muller)
  - **Nuculana permula* (Muller)
  - *Pseudamussium septemradiatum* (Muller)
  - *Thyasira flexuosa* (Montagu)

Arthropoda
- Pycnogonida
  - *Nymphon rubrum* Hodge
- Ostracoda
  - *Philomedes globosus* Lilljeborg
- Cumacea
  - *Diastylis rathkei* Kroyer
  - **Diastylis tumida* Lilljeborg
  - *Leptostylis ampullacea* Lilljeborg
  - *Leucon nasicoides* Lilljeborg
- Amphipoda
  - *Ampelisca spinipes* Boeck
  - *Dulichia porrecta* Bate
  - *Haploops tenuis* Kanneworff
  - *Haploops tubicola* Lilljeborg
  - *Maera loveni* Bruzelius
  - *Phthisica marina* Slabber
  - Photidae - 2 spp.
- Decapoda
  - **Pagurus bemhardus* L.

Echinodermata

Table 2. (Continued)

Ophiuroidea
Amphiura chiajei Forbes
Amphiura filiformis (Muller)
Ophiocomina nigra (Abildgaard)
Ophiopholis aculeata (L.)
Ophiura albida Forbes
Ophiura robusta Ayres
Asteroidea
Asterias rubens L.
Echinoidea
Echinocardium cordatum (Pennant)
Psammechinus miliaris (Gmelin)
**Strongylocentrotus droebachiensis* (Muller)
Holothuroidea
**Psolus phantapus* (Strussenfelt)
Hemichordata
Harrimania kupfferi Willemoes-Suhm

The following taxonomic references were used in some of the above identifications: crustaceans Enckell (1980); polychaetes Fauchald (1977); nemerteans Gibson (1982); Hartmann-Schröder (1971); molluscs Høisaeter (1986).

The major potential food sources for heteronemerteans are probably polychaetes and bivalves (McDermott & Roe, 1985). Polychaetes were relatively numerous (Table 1) and certainly the most diverse (Table 2). The numbers of carnivorous versus non-carnivorous species of polychaetes are 11 and 19, respectively (this trophic categorization is based on Fauchald & Jumars (1979) and my own experience); the non-carnivores were at least 2.4 times more numerous in the cores. Bivalves were much less diverse, and *Nuculana* spp. were the most numerous (55% of all species recovered from the cores). The marine hoplonemerteans are crustacean feeders (McDermott & Roe, 1985), and it appears that amphipods, being relatively numerous but not very diverse (Table 2), are the major potential prey for this group. *Haploops* spp. outnumbered all others by about 12 to 1 (361 to 31 individuals in the 47 cores), a distinct decrease from 1982 when the ratio was 46 to 1 (369 to 8 in the 14 cores). This decrease in amphipods, especially *Haploops* spp., appears to have developed in the Øresund in the last few years at least partially as a result of prevailing low dissolved oxygen levels in the bottom waters (Hagerman & Baden, 1988; E. Kanneworff, pers.

Table 3. Summary of nemerteans taken from the *Haploops* community in the Øresund, employing the 'Haps' corer; comparison of 14 cores from 1982 (McDermott, 1984) with 47 from 1989.

Species	1982		1989	
	No.	No. m^{-2}	No.	No. m^{-2}
Palaeonemertea				
Callinera – like sp.			1	1.6
Heteronemertea				
Cerebratulus fuscus	1	5.3	4	6.3
Micrura fasciolata			1	1.6
unidentified	5	26.5	1	1.6
Hoplonemertea				
Amphiporus bioculatus	2	10.6		
Amphiporus dissimulans	1	5.3		
Nipponnemertes pulcher	14	74.1	43	67.8
unidentified			2*	3.2
Totals	23	121.7	52	82.0

* Damaged, but one was probably *N. pulcher* and other an *Amphiporus* sp.
Other heteronemerteans collected in the same locality but not in the 'Haps' cores: *Cerebratulus marginatus*, *Lineus bilineatus* and *Micrura purpurea*.

comm.)The numbers of the most abundant crustacean, *Philomedes globosus*, were about the same as in 1982, and the cumacean fauna was not common or diverse.

Overall, the macrofaunal species diversity (91) (Table 2) and the abundance of individuals (approximately 3000 m^{-2}) in this deep-water region of the Øresund are moderate compared to a variety of other soft-bottom marine habitats.

Laboratory feeding experiments

Studies with Nipponnemertes pulcher

In a series of four similar 48 h experiments, each with two replicates, the mean feeding rate (prey killed $worm^{-1}$ d^{-1}) of starved *Nipponnemertes pulcher* during the first 24 h was more than 4 times the rates during the next 24 h (Table 4). The mean lengths and wet weights of worms and mean length of amphipods for each replicate in these experiments are given in Table 5. In two of these experiments which were continued beyond

Table 4. Rates of predation by *Nipponnemertes pulcher* (number prey killed worm^{-1} d^{-1}) at 12-hour intervals in laboratory experiments with *Haploops* spp. as prey; four experiments, each replication with 5 worms and 100 amphipods; x̄ (S.D.) given for each time interval.

Expt. no.	Replication no.	Predation at each of four 12 h intervals							
		1		2		3		4	
		No. killed	Rate	No. killed	Rate	No. killed	Rate	No. killed	Rate
1	1	5	2.0	2	0.8	2	0.8	0	0.0
	2	7	2.8	1	0.4	0	0.0	1	0.4
2	1	7	2.8	2	0.8	3	1.2	0	0.0
	2	8	3.2	2	0.8	1	0.4	1	0.4
3	1	9	3.6	0	0.0	1	0.4	0	0.0
	2	5	2.0	0	0.0	0	0.0	0	0.0
4	1	4	1.6	1	0.4	3	1.2	2	0.8
	2	7	2.8	2	0.8	0	0.0	0	0.0
	8	52	x̄ 2.60 (0.68)	10	x̄ 0.50 (0.35)	10	x̄ 0.50 (0.51)	4	x̄ 0.20 (0.30)

48 h (up to 192 h) a mean feeding rate of 0.21 worm^{-1} d^{-1} was maintained. Two other tests, employing 10 worms in an aquarium of the same dimensions with 100 amphipods and run for 72 h and 144 h beyond the 48 h period, gave rates of 0.23 and 0.22 worm^{-1} d^{-1}, respectively. Thus it appears that, in the laboratory, starved *N. pulcher* will at first feed on *Haploops* spp. at a high rate and then the feeding stabilizes to a rate of ± 0.2 amphipods worm^{-1} d^{-1}.

During the first 12 h of the above-mentioned four experiments (Table 4), when the predation

Table 5. Mean length and wet weight (S.D.) of *Nipponnemertes pulcher* and the length (S.D.) of *Haploops* spp. killed by this nemertean in the feeding rate experiments (Table 4).

Expt. no.	Replication no.	Worms*		Prey**		
		Length mm	Wet wt mg	Number***	Length mm	Range mm
1	1	18.0 (2.8)	24.8 (9.4)	9	7.32 (1.29)	5.21–9.07
	2	17.8 (1.9)	28.5 (6.0)	9	8.01 (1.30)	5.88–9.58
2	1	22.2 (1.8)	42.2 (16.9)	13	7.66 (1.00)	5.46–8.90
	2	21.0 (2.6)	35.2 (10.6)	13	7.41 (1.26)	5.38–10.00
3	1	21.6 (0.9)	40.1 (4.1)	10	8.18 (1.32)	6.13–10.08
	2	21.2 (1.1)	36.8 (6.3)	5	7.17 (0.80)	6.47–8.40
4	1	22.6 (4.6)	36.4 (12.3)	13	6.42 (1.80)	3.11–9.24
	2	22.2 (3.6)	36.9 (8.7)	19	6.03 (1.63)	3.02–8.90

* 5 worms in each group.
** *H. tenuis* and *H. tubicola* combined.
*** The numbers here also include those amphipods killed beyond the 48-hour time limit given in Table 4.

rate was high, 60% of the worms completely evacuated the exoskeletons of amphipods, while only 29% fed partially and 11% apparently did not feed after killing the prey (Table 6). During the next 36 h, as the feeding rate decreased, complete evacuation occurred in 33.3% of the cases while partial and no apparent feeding occurred in 66.6% of the cases. In other experiments (not presented here) a similar trend was also usually seen.

On a few occasions *N. pulcher* was tested with other potential prey species, *i.e.* sympatric crustaceans (Table 2). Two tests with the large burrowing amphipod *Maera loveni*, 21 and 22 mm long, were negative with starved worms of 20 and 25 mm length, respectively. One test with the cumacean *Diastylis tumida* (12.5 mm long, rostrum to telson) was positive; the prey was killed and evacuated by a 28 mm worm at 12 °C. A common crustacean in the soft-bottom community of the Øresund is the relatively large ostracod *Philomedes globosus* (see benthic data). Two *N. pulcher*, each 18 mm long, were placed in a standard (100 mm dia) petri dish with 10 *Philomedes* ($\bar{x}$ length 1.99 ± 0.28 mm). The dish was kept at 12 °C and the seawater was changed daily. Another dish with 9 ostracods ($\bar{x}$ length 2.02 ± 0.32 mm) served as a control. One control ostracod died on the fourth day, but there was no predation by the nemerteans during the eight-day observation period.

After *N. pulcher* strikes and kills an amphipod with toxin injected via its stylet, is there a necessary period of recovery for resynthesis of more toxin before it is able to kill another amphipod? Or is there sufficient toxin available to immediately kill additional prey? An experiment was performed in an attempt to resolve this problem. Two worms (both 16 mm in length), starved for 15 days, were placed in small (50 mm dia) glass evaporating dishes. One amphipod was then added to each dish. After a worm struck an amphipod, the latter was immediately and carefully dislodged from the worm with fine forceps and placed in a separate container of seawater to determine if it had been killed by the proboscis strike. A fresh amphipod was added at once, and the process was repeated until the worm could no longer be induced to strike an introduced prey. These animals were observed continuously at 3 or 10 × magnification. Amphipods were later preserved in 10% seawater formalin, then measured and the wet and dry weights determined. The nemerteans were also preserved and weighed.

Nemertean 1 (15.3 mg wet wt., 3.5 mg dry wt.) struck seven amphipods in 69 min, six of which were killed (Table 7). The fourth amphipod in the series survived because it was apparently not struck with a completely everted proboscis (noted

Table 6. The degree of feeding by *Nipponnemertes pulcher* on species of *Haploops* during 12-hour intervals of 48-hour experiments (Table 4).

Time after start (hours)	Number of amphipods killed	Feeding-number (%)		
		Complete	Partial	None
12	52	31 (60)	15 (29)	6 (11)
24	10	4 (40)	6 (60)	0
36	10	3 (30)	5 (50)	2 (20)
48	4	1 (25)	3 (75)	0
Totals	76	39	29	8

Table 7. Dimensions of *Haploops tenuis* attacked by *Nipponnemertes pulcher* (N = 2) in experiments to evaluate the toxin capacity of the nemertean. The amphipods are numbered as they were successively attacked by the nemerteans.

Amphipods attacked	Length mm	Weight mg	
		Wet	Dry
Worm 1			
1	6.30	7.4	0.6
2	5.88	7.3	0.9
3	7.06	11.8	0.9
4*	6.72	10.0	1.2
5**	6.64	11.4	0.6
6	6.80	11.6	2.2
7	6.22	7.4	1.1
$\bar{x}$ (S.D.)	6.48 (0.43)	9.5 (2.3)	1.1 (0.6)
Worm 2			
1	7.90	15.1	1.8
2*	5.88	7.3	0.9

* Amphipod attacked but not killed.
** Ovigerous amphipod.

under the microscope). All of the amphipods were *Haploops tenuis*, and there was nothing unique in the size or weight of amphipod no. 4 (Table 7). The second nemertean (16.2 mg wet wt., 3.7 mg dry wt.) struck two amphipods in 16 minutes. The second amphipod, which was struck dorsally was not killed. Amphipods are usually struck on their more vulnerable ventral side (McDermott, 1976, 1984). This worm did not attack additional amphipods.

Results with the first nemertean showed that the supply of toxin in the proboscis was sufficient in this case to kill six normal-sized amphipods in rapid succession. These six amphipods were equal to 3.7 × the worm's own body wet wt and 1.8 × its dry wt. As the toxin is contained only in the proboscis, the weight ratios may be expressed in another manner. The proboscis of this worm had a wet wt. of 1.4 mg, equivalent to 9.15% of the total body wt. The dry wt. of the proboscis then is $3.5\ \text{mg} \times 0.0915 = 0.32\ \text{mg}$. Thus this nemertean killed six amphipods, equivalent to 41 × and 20 × its proboscis wet wt. and dry wt., respectively, presumably with its stored complement of toxin that had little time for synthetic replenishment.

The sexes, lengths of the dorsal shield and wet weights of the four hermit crabs used to test their reactions to *N. pulcher* were as follows: F 9.9 mm, 5.4 g; F 12.0 mm, 10.8 g; M 14.0 mm, 19.8 g; M 16.8 mm, 39.1 g. Crabs were kept separately in standing seawater (12 °C) which was changed every other day. They were tested on three dates for a total of four trials. The procedure for each trial involved dropping the nemertean near the mouth of the crab, noting its reaction, presenting it with a fragment of a fresh polychaete five minutes later, noting the reaction, and then finally presenting it with the nemertean a second time. Fragments of *Lumbrinereis fragilis* were employed in three trials and *Glycera alba* in the other. In all cases the nemertean was rejected violently, whereas there was no such reaction with the polychaetes, and in 10 of 16 cases they were eaten. The behavioral reactions of the crabs to the nemertean were quite stereotyped. As the nemertean touched appendages such as the third maxillipeds or the chelipeds, the crab jerked backward rapidly by pushing forward with the thoracic appendages. Often the worm was grasped by the minor cheliped and then quickly flicked off followed by vigorous wiping of this appendage in the mouthparts, which in turn became particularly active. If the nemertean stuck to the minor claw the crab would try to dislodge it with the tips of the major cheliped. This appendage was then also rubbed with the mouthparts. Most of the external appendages (*i.e.* those outside of the shell), including the base of the antennae, appeared to be irritated if touched by the worm.

Studies with other nemerteans

Limited observations were made on feeding in nemerteans sympatric with *Nipponnemertes*. The two species of hoplonemerteans found previously in the same area in 1982, *Amphiporus bioculatus* and *A. dissimulans*, were present but still scarce. *A. dissimulans* did not attack *Maera loveni* or *Diastylis tumida* in the laboratory. Both species of *Amphiporus* will, however, attack and feed on *Haploops* spp., as was determined before (McDermott, 1984) and confirmed in the present study.

Among the heteronemerteans in the Øresund community, only *Micrura fasciolata* was relatively common (in the dredged samples). Specimens collected were 15–61 mm in length (11 collected 10 Aug. 1989 had an average length of 38 mm). What little information exists in the literature on feeding by *Micrura* spp. (McDermott & Roe, 1985) indicates that polychaetes might be their major prey. Specifically, Cantell (1975) found the remains of polynoid worms in the intestinal tract of *M. fasciolata*. I tested this species in the laboratory with the following seven species of polychaetes: *Anobothrus gracilis* (the most abundant species of polychaete in the community, about 42% of all species collected), *Artacama proboscidea*, *Glycera alba*, *Lumbrinereis fragilis*, *Ophelina acuminata*, *Sige fusigera* and *Sphaerodorum gracilis*. There was a positive reaction only to a 30 mm *Glycera* by a 61 mm worm. The polychaete was ingested head first for about two-thirds of it length, and found regurgitated the next day in a partly digested condition. Other *Glycera* presented to this and other worms were not ingested.

Cerebratulus fuscus (2 worms, 40 and 45 mm long), the next most common heteronemertean, was tested with *Amphitrite cirrata*, *Glycera* and *Lumbrinereis*, with negative results.

Discussion

Quantitative studies on the incidence and density of nemerteans in marine communities throughout the world are scarce. In the soft-bottom *Haploops* community of the deeper parts of the Øresund, Brunberg (1964) recorded about 16 species of nemerteans, but gave no quantitative information on their density. None of the species she listed was considered abundant except for *Nipponnemertes pulcher* which was characterized as occurring in 'great numbers'. At the turn of the century Lönnberg (1903) had essentially the same comment for this species. Petersen (1918) estimated 12 nemerteans m^{-2}, and although he did not identify the species, both Brunberg (1964) and I assumed that he was referring to *N. pulcher*. My preliminary estimate of the density of *N. pulcher* in the Øresund was 74 m^{-2}, which was based on a total 14 cores collected 7 and 10 December 1982 (McDermott, 1984). In the present analysis, a total of 47 cores taken on five separate dates from September to December 1989, gave a mean of 68 *N. pulcher* m^{-2}. This figure lends credence to the original estimate. In 1982 *Amphiporus bioculatus* and *A. dissimulans* were also found in the cores, and it was shown that both feed on *Haploops* in the laboratory. Thus when these species were added to the more numerous *N. pulcher*, the estimate became 90 suctorial hoplonemerteans m^{-2}. Including the two unidentified hoplonemerteans, from the 1989 cores, with *N. pulcher* gives 71 m^{-2}.

Unfortunately there are few studies of other hoplonemertean species with which to compare the densities mentioned above. Bartsch (1973) estimated a density of 116 *Tetrastemma melanocephalum* m^{-2} in the intertidal of the German Wadden Sea. Reise (1985b) recorded 19 m^{-2} for the same species and 6 m^{-2} for *Amphiporus lactifloreus* from the same region. I have shown that several species of suctorial hoplonemerteans living in eelgrass communities had a total density of 40 m^{-2} (mean from monthly sampling), but in June they were >300 m^{-2} (McDermott, 1988). Roe (1976) found that the macrophagous hoplonemertean *Paranemertes peregrina*, living on intertidal mud flats, varied in density over the year from 1 to 9 m^{-2}, but averaged about 3 m^{-2}; it showed peaks during periods of recruitment in the fall. These lower values compared to the suctorial species may be related to the much larger size of *Paranemertes* (adults mainly 100 to 300 mm in length and ranging up to >600 mm).

Brunberg (1964) found nine species of heteronemerteans in the Øresund, and I have recorded all except *Lineus albocinctus*, *Cerebratulus pantherinus* and *C. roseus*. She considered only *Micrura fasciolata* to be a widespread and abundant species, which I have confirmed. We still have poor information, however, on the densities of the heteronemerteans in the Øresund. Fourteen 'Haps' cores in 1982 contained 1 *C. fuscus* and 5 unidentified partial specimens (none of which was *M. fasciolata*), or 32 heteronemerteans m^{-2}. Of the 6 heteronemerteans recovered from the 47 cores in 1989, 4 were *C. fuscus*, 1 was *M. fasciolata* and the other was an unidentified fragment; this gives an estimate of only 10 m^{-2}.

Again, few studies are available with which to compare densities of heteronemerteans. Sanders *et al.* (1962) found *Micrura albida* and *M. leidyi* to have maximum densities of 4 m^{-2} and 7 m^{-2}, respectively, in the intertidal of Barnstable Harbor, Massachusetts. On a soft muddy subtidal bottom in upper Chesapeake Bay, Holland *et al.* (1977) recorded a maximum of 13 *Micrura leidyi* m^{-2}. Withers (1977) found an unidentified species of *Cerebratulus* from intertidal areas of Wales to have densities that ranged from 30–100 m^{-2}. Commito (1982) calculated approximately 4 *Cerebratulus lacteus* m^{-2} in a soft-bottom intertidal community in Maine. Reise (1985a) found *Lineus viridis* in the German Wadden Sea to have densities of 6 m^{-2} and 88 m^{-2} on an intertidal flat and eelgrass bed, respectively. Nordhausen (1988), who studied the same species in the same general area, found a mean density of 14 m^{-2} on sandy tidal flats and 38 m^{-2} in an eelgrass bed.

The experimental laboratory data on predation of *Haploops* spp. by *N. pulcher* suggested that a feeding rate of 0.2 amphipods $worm^{-1}$ d^{-1}, which occurred after the first 48 hours, may be a realistic figure to use in estimating potential predation in the natural environment. This value is lower than obtained in previous laboratory experiments employing isolated worms fed *Haploops* spp. (McDermott, 1984), but it confirmed that worms feed at a significantly reduced rate after the first 24 hours of more intensive feeding. If ± 70 suctorial hoplonemerteans m^{-2} consume a maximum of 0.2 amphipods $worm^{-1}$ d^{-1}, this gives 73 prey $worm^{-1}$ yr^{-1} or 5110 prey consumed m^{-2} yr^{-1}. This is only about 40% of the estimate suggested earlier employing 90 suctorial hoplonemerteans m^{-2} (McDermott, 1984). The present estimate is still far beyond the yearly production of *Haploops* spp. which appears to be equal to the biomass (E. Kanneworff, pers. comm.), *i.e.* the production may be equal to the standing crop of amphipods (361 *Haploops* in 47 cores = 570 m^{-2}). According to Wildish (1984), deep cold-water populations of *Haploops* spp. generally have relatively low P/B ratios. The nemertean feeding rate used here was derived from experiments employing relatively large worms (Table 5), as compared to the mean length of the worms collected in the 47 cores ($\bar{x} = 13$ mm, page 18); nearly half of the worms were below this mean, and most of these were in the 5–8 mm size range. Thus the 0.2 rate might only be realistic as a maximum value, and would have to be scaled down considerably to account for lower rates to be expected in worms with the overall smaller mean size found in the cores. To be compatible with the standing *Haploops* population and its presumed low P/B ratio, a value < 0.02 would have to be postulated. Another consideration is that the bottom water temperature in the Øresund for most of the year (Christensen & Kanneworff, 1985) is well below the 12–13 °C temperature used in the feeding experiments, and this presumably would affect on the feeding rates in nature. Although the results of laboratory experiments are tentative, they do indicate that nemerteans can exert a *potentially* significant effect on benthic communities.

Hoplonemertean predation may be directed to other species of amphipods which are relatively uncommon in the Øresund (49 m^{-2}), and possibly to the cumaceans, also not very abundant or diverse (30 m^{-2} in 1989 and 85 m^{-2} in 1982). One positive laboratory test with *Diastylis tumida* suggests that this group may be vulnerable to hoplonemerteans in nature. Negative observations suggest that the relatively numerous ostracod *Philomedes globosus* (Table 1) may not be a prey candidate. Even if attacked, it is unlikely that the nemertean stylet could penetrate the ostracod's hard valves. The meiofaunal harpacticoid copepods, which were not accounted for with my sampling methods, could possibly be a food source, especially for small developing worms. However, gross examination of sediment passed through a 0.5 mm screen did not reveal any indication of harpacticoids. It is possible that the 'Haps' corer disturbs the fine detritus surface layer and displaces some of the small crustaceans as suggested by Jensen (1983).

It was shown in earlier laboratory studies that some *N. pulcher* may consume up to 5 amphipods in 24 hours (McDermott, 1984), an extraordinary rate which demonstrated that enough toxin was available to the worm in a 24-hour period to kill this many prey. I presumed that after each proboscis strike and release of toxin into the prey, more toxin was synthesized in preparation for the next successful attack and feeding. The experiment presented here in which a nemertean was allowed to strike an amphipod but not feed on it, showed that the worm had enough toxin stored in its proboscis to kill six amphipods in slightly over one hour. Thus in nature, if for some reason a worm is unsuccessful in one or more strikes with its proboscis, it may still have sufficient toxin to kill an amphipod on subsequent strikes. That is, there is no need for a dormant period requiring resynthesis.

It is probable that only a miniscule amount of the toxin released from the anterior proboscis during a strike ever gets beyond the exoskeleton,

polychaetes in controlling the structure of a soft-bottom community in Maine, USA. Mar. Biol. 68: 77–81.

Commito, J. A. & W. G. Ambrose, Jr., 1985. Predatory infauna and trophic complexity in soft-bottom communities. In P. E. Gibbs (ed.), Proceedings of the 19th European Marine Biology Symposium. Cambridge University Press, Cambridge: 323–333.

Commito, J. A. & P. B. Shrader, 1985. Benthic community response to experimental additions of the polychaete *Nereis virens*. Mar. Biol. 86: 101–107.

Enckell, P. H., 1980. Kräftdjur. Bokförlaget Signum I, Lund. 685 pp.

Fauchald, K., 1977. The Polychaete Worms. Definitions and keys to orders, families and genera. Natural History Museum of Los Angeles County, California, 188 pp.

Fauchald, K. & P. A. Jumars, 1979. The diet of worms: a study of polychaete feeding guilds. Oceanogr. Mar. Biol. annu. Rev. 17: 193–284.

Gibson, R., 1982. British Nemerteans. Cambridge University Press, Cambridge, 212 pp.

Hagerman, L. & S. P. Baden, 1988. *Nephrops norvegicus*: field study of effects of oxygen deficiency on haemocyanin concentration. J. exp. mar. Biol. Ecol. 116: 135–142.

Hartmann-Schröder, G., 1971. Annelida, Borstenwurmer, Polychaeta. Volume 58. In Die Tierwelt Deutschlands, Fischer Verlag, Jena. 594 pp.

Holland, A. F., N. K. Mountford & J. A. Mihursky, 1977. Temporal variation in upper bay mesohaline benthic communities: I. The 9-m mud habitat. Chesapeake Sci. 18: 370–378.

Høisaeter, T., 1986. An annotated check-list of marine molluscs of the Norwegian coast and adjacent waters. Sarsia 71: 73–145.

Jennings, J. B. & R. Gibson, 1969. Observations on the nutrition of seven species of rhynchocoelan worms. Biol. Bull. 136: 405–433.

Jensen, P., 1983. Meiofaunal abundance and vertical zonation in a sublittoral soft bottom, with a test of the Haps corer. Mar. Biol. 74: 319–326.

Kanneworff, E., 1966. On some amphipod species of the genus *Haploops*, with special reference to *H. tubicola* Liljeborg and *H. tenuis* sp. nov. from the Øresund. Ophelia 3: 183–207.

Kanneworff, E. & H. Christensen, 1986. Benthic community respiration in relation to sedimentation of phytoplankton in the Øresund. Ophelia 26: 269–284.

Kanneworff, E. & W. Nicolaisen, 1973. The 'Haps' a frame-supported bottom corer. Ophelia 10: 119–129.

Kem, W. R., 1971. A study of the occurrence of anabaseine in *Paranemertes* and other nemertines. Toxicon 9: 23–32.

Kem, W. R., K. N. Scott & J. H. Duncan, 1976. Hoplonemertine worms – a new source of pyridine neurotoxins. Experientia 32: 684–686.

Lönnberg, E., 1903. Undersökningar rörande Skeldervikens och angränsande Kattegat-Områdes Djurlif. Meddn. K. Lantbr. Styr. 2: 1–70.

McDermott, J. J., 1976. Observations on the food and feeding behavior of estuarine nemertean worms belonging to the order Hoplonemertea. Biol. Bull. 150: 57–68.

McDermott, J. J., 1984. The feeding biology of *Nipponnemertes pulcher* (Johnston) (Hoplonemertea), with some ecological implications. Ophelia 23: 1–21.

McDermott, J. J., 1988. The role of hoplonemerteans in the ecology of seagrass communities. Hydrobiologia 156: 1–11.

McDermott, J. J. & P. Roe, 1985. Food, feeding behavior and feeding ecology of nemerteans. Am. Zool. 25: 113–125.

McDermott, J. J. & R. L. Snyder, 1988. Food and feeding behavior of the hoplonemertean *Oerstedia dorsalis*. Hydrobiologia 156: 47–51.

Nordhausen, W., 1988. Impact of the nemertean *Lineus viridis* on its polychaete prey on an intertidal sandflat. Hydrobiologia 156: 39–46.

Ockelmann, K. W. & O. Vahl, 1970. On the biology of the polychaete *Glycera alba*, especially its burrowing and feeding. Ophelia 8: 275–294.

Petersen, C. G. J., 1918. The sea bottom and its production of fish-food. Rep. Dan. Biol. Stn. 25: 1–62.

Reise, K., 1977. Predation pressure and community structure of an intertidal soft-bottom fauna. In B. F. Keegan, P. O'Ceidigh & P. J. S. Boaden (eds), Biology of Benthic Organisms. Pergamon Press, New York: 513–519.

Reise, K., 1985a. Predator control in marine tidal sediments. In P. E. Gibbs (ed.), Proceedings of the 19th European Marine Biology Symposium. Cambridge University Press, Cambridge: 311–321.

Reise, K., 1985b. Tidal Flat Ecology. An experimental approach to species interactions. Springer-Verlag, Berlin, 191 pp.

Riches, T. H., 1893. A list of the nemertines of Plymouth Sound. J. mar. biol. Ass. U.K. 3: 1–29.

Roe, P., 1970. The nutrition of *Paranemertes peregrina* (Rhynchocoela: Hoplonemertea). I. Studies on food and feeding behavior. Biol. Bull. 139: 80–91.

Roe, P., 1976. Life history and predator-prey interactions of the nemertean *Paranemertes peregrina* Coe. Biol. Bull. 150: 80–106.

Roe, P., 1979. A comparison of aspects of the biology of *Paranemertes peregrina* (Nemertea) from Bodega Harbor, California, and Washington State. Pacific Sci. 33: 281–287.

Sanders, H. L., E. M. Goudsmit, E. L. Mills & G. E. Hampson, 1962. A study of the intertidal fauna of Barnstable Harbor, Massachusetts. Limnol. Oceanogr. 7: 63–79.

Sundberg, P., 1979. *Tubulanus annulatus*, an aposematic nemertean? Biol. J. linn. Soc. 12: 177–179.

Wildish, D. J., 1984. Secondary production of four sublittoral, soft-sediment amphipod populations in the Bay of Fundy. Can. J. Zool. 62: 1027–1033.

Wilson, W. H., 1990. Competition and predation in marine soft-sediment communities. Annu. Rev. Ecol. Syst. 21: 221–241.

Withers, R. G., 1977. Soft-shore macrobenthos along the south-west coast of Wales. Estuarine coast. mar. Sci. 5: 467–484.

Hydrobiologia **266**: 29–44, 1993.
R. Gibson, J. Moore & P. Sundberg (eds), Advances in Nemertean Biology

Aspects of the biology of *Pantinonemertes californiensis*, a high intertidal nemertean

Pamela Roe
Department of Biological Sciences, California State University Stanislaus, Turlock, CA 95380, USA

Key words: semi-terrestrial, feeding, reproduction, population parameters

Abstract

I studied the distribution, feeding biology, and reproductive biology of *Pantinonemertes californiensis*, described as a semi-terrestrial nemertean, along the central California coast. At the sites used in this study, maximal tidal height is about 2 m, and *P. californiensis* typically occurred under boulders between 1.3 and 1.7 m tidal height. Worms fed primarily on the semi-terrestrial amphipod *Traskorchestia traskiana*. Distribution of nemerteans was similar to that of the prey, although prey extended higher on the beach than did the worms. Nemerteans were largest and most abundant at the site with highest abundance of *T. traskiana* and smallest and least abundant at the lowest prey abundance site. In laboratory feeding trials, nemerteans from the site with lowest prey abundance fed most readily. Non-reproductive nemerteans lived for at least a week when submerged in sea water; some prey died within a week of being submerged. Nemerteans only lived minutes when submerged in fresh water; 50% of prey lived 4.5 h. Eggs are approximately 90–100 μm in diameter and hundreds to thousands are shed per female. Larvae are planktonic and apparently planktotrophic, and are morphologically similar to other marine hoplonemertean larvae. At the sites studied life history characteristics of *P. californiensis* provided little evidence of adaptations to terrestrial life in these worms and were not helpful in elucidating the role of semi-terrestrial nemerteans in the evolution of terrestrial nemerteans.

Introduction

Most of the approximately 1400 (Gibson, pers. comm.) species of nemerteans are marine, occurring in both subtidal and intertidal habitats (Gibson, 1982a). About 20 terrestrial nemertean species have been described, primarily from oceanic islands (Moore & Gibson, 1981, 1985; Moore, 1985). Included with terrestrial nemerteans are a few species described as semi-terrestrial; these are considered transitional between marine and terrestrial species and have been postulated to be ancestral to terrestrial species (Moore & Gibson, 1981, 1985). Sundberg (1989) suggested, however, that terrestrial forms evolved directly from a marine form without passing through semi-terrestrial forms, and that upper littoral and marine environments have been recolonized by descendants of a terrestrial ancestor.

Among the species described as semi-terrestrial, *Pantinonemertes californiensis* is of special interest as a transitional species in the Group I series of Moore & Gibson (1981) (Gibson *et al.*, 1982), because it is in the only known genus with marine, transitional and terrestrial species (Moore, 1985). *Partinonemertes californiensis* has been found between Huntington Beach, south of Los Angeles, CA (Gibson *et al.*, 1982) north to

Argyle Lagoon on San Juan Island, Washington (Kozloff, pers. comm.). It occurs beneath rotting logs and stones in marshes or sandy intertidal areas 'where the ground is dampened by only an occasional extremely high tide, perhaps half a dozen times a year' (Gibson *et al.*, 1982). It is tolerant of a wide salinity range but if forced to remain submerged, worms rarely survive more than 24–36 hours (Gibson *et al.*, 1982). No extensive studies have been done on the biology of any species of semi-terrestrial nemertean. The present study was undertaken to elucidate aspects of the distribution, feeding and reproductive biology of *Pantinonemertes californiensis* so that such life history information would be available for at least one semi-terrestrial nemertean. In addition I wanted to determine whether life history parameters of a semi-terrestrial nemertean could be used to help elucidate pathways of evolution of terrestrial nemerteans.

Materials and methods

The study was conducted between July 1989 and July 1991 at three intertidal localities of the central California coast. One study site was south of Blakes Landing along the northeastern shore of Tomales Bay. The other two sites were at Moss Landing, California, east of the Highway 1 bridge over Elkhorn Slough and along the bank of Jetty Road in the North Harbor area of Elkhorn Slough.

Field studies were conducted at least each season. At field sites, water and air temperatures were measured and the area was searched for worms and associated organisms by turning over rocks.

Tidal height (as height above mean lower low water) of areas inhabited by *P. californiensis* was correlated with height of the tide if high water covered part of the worm zone while I was at the site. If water did not reach to the nemertean zone, height above water was measured, and later the height of the worm zone was calculated. Correction factors for Tomales Bay were calculated using U.S. Dept. Commerce Tide Tables (1989, 1990, 1991). Tidal heights at times of measurements of height above water of worm zones for Elkhorn Slough were determined by use of a computer simulation for tidal data at Moss Landing (Broenkow, pers. comm.).

In the laboratory worms were maintained individually or in groups in fingerbowls or small jars. Each container was supplied with a paper towel dampened with sea water and usually a small rock for worms to crawl onto. During the first part of the study worms were maintained at 8.5–12 °C but were later maintained at 16.5 °C.

Observations of worm activity at 'high' and 'low' tides were made using a small tide chamber in the laboratory. The chamber was made from two plastic bowls, the larger of which had a hole cut in the bottom into which fit a cork with a glass stand pipe. The smaller bowl had holes in the bottom, and fit within the larger dish. High tide was made by adding water to the larger bowl; the water would seep up into the mud in the small bowl and would soon cover the mud surface. Low tide occurred when the standpipe was removed from the outer bowl.

To determine whether worms seek out a particular amount of wetness of substratum, a galvanized steel tray 111 cm × 18.7 cm × 13.6 cm partly filled with sediment from Elkhorn Slough was used. Water was added to cover the sediment and 11 or 14 worms were added (two experiments, one in dark and one in light). One end of the tray was then inclined so that a moisture gradient was made over the sediment. Distribution of worms with respect to moisture gradient was recorded several hours after the gradient was established. Worms used in these experiments were post reproductive individuals that had been collected from Tomales Bay approximately one month prior to the experiments and that had spawned in the laboratory.

To determine whether *P. californiensis* could tolerate submersion in sea water, seven worms were held for one week at 8.5–11 °C and six worms were held at 16.5 °C for one week. In addition, a relatively mature male and two moderately mature females were held in pint jars under water for 3–4 days at 10 °C (one female) and

16.5 °C. Jars were filled with sea water; however a small air space was sometimes trapped under the lid of some jars even when they were filled with water. To determine whether worms are tolerant of fresh water, nine of the specimens that had been held under sea water were placed into pint jars containing fresh city well water from Turlock, California, after they had been removed from sea water and held on paper towels dampened with sea water for three days. Fresh water containers were kept at 16.5 °C. Specimens were wet weighed before and after submersion in both sea water and fresh water.

All feeding observations were made under laboratory conditions. Potential prey were held at the anterior edge of a worm. If the worm struck the prey with its proboscis the prey was considered to be a potential food species. If the worm continued the feeding process, the time it took for feeding to be completed was usually recorded. To determine whether associated fauna fed on *Pantinonemertes californiensis*, worms were held before shore crabs and glycerids. Responses of these animals was recorded.

During fall 1989 specimens were held in groups at 10–12 °C and were offered food once or twice/week over a several week time period. In spring, 1990, worms were held individually at 16.5 °C for varying time lengths (8–171 days). They were offered food approximately twice/week. Prey wet weight was determined before and after feeding and worm wet weight was determined weekly.

Relative prey densities in nature were determined by two methods: 1) the number of amphipods under each rock that was turned over within the worm habitat was counted and 2) transects 37 m in length (Tomales Bay) and 17 m length (Elkhorn Slough Bridge site) were laid parallel to beach contours above, at the top of, through the middle of (2) at the lower edge of, and below the worm habitat. Numbers of prey under the nearest rock, eel grass wrack or both, to the beginning of each of 10 random meters along each transect were counted. Areas counted were approximately 625 cm^2.

To determine whether amphipod prey could tolerate submersion in sea water, eight amphipods were held for one week at 8.5–10 °C and eight for one week at 16.5 °C in plastic jars 5 cm diameter by 4 cm height filled with sea water. To determine amphipod prey tolerance to fresh water, 10 specimens were placed into plastic jars filled with fresh Turlock city well water at 16.5 °C. Time to death was recorded for each amphipod.

In the field, nemerteans were counted, sexed (by observing appearance of gonads through the body wall) and measured (length and width), or collected and measured (length, width and sometimes weight) upon return to the laboratory. Only worms that could be seen when rocks or wrack were removed were counted. No sediment was sieved for possible buried worms.

Spawning was observed and induced in the laboratory and was induced in a few specimens in nature. Developing larvae were maintained in fingerbowls at temperatures between 10 °C and 23 °C, and even to 30 °C for short time periods. When larvae started dying, juice from ground up krill was added as nutrients to some cultures. In other cultures, approximately 2.5 ml of a qualitative mixture of glucose, isoleucine, leucine, lysine, methionine, phenylalanine, threonine, tryptophan and valine diluted in sea water was added to sea water in fingerbowls with larvae.

Data were statistically analyzed with the use of the statistical computer program Minitab.

Description of study sites

Worms were found under rocks imbedded in sand and gravel in high intertidal areas where there was at least some beach wrack vegetation.

A thick, wide band of washed-in *Zostera marina* carpeted the high intertidal to the lower edge of cliffs that marked the top of the intertidal at the Tomales Bay study site. The *Zostera* wrack varied in thickness and patchiness but was present throughout the study area and duration, and covered at least the rocks in the upper one-half of the worm zone every time I visited the area. *Pantinonemertes californiensis* was rarely observed at Tomales Bay in areas without this wrack, and occasionally was found crawling among the wetter leaves. The wrack kept the high intertidal cool and damp, and probably increased the effective height of moisture from high tide. Organisms associated with *P. californiensis* at the Tomales Bay site included the snails *Littorina* sp. and *Ovatella myosotis*, the centipede *Lionyx hedgpethi*, small red mites, a flatworm, an oligochaete, pseudoscorpions, the isopods *Armadilloniscus holmesi* (at the highest part of the worm zone) and *Ligia occidentalis*, the amphipod *Traskorchestia traskiana*, and shore crabs *Pachygrapsus crassipes* and *Hemigrapsus nudus*. The nemerteans *Lineus vegetus* and *Amphiporus* sp. were found barely below and within the lower *P. californiensis* zone, respectively.

At the Elkhorn Slough site east of Highway 1 bridge (called the Elkhorn Slough Bridge site) rocks were imbedded in a more sandy substratum, and vegetational wrack, comprised mainly of ulvoid algae, was usually patchy and often not present at all. (It did form a solid band in June 1991). Few associates were found with *P. californiensis*, including *T. traskiana* and a few specimens of the oligochaete and *Littorina* sp. The shore crab *P. crassipes* occurred throughout the worm zone and a species of isopod occurred at the highest edge of the worm zone. Under rocks at the lower edge of, and slightly below *P. californiensis*, the nemerteans *Emplectonema gracile* and *Nemertopsis gracilis* were fairly common, as were the shore crabs *H. oregonensis* and *H. nudus* and the glycerid polychaete *Hemipodus borealis*.

The study site in the North Harbor region of Elkhorn Slough (called the Jetty Road site) consisted of rocks on mud within patches of *Salicornia* and *Distichlis* that grew on the intertidal part of the road bank of Jetty Road. This area housed the fewest and smallest of the *P. californiensis* specimens and the fewest *T. traskiana* specimens of the 3 study sites. Red mites, oligochaetes, *P. crassipes*, dipteran larvae and a few centipedes were found under these rocks during the study. The snail *Batillaria attramentaria* was extremely abundant in and below the worm zone, and a few *Ovatella myosotis* were found at the upper edge of the worm zone. *Lineus vegetus* was found in the same zone as *P. californiensis* as were three specimens of an undescribed species of probably *Prosorhochmus*. Rocks were sometimes covered by a mat of *Enteromorpha* but vegetational cover was sparse and ephemeral.

During the Loma Prieta earthquake October 17, 1989, Jetty Road cracked and fell into the Slough where the road crossed a culvert that connected the North Harbor and Bennett Slough. The cave-in was approximately 27 meters from the original area of high worm density. Pressure from the settling road increased liquefaction and lateral spreading or slippage of the road bed and beach. The worm zone became a trench about a meter wide and deeper by 28 cm on the lower edge to 18 cm on the higher intertidal edge; the entire bank and surrounding area sank approximately 15 cm within seconds. Tensional cracks parallel to the road formed, which remained visible for at least five months. Mud lost its cohesiveness due to dewatering, many small rocks that had been buried, appeared on the mud surface, and *P. californiensis*, oligochaetes and other under-rock fauna associated with *P. californiensis* disappeared except higher on the road bank where rocks were not so disturbed. (Descriptions of causes of post-earthquake observations are from Greene (pers. comm.)). Even 20 months after the earthquake, the condition of few organisms under rocks pertained. Most rocks lower on the beach covered bare sand or an occasional tiny *H. oregonensis*, and few organisms were found in association with *P. californiensis* as well.

Water and air temperatures at the three study sites during times I did field sampling were 11.5 to 22.5 °C and 12 to 19 °C. respectively. Although I was not there at the time, sea ice formed in parts of Elkhorn Slough (Heine, pers. comm.) and all of Central California was below freezing for several days in December 1990.

Results

Tidal height and moisture gradient of nemertean distribution

At Tomales Bay *Pantinonemertes californiensis* occurred between + 1.22 m and + 1.58–1.68 m tidal height. Most individuals occurred close to 1.40 m. At Elkhorn Slough Bridge nemerteans ranged between 1.37 and 1.71 m. At Jetty Road, prior to the Loma Prieta earthquake, *P. californiensis* was found between 1.31 m and 1.49 m. After October 17, 1989, worms were found between 1.46 m and 19.2 m (one individual) with most concentrated between 1.62 m and 1.71 m. Of the three sites, *P. californiensis* was found at highest tidal elevation at Jetty Road. However, if that entire area subsided approximately 15 cm during the Loma Prieta earthquake, as suggested, then the animals at that site are not higher than those at the other sites. In summary, *P. californiensis* typically occurred between 1.3 and 1.7 m tidal height at all three study sites.

The highest tide of the year for 1989–1991 at the study sites was 1.98–2.04 m. At the lower edge of *P. californiensis*, *i.e.* 1.22 m, high water would cover worms nearly every day: at 1.40 m, the center of the zone at Tomales Bay, worms would be covered by water some time during the day 84% of the days of a year. At 1.68 m, the upper limit to Tomales Bay worms and within the upper part of the two Moss Landing populations, water would still cover worms approximately 84 days, or 23% of the days per year.

In laboratory experiments using the inclined metal tray under dark conditions, 8/11 individuals were distributed under water, 1/11 was partly in and partly out of water and 2/11 were under a rock in damp sand at the end of the experiment. In the experiment run with the tray exposed to indoor lights, 9/14 were under water, 2/14 were partly in and partly out of water, 3/14 were under the rock in damp sand and 2/14 were buried deep within the sediments at the end of the experiment. Several mucus trails seen on the sediment surface indicated that the nemerteans moved around extensively during both experiments. Results of these two experiments indicate that *P. californiensis* might actually prefer somewhat wetter conditions than those found at the tidal heights in which it occurs in nature.

Tolerance to physical conditions by Pantinonemertes californiensis

When placed in pint jars filled with sea water (32 ppt) from Elkhorn Slough, sexually immature individuals from Tomales Bay withstood submersion for at least one week. Two immature females, two immature males and three individuals whose sex could not be determined held one week in sea water at 8.5–10 °C all survived. One male was found on the jar lid with only a film of water surrounding him on one of the seven days, one immature individual was found on the jar lid three days, and one immature individual was found on the jar lid two days, and also broke into two pieces upon handling on the seventh day of the experiment. The front end continued to live many days post-experiment. One immature female, two immature males and three immature individuals held under sea water for one week at 16.5 °C also survived the seven day submersion. The female was found on the jar lid five days, two immatures and one male were found on the jar lid 1 day each.

Combining all individuals at both temperatures that were completely submerged seven full days, six survived and none died. Including all individuals that may have found slightly dryer conditions on lids of jars for varying time lengths, all 13 animals survived. In feeding trials, none ate, either under water or when removed from water and placed on paper towels dampened with sea water immediately after termination of submersion experiments, but all tested responded to presence of food normally. (One male broke into two pieces between the two feeding experiments, so was not tested for feeding in damp condition.) Wet weights before submersion and post submersion for 11 individuals averaged 121 mg and 132 mg, respectively.

One near-mature female anterior end placed in sea water for four days survived, at 10 °C. One moderately mature female in seawater for four days at 16.5 °C survived (and wet weight changed from 190 mg to 177 mg). This female broke into two pieces with handling at the end of the experiment, but the anterior end continued to live. One nearly mature male placed in sea water at 16.5 °C appeared moribund, with head bent to the side and body barely moving upon prodding, after submersion for three days, and his wet weight increased from 549 to 613 mg. He was removed from the sea water jar, placed on paper toweling dampened with sea water, and by the next day had recovered and wet weight was 542 mg.

All nine individuals from the above experiments, when submerged in fresh water (Turlock, city well water), died within minutes of submersion. After 1 1/2 hour all had produced a thick layer of mucus, and their bodies were swollen and fell apart very easily. Average weights increased from 85 to 240 mg.

Pantinonemertes californiensis was not tolerant of dry conditions. It fell apart easily if handled with dry forceps or if an individual was exposed to air from an overturned rock and was not dampened with a few drops of sea water before handling.

Nemertean densities

Pantinonemertes californiensis occurrence was highly patchy and variable within its zone (Table 1). It was usually found under rocks, but occasionally worms at Tomales Bay were seen crawling through the thick *Zostera* wrack. The number of worms found under individual rocks ranged from 0 (often) to 26 (once, Tomales Bay). Densities of *P. californiensis* at Tomales Bay were higher than at Elkhom Slough Bridge, and at Elkhom Slough Bridge than at Jetty Road. It is not known if most of the population can be seen on the sediment surface when a rock is removed at low tide, or if an unknown portion of worms were buried within sediments during low tides when sampling for this study occurred. When individuals of *P. calforniensis* placed in a tide chamber in the laboratory were covered with water to simulate high tide, most buried into the sediment, and returned to the surface at simulated low tide. These observations indicate that *P. californiensis* is active at high tide. When the number of worms under marked rocks at Tomales Bay were counted before high tides covered them and again after water had receded, 78% and 62% of the pre-wetting numbers remained after water receded. When the numbers of worms under or very near to marked rocks at Tomales Bay were compared over successive summer months, numbers were 68 (June), 40 (July) and 20 (August).

Table 1. Densities of *Pantinonemertes californiensis* and number of square meters sampled, over time, for three study sites.

Date	Density, number of M^2 sampled		
	Tomales Bay	Elkhorn Slough Bridge	Jetty Road
Jul 89	4.7, 3	–	1.8, 16
Aug 89	3.4, 10	–	–
Oct 89	2.8, 4	–	3.5, 2
Nov 89	–	0.82, 17	0.05, 40
Jan 90	1.2, 28	4.7, 3	0.8, 5
Mar 90	7.75, 4	1.6, 3	1.7, 10
Jun 90	4.5, 30	–	–
Jul 90	–	0.88, 17	–
Jun 91	–	–	0.82, 11

Laboratory observations on feeding

Individuals of *P. californiensis* are suctorial feeders, primarily on the amphipod *Traskorchestia traskiana*. They also readily feed on a variety of isopods. *Ligia occidentalis* was eaten readily by individuals from Jetty Road and rarely by Elkhorn Slough Bridge individuals. *Armadilloniscus holmesi* was readily accepted by Jetty Road individuals, and was accepted by one Tomales Bay worm on one occasion. Jetty Road individuals also accepted the isopod found occasionally at the upper edge of, and above, worms at Elkhorn Slough Bridge (3 of 6 trials) and a small isopod from San Francisco Bay (4 of 5 trials). The last two prey were not tested with Elkhorn Slough Bridge or Tomales Bay worms, nor was *A. holmesi* with Elkhorn Slough Bridge worms. No other animals found in association with *P. californiensis* was ever eaten: six worms never ate 18 oligochaetes; four specimens of *Ovatella myosotis* lived with Jetty

Road worms many days with no recognition by either species of the other, two other *O. myosotis* individuals were rejected by Jetty Road individuals in feeding trials, as well as three rejected by Tomales Bay worms.; tiny *Hemigraphus oregonensis* were rejected in 20 trials by both Jetty Road and Elkhorn Slough Bridge individuals; *Lionyx hedgpethi* was rejected in six trials; red mites were rejected in 11 trials, and fly larvae were rejected in three trials. Worms also did not feed on dead *T. traskiana*, dead *L. occidentalis* or brooded embryos of *T. traskiana*.

In feeding trials to determine if other animals would feed on *P. californiensis*, no worm predators were found. Four individuals of *Pachygrapsus crassipes* showed no response to worms, even when one was placed into the claws of two crabs. In another experiment nine pieces of worms put into a small aquarium with six freshly caught *P. crassipes* were quickly grabbed by four crabs. Almost immediately the crabs pushed the worm pieces away and two crabs appeared to try to clean off their claws. Three *Hemigrapsus oregonensis* did not respond to worms placed before them. The glycerid polychaete *Hemipodus borealis* showed no response to presence of *P. californiensis* in five trials. In another trial, a glycerid everted its proboscis, but not on the nemertean, and did not bite the nemertean. In one trial the glycerid's proboscis was everted and the teeth even stayed on the nemertean, but no biting occurred. The polychaetes appeared to be more agitated when *P. californiensis* touched them; however, this could have been simply due to the mechanical stimulus of contact. In one feeding trial with a large *T. traskiana*, the amphipod prey pinched the nemertean; the worm's proboscis came out a rupture in the neck region due to the pinching, and later the worm died.

Feeding took an average of 29.5 ± 8.3 ($n = 46$) minutes when *P. californiensis* fed on *T. traskiana*, 23.5 ± 5.4 ($n = 11$) minutes for *L. occidentalis*, 33.3 ± 17 ($n = 4$) for *A. holmesi* 22 ± 8.5 ($n = 2$) minutes for the Elkhorn Slough Bridge isopod and 32.5 ± 7.6 ($n = 4$) for San Francisco Bay isopods. The overall average time to complete feeding was 28.7 ± 8.8 minutes ($n = 67$).

During fall, when worms were held in groups at 11 °C, it was not possible to determine feeding rates of individuals. Eight Jetty Road individuals fed 111 of 223 opportunities (44% or once every 10.6 days). Five Elkhorn Slough Bridge worms ate 3 of 30 possible times (10% per opportunity or every 70.7 days). Several Tomales Bay worms fed only 3 times between Sept and Jan, with 104 opportunities (2.9%).

In spring, when individuals were held separately at 16.5 °C, 13 Jetty Road worms ate 153/294 (52%) of trials. (Accepted prey were 1 *L. occidentalis*, 4 *A. holmesi*, 3 Elkhorn Slough Bridge isopods and 145 *T. traskiana*). Eight Elkhorn Slough Bridge worms ate 25 *T. traskiana*, and 2 *L. occidentalis* in 83 feeding trials (32.5%). Twenty-seven Tomales Bay worms ate 11 *T. traskiana* in 168 (6.5%) feeding trials. Individual feeding rates ranged from 85.7% to 0% for 44 worms. Nine individuals (7 from Jetty Road) fed 50% (once/8 days) to 85.7% (once/6.4 days) of trials, 12 from all three study sites fed between 8.3 (once/108 days) and 33.3% (once/11 days) of trials and 23 from Jetty Road (1), Elkhorn Slough Bridge (4) and Tomales Bay (18) never fed. (Four animals broke or died so percent feeding and weight change were unavailable.) All individuals that fed in > 50% of the feeding trials had gained weight by the end of the experiments. Three individuals gained weight (33.3%, 20%, 17.6%) and all others lost weight at feeding ranges of 33.3–8.3%. Table 2 shows growth of 3 individuals through duration of experiments. Increase in weight during one feeding of two worms was 61 to 66 mg and 89 to 103 mg. One other animal went from 20 to 19 mg during one feeding, but its posterior end was previously broken off, and such damaged individuals often presented non-typical results. Average intake of *Traskorchestia traskiana* per feeding by *P. californiensis* was 7.56 ± 13.19 mg ($n = 66$). Average weight loss of the 23 worms that did not feed plus 9 additional Tomales Bay worms held 82 days without feeding was 2.13 mg/d ± 2.37, or 0.82% of body wt/d ± 1.11.

Table 2. Size increase during laboratory feeding experiments of 131 day duration with animals held at 16.5 °C. Original weight = Orig Wt (mg), original length and width = Orig L, W (mm), final weight = Fin Wt (mg), final length and width = Fin L, W (mm), percent feeding = % Feed, average number of days between feedings = F/#d, percent weight gain = % Wt. Gain (mg), increase in length and width = Inc L, W (mm).

Orig Wt (mg)	Orig L, W (mm)	Fin Wt (mg)	Fin L, W (mm)	% Feed	F/#d	% Wt Gain (mg)	Inc L, W (mm)
35	48, 1.5	91	65, 1.5	77.8	6.2	160	17, 0.0
12	22, 1.0	62	60, 1.5	69.2	7.3	417	38, 0.5
8	15, 0.9	31	38, 1.5	61.5	8.2	286	23, 0.6

Distribution and abundance of prey amphipods

The upper limit to the distribution of *Traskorchestia traskiana* is higher in the intertidal than the upper limit of *Pantinonemertes californiensis*, it becomes rare in the lower levels of the *P. californiensis* zone, and rarely occurs below the nemertean zone. At Tomales Bay in June 1991 average numbers of *T. traskiana* in 10 random 625 cm^2 quadrats in each of six transects (transect A approximately 60 cm above highest *P. californiensis* distribution, transect B at upper limit of *P. californiensis*, transect C approx. 150 cm below transect B and at 1.40 m tidal height, transect D approx. 45 cm above lower edge of worm zone and about 35 cm below transect C, transect E at lower edge of worms and quadrats 30 cm below the transect, transect approximately 1 m below transect 3 and below worm zone) were: transect A, 12.6 ± 9.8; transect B, 9.4 ± 9.2; transect C, 22.3 ± 23.9; transect D, 13.4 ± 19.7; transect E, 1.6 ± 2.6; transect F, 0.1 ± 0.32. Kruskal-Wallis analysis showed significant differences ($P = 0.001$) in median numbers of *T. traskiana* in the six transects. Mann Whitney U comparisons of median numbers showed significantly more *T. traskiana* within the *P. californiensis* zone (transects B–D, $\bar{x} = 15.03$) than below the nemertean zone (transects E + F, $\bar{x} = 0.85$) ($P < 0.0001$). Transects A and B were not significantly different. Transects A + B ($\bar{x} = 11.0$) and transects C + D ($\bar{x} = 17.85$) were not significantly different while transects A and F were significantly different. At Elkhorn Slough Bridge in June 1991 average numbers of *T. traskiana* in 10 random 625 cm^2 quadrats in each of six transects (transect A, 50 cm above highest *P. californiensis* distribution; transect B at upper limit of *P. californiensis*; transect C, 50 cm below transect B; transect D, 50 cm above lower limit of *P. californiensis* distribution; transect E at lower edge of *P. californiensis* zone; transect F, 50 cm below lower *P. californiensis* limit) were: transect A, 5.3 ± 6.8; transect B, 5.3 ± 4.4; transect C, 3.2 ± 3.7; transect D, 2.2 ± 6.3; transect E, 0.0 ± 0.0; transect F 0.4 ± 0.97. Kruskal-Wallis analysis showed significant differences ($P < 0.0001$) in median numbers of *T. traskiana* in the six transects. Mann Whitney U comparisons of median numbers showed significantly more *T. traskiana* in the top three transects ($\bar{x} = 4.6$) than in the bottom three ($\bar{x} = 0.87$) ($P < 0.0001$), no significant difference between transect A and transects B–E within the worm zone ($\bar{x} = 2.7$), significantly more *T. traskiana* in transect A than transect F ($P = 0.0404$), significantly more *T. traskiana* in transects A + B ($\bar{x} = 5.3$) than in transects C + D ($\bar{x} = 2.7$) ($P = 0.0468$), significantly more *T. traskiana* in transects C + D than transects E + F ($\bar{x} = 0.2$) ($P = 0.0120$) and no significant difference between the worm zone (transects B–E, $\bar{x} = 2.7$) and transect F. So few individuals of *T. traskiana* were seen at Jetty Road that similar analysis was not done for that site.

Traskorchestia traskiana was more abundant at Tomales Bay than at Elkhorn Slough Bridge, and at Elkhorn Slough Bridge than at Jetty Road. Although variation in abundances occurred within areas, this pattern was always apparent even from casual observation. In June 1991 the average number of *T. traskiana* at Tomales Bay was 9.9 ± 15.24 ($n = 60$) throughout its distributional zone; at Elkhorn Slough Bridge the average number was 2.7 ± 4.76 ($n = 60$) and at Jetty Road four individuals were seen in 11 m^2. Averages for other dates were: at Tomales Bay in July 1990, 67 *T. traskiana* and 71 *P. californiensis* were found under 159 rocks; however, *Traskorchestia traskiana* was far more abundant under the *Zostera* wrack than under rocks. In August, 1990, densities of *T. traskiana* above and below 1.43 m tidal height (which corresponded to the bottom of the *Zostera* wrack) were 315 *T. traskiana* and 29 *P. californiensis* under 72 rocks (4.4 amphipods/rock) above 1.43 m and covered with *Zostera* and 116 *T. traskiana* and 14 *P. californiensis* under 73 rocks (1.6 amphipods/rock) below 1.43 m where rocks were not covered with *Zostera*. At Elkhorn Slough Bridge, July 1990, at the top of the worm zone 103 amphipods and 1 *P. californiensis* occurred under 34 rocks (3.03 amphipod/rock). In the middle of the worm zone, 47 *T. traskiana* and 11 *P. californiensis* occurred under 80 rocks (0.59 amphipod/rock). In the lower worm zone two amphipods and three worms occurred under nine rocks (0.22 amphipod/rock). Other *T. traskiana* abundances at Jetty Road were: July 1989, several; Nov, 1989 0/20 m^2: March 1990, 1/10 m^2; Oct 1990, 5/20 m^2.

Amphipod prey tolerances to submersion in sea and fresh water

When submerged in sea water for one week all individuals of *Traskorchestia traskiana* held at 8.5–10 °C survived ($n = 8$). Three of eight held at 16.5 °C died, one on day 4, one on day 5, and one on day 6. One animal molted while submerged and lived, and one brooding female released live young while submerged. When 10 individuals were submerged in fresh Turlock city well water, all were alive at 2.5 h, 6 at 3.5 h, 5 at 4.5 h, 3 at 5.5 h and 6.5 h, 2 at 7.5 h, 1 at 8.5 h, 10 h, 12 h and died between 12 and 16 h post-submersion.

Nemertean sizes

It is difficult to obtain accurate measurements of living nemerteans because they do not possess external hard parts. However, when a worm is fully extended, it tends to be more narrow in width, and wider if not completely extended. Area (length × width), therefore, is usually a more accurate measurement than either length or width. Wet weight is also variable due to the amount of water on the worm's surface or leaked during weighing. Length to wet weight and area to wet weight were both strongly correlated; width was not strongly correlated to wet weight (r, length to wet weight = 0.878, $r^2 = 77\%$; r, area to wet weight = 0.917, $r^2 = 84.1\%$; r, width to wet weight = 0.379, $r^2 = 14.4\%$, $n = 59$).

Average sizes of all accessible worms observed during field studies, or of worms selected haphazardly from field samples, are shown on Table 3. The largest animal measured in this study was a Tomales Bay female, 38 cm long and 2 mm wide. The average size of individuals from Tomales Bay was both longer and wider than average size at Elkhorn Slough Bridge, and Elkhorn Slough Bridge worms averaged longer and wider than Jetty Road individuals. Few small animals were seen during the two year study at Tomales Bay; no individual longer than 7.5 cm was seen at Jetty Road. It is not known if the Jetty Road individuals were young or small for other reasons.

Mann Whitney U comparisons of median lengths or worm areas showed significant differences between worm sizes at Tomales Bay in July 1989, June 1990, June 1991, August 1990 and March 1990 compared to January 1990, and between June 1990 and July 1990, June 1990 and August 1990, July 1990 and Aug 1990, July 1989 and July 1990, and June 1990 and June 1991.

At Elkhorn Slough Bridge, significant differences in worm lengths occurred between January 1990 and March 1990 and between January 1990 and July 1990.

The average sizes of all measured Tomales Bay females, males and worms whose sex could not be determined are shown on Table 4. No significant differences occurred between males and females during any individual sampling periods or between overall average lengths, widths or areas. During some individual sampling periods there were significant differences when lengths or areas

Table 3. Average length (cm) ± s.d., width (mm) ± s.d. and area (mm^2) ± s.d., and number in sample of *Pantinonemertes californiensis* at three study sites, July 1989–July 1991.

Date	L (cm) ± s.d.	W (mm) ± s.d.	Area (mm^2) ± s.d.	N
Tomales Bay				
Jul 89	18.3 ± 8.1	1.6 ± 0.56	333 ± 221.3	35
Jan 90	17.4 ± 5.5	1.5 ± 0.34	250 ± 88.7	28
Mar 90	19.4 ± 5.8	2.3 ± 0.68	454 ± 175.5	25
Jun 90	17.5 ± 5.8	2.3 ± 0.66	403 ± 142.7	55
Jul 90	15.3 ± 4.7	1.4 ± 0.51	225 ± 119.1	18
Aug 90	11.9 ± 4.7	1.2 ± 0.30	137 ± 65.1	28
Jun 91	13.0 ± 5.7	1.4 ± 0.61	203 ± 139.7	20
Overall	13.7 ± 7.6	1.78 ± 0.72	257 ± 185.7	209
Elkhorn Slough Bridge				
Nov 89	7.6 ± 2.7	1.1 ± 0.12	86.4 ± 41.4	5
Jan 90	8.6 ± 3.9	1.2 ± 0.35	105 ± 58.3	14
Mar 90	5.1 ± 1.4	1.4 ± 0.38	72 ± 32.6	16
Jul 90	5.9 ± 1.8	1.2 ± 0.22	66 ± 13.9	12
Jun 91	7.0 ± 1.4	1.5 ± 0.00	105 ± 21.2	2
Overall	6.6 ± 2.9	1.25 ± 0.32	82.7 ± 41.5	49
Jetty Road				
Jul 89	3.7 ± 1.5	0.8 ± 0.20	30.0 ± 15.9	19
Oct 89	2.8 ± 1.1	0.3 ± 0.28	9.8 ± 11.0	2
Nov 89	1.7	0.9	15.3	1
Jan 90	2.8 ± 1.6	1.1 ± 0.47	35.6 ± 31.7	20
Mar 90	2.9 ± 1.6	1.2 ± 0.47	40.1 ± 32.7	16
Jun 91	2.8 ± 1.8	0.8 ± 0.29	26.6 ± 22.1	8
Overall	3.1 ± 1.6	0.97 ± 0.43	32.9 ± 26.6	66

Table 4. Average sizes (lengths, widths, areas ± s.d.) and number in sample (N) of Tomales Bay non-ripe individuals, females and males of *Pantinonemertes californiensis*.

	N	L (cm) ± s.d.	W (mm) ± s.d.	Area (mm^2) ± s.d.
Non-ripe	116	14.9 ± 6.8	1.56 ± 0.60	245.3 ± 150.4
Females	68	13.7 ± 9.2	1.96 ± 0.82	280.2 ± 227.2
Males	45	13.3 ± 8.6	2.04 ± 0.54	289.7 ± 209.5

of combined males and females were compared to non-ripe worms. The overall average width of non-ripe worms was significantly smaller than the overall width of females and the overall width of males. The overall smaller width of non-reproductive worms compared to male or female widths reflects in part a width decrease upon spawning (see section on sex ratios and spawning below).

Sex ratios and spawning

I was able to recognize males and females from external appearances, as early as March, when gonads began to show color of developing gametes, and as late as October. The pink of ovaries was more apparent than the white of testes throughout the reproductive season, and many of the individuals considered as non-ripe (Table 4) could have been immature or post-reproductive males. Sex ratios are probably 1:1, although I did not ever see a 1:1 ratio (Table 5). Sex ratios (Table 5) from nature indicate that the major spawning period is late June–early July, (see especially June–August 1990) with some individuals spawning as late as October.

At Elkhorn Slough Bridge site, females were seen in March (1/17 individuals observed), June (3/3 individuals observed), July (9/15 individuals observed) and October (1 ripe individual). Only one animal was tentatively recognized as a developing male, in March. Females were all approximately 7 cm long by 1.2 mm wide.

At Jetty Road, one individual, 6–7 cm long, appeared to be very ripe in July, and one male contained mature sperm in September. However, all other in individuals recognized as male or females had immature gonads (March) or were recognized by sex only upon observation with a compound microscope (July).

Both in nature and in the laboratory, if worms were ripe, they would spawn when submerged for a few minutes in sea water. Especially in nature, but occasionally in the lab, individuals would also spawn when exposed to air, drying conditions, sun or after submerged worms were removed from water. Spawning was observed in nature at Tomales Bay June 22, 1990, July 9, 1990, Aug 28, 1989, and already-spawned eggs were seen under one rock Aug 21, 1990. In early June, 1991 (Jn 6–8) no worms (9 females, 7 males) spawned when submerged in sea water, nor did any spawn when exposed to air.

Table 5. Sex ratios of *Pantinonemertes californiensis* observed at Tomales Bay, 1989–1991.

Date	Total # in sample	Males	Females	Non-ripe individuals
Jul 1989	47	0.40	0.51	0.09
Aug 1989	34	0.38	0.44	0.18
Oct 1989	11	0.36	0.55	0.09
Dec 1989	23	?0.22	?0.04	?1.00
Jan 1990	40	0	0	1.00
Mar 1990	26	0.08	0.15	0.77
Jun 1990	133	0.30	0.65	0.05
Jul 1990	73	0.11	0.38	0.51
Aug 1990	44	0.05	0.16	0.80
Jun 1991	37	0.27	0.51	0.22

In the laboratory, a female collected June 8, 1991 spawned June 27 when submerged in sea water; she did not spawn June 12, 17 or 24 when submerged. Six mature worms collected June 22, 1990 did not spawn when submerged in sea water until July 4 (1) July 7 (4) and July 12 (1 plus another section of the one that spawned July 4). Worms collected July 19, 1989 spawned when submerged in sea water Aug 4–16 and ones collected July 9, 1990 spawned Aug 21; worms collected October 7 spawned 10 days later when submerged, and some individuals continued to spawn sporadically until Nov 21.

From sex ratios, observations of spawning in nature, observations of no spawning with submersion in sea water both in nature and laboratory, and dates of spawning of animals in the laboratory, it appears that most spawning occurs late June–early July, with some spawning occurring throughout late summer and fall. Both in nature and in the laboratory, individuals often spawned gametes from only part of the body at one time. Gametes at the posterior end of the body were always spawned prior to those in more anterior gonads.

When animals spawned, they became much narrower in diameter in the post-reproductive part of the body. Four animals measured anterior to and in the spawned region had an average pre-spawning width of 3 mm and post-spawning width of 1.28 mm. It appears that ripe animals usually rapidly absorb water when submerged, so that immediately after spawning, they weigh more than before spawning. By the following day, their decreased width is reflected in decreased weight. Five of 6 animals went from a total of 1.378 g pre-spawning wt to 1.575 g immediately after submersion in sea water and spawning. One animal, however, lost from 449 mg to 166 mg. The day after spawning, four animals decreased from 1.065 g to 0.641 g.

In the laboratory and in nature individuals did not die upon, or soon after, spawning. Reproductively mature and immediate post-reproductive individuals were, however, much more fragile than animals in a non-reproductive condition, and often lost posterior sections of the body.

Larval development

Females shed hundreds to thousands of eggs at a time. At least under laboratory conditions, eggs were shed as primary oocytes. Each egg was surrounded by a thick jelly coat. Eggs were usually initially held together in poorly formed mucus strings or clumps and were easily dispersed individually into sea water. Average egg diameter without the jelly coat was 95.8 ± 8.75 μm ($n = 20$), and 247.2 ± 34.09 μm ($n = 18$) with the jelly covering. A thin fertilization membrane formed between the egg and the jelly coat. Eggs held at 11 °C developed into swimming larvae within 22–48 hours; those held at 16.5 °C hatched in 15–18 hours, and eggs held at 22–23 °C became swimming larvae in 12–16 hours. Larvae lived when maintained for short periods (hours) at temperatures as high as 30 °C. Ciliated larvae began to rotate slowly within the fertilization membrane and jelly coat, and rotated faster and faster until they hatched. Prior to hatching the inner membrane sometimes became misshapen; larvae would break through it and rotate within the outer, broad jelly coat until they came to the coat's outer edge. Then they would not only rotate, but also rock back and forth, gradually forming a bulge in the coat. Finally the coat would break; the larval anterior end would be free, and the rest of the larva was freed as it swam forward. Upon hatching larvae swam anteriorly in a rotating pattern. Just hatched larvae were approximately the size of eggs without jelly coats, *i.e.* about 100–110 μm diameter. They were completely covered with cilia and had a group of flagella both anteriorly and posteriorly. Within a week of fertilization larvae could lengthen and shorten with development of muscles, became elongate in shape, and development of gut, nervous system and eyes were visible with compound microscope. I was never able to see development of stylet or basis. Morphology visible with a compound microscope was similar to that of other hoplonemertean larvae.

Throughout the time I was able to keep larvae alive they appeared to be photopositive. Addition of sand or adult *P. californiensis* to fingerbowls containing larvae did not stimulate them to settle out of the plankton.

All attempts to maintain eggs and larvae in damp conditions to determine if any development pattern other than planktonic existed, failed. In one such experiment, larvae were found swimming in the thin water column above wetted filter paper.

Larvae held at 11 °C lived up to 18 days post-fertilization. Addition of juice from ground up krill to water containing larvae resulted in larger, more elongate, and faster moving larvae compared to those without food. Krill juice may have extended time to death by approximately 2 days. Average size of 11 °C larvae between day 6–18 of development was 150.6 ± 32.8 μm length and 70.6 ± 25.97 μm width.

Larvae held at 16.5–23 °C lived up to 16 days without added food. Larvae without added food invariably became misshapen and the digestive tract accumulated many brown irregular inclusions. When I added a mixture of glucose and amino acids to the water in some fingerbowls with larvae on day 8 after fertilization, larvae with this 'food' continued to live 18–20 days longer than those without, for a total of 34–36 days. Average size of 16.5–23 °C larvae between day 1–8 was 158.8 ± 36.1 μm by 56.25 ± 9.46 μm. Table 6 shows comparisons of sizes of fed and unfed 16.5–23 °C larvae, and sizes of fed larvae after unfed ones had died. The mixtures of amino acids and glucose did not supply requirements for further development or settlement from plankton, but longer life of larvae with these dissolved organics in the water did show that these larvae are planktotrophic.

Discussion and conclusions

Terrestrial nemerteans share several characteristics in addition to those common for all mono-

Table 6. Sizes (length ± s.d., μm and width ± s.d., μm) of fed and unfed larvae of *Pantinonemertes californiensis* held at 16.5–23 °C. Age is day post-fertilization. N is number in sample measured. Median length of fed and unfed larvae compared with Mann Whitney U test (significantly different, $P < 0.05$).

Age	Fed larvae			Unfed larvae			Sig. diff.
	L ± s.d.	W ± s.d.	N	L ± s.d.	W ± s.d.	N	
12	180 ± 28	70 ± 28	2	139 ± 19	68 ± 15	5	No
13	197 ± 21	75 ± 22	3	118 ± 21	59 ± 16	9	Yes
14	146 ± 26	58 ± 11	11	120 ± 32	54 ± 13	14	Yes
16	150 ± 20	60 ± 16	9	97 ± 15	63 ± 15	3	Yes
17	143 ± 20	51 ± 13	5				
20	130 ± 16	57 ± 6.7	5				
21	139 ± 14	51 ± 7.4	5				
22	149 ± 19	56 ± 5.5	5				
23	137 ± 17	56 ± 4.9	6				
24	133 ± 21	54 ± 9.5	7				
27	118 ± 8.4	53 ± 4.5	5				
28	122 ± 14	49 ± 10	7				
29	138 ± 25	57 ± 8.4	5				
32	120	55	1				

stiliferous hoplonemerteans, including a) small size; b) reduction of the cephalic longitudinal furrows; c) proboscis large and used in many species for rapid escape locomotion; d) cephalic glands strongly developed for secretion of copious amounts of mucus; e) excretory system especially well-developed, with many flame bulbs and thousands of nephridiopores; f) circulatory systems very well-developed, with well-developed vascular plugs (Moore & Gibson, 1985). The genus *Pantinonemertes* contains fully marine *P. enalios* (Moore & Gibson, 1981), *P. mooreae* (Gibson, 1982b), *P. mortoni* (Gibson, 1986), *P. spectaculum* (Gibson, 1986), semi-terrestrial *P. californiensis* (Gibson *et al.*, 1982), *P. winsori* (Moore & Gibson, 1981), also considered a brackish water species and terrestrial species *P. agricola* (Willemoes-Suhm, 1874, in Moore & Gibson, 1981).

Members of this genus are characterized by the following characteristics that relate to the ones above for terrestrial nemerteans: proboscis large and often used for locomotion, cephalic glands extensive, excretory system very well-developed and opening via numerous nephridiopores and flame cells which are fused in pairs and reinforced with transverse bars, blood system thick-walled with a large number of extravascular pouches and few valves, mid-dorsal blood vessel with one vascular plug (Moore & Gibson, 1981). Of these characteristics of the genus, *P. californiensis* shares a large proboscis with all but *P. enalios*; it does not use its proboscis for locomotion, shared with marine relatives only; its multiplication of flame cells, excretory ducts and nephridiopores is shared with all relatives except *P. enalios* (Gibson, 1986). *Pantinonemertes californiensis* therefore shares some typically terrestrial characteristics with strictly marine forms within its genus.

In addition, *P. californiensis* is large, in contrast to terrestrial nemerteans (Gibson *et al.*, 1982; pers. obs.; Moore & Gibson, 1985). *Pantinonemertes californiensis* has well-developed cephalic glands which discharge via many improvised ducts, which is not shared with other members (including *P. agricola*) of the genus (Gibson *et al.*, 1982). Well-developed cephalic glands are characteristic of terrestrial nemerteans (Moore & Gibson, 1985); however, I have not seen that *P. californiensis* produces copious amounts of mucus. Production of much mucus and well-developed cephalic glands also occurs in some marine nemerteans, for example, *Carcinonemertes epialti* (Coe, 1902). *Pantinonemertes californiensis* does share a capillary network with terrestrial nemerteans in the genera *Argonemertes*, *Antiponemertes*, and *Acteonemertes* (Moore & Gibson, 1985) plus the supra-littoral *Notogaeanemertes folzae* (Riser, 1988). This is not found in any other member of its genus (Gibson, 1986). Of characteristics typical for terrestrial nemerteans, therefore, *P. californiensis* has no characteristics that are not also present in marine species within its genus or found in other genera of marine nemerteans.

Distribution of *P. californiensis* with respect to tidal height at areas I studied was at upper levels of the intertidal. Virtually all specimens I found, however, were distributed low enough to be wetted by high tides at least 23% of the days/year. Nemertean distribution did not reach as high intertidally as did that of their primary prey, *Traskorchestia traskiana*, and nemerteans easily broke into pieces if allowed to become dry. Experiments of worms in a moisture gradient tray in the laboratory suggested that *P. californiensis* might sometimes seek out wetter conditions than those in which it is usually found in nature.

Pantinonemertes californiensis is distributed in nature from at least south of Los Angeles, CA (Gibson *et al.*, 1982) north to San Juan Isl., WA (Kozloff, pers. comm.). In addition, Scott (pers. comm.) found that when 10 specimens of *P. californiensis* from Tomales Bay were placed into a Gilson Differential Respirometer and monitored for 1 hour at each temperature, 6–7°, 11–12°, 16–17°, 22–23°, 26–27°, 31–32° and 36–37 °C, respiration rate increased with temperature through 31–32 °C. At higher temperatures respiration decreased due to death of several worms. Six months after a several-day severe freeze throughout central California, I found *P. californiensis* at all three study sites. Since December is not within the reproductive season of *P. californiensis*, these worms had to survive the cold; they were not new recruits. *Pantinonemertes californiensis* therefore appears to have wide tolerance limits for temperature changes.

According to Gibson *et al.* (1982), most specimens of *P. californiensis* die within 24–36 hours if forced to remain submerged, although specimens can withstand salinity changes from hypersaline to nearly fresh water, providing they can avoid prolonged immersion. I found that immature specimens of *P. californiensis* could withstand submersion in sea water for at least a week, a much longer continuous time than any intertidal organism is covered by water in nature. Nearly mature adults broke apart or became moribund within 3–4 days; however all recovered when removed from submersion, and even 3 days continuous submersion is longer continuous submersion time than occurs intertidally in nature.

Completely terrestrial animals should have adaptations to handle fresh water. Specimens of *P. californiensis* that I submerged in fresh water died almost immediately. Even though *P. californiensis* has many binucleate, reinforced flame bulbs and many nephridiopores (Gibson *et al.*, 1982), it does not appear to be able to adjust to fresh water at all. One other species of *Pantinonemertes*, the terrestrial *P. agricola*, is also killed by fresh water (Moore, 1985), unlike most other terrestrial nemerteans (Moore & Gibson, 1985).

Traskorchestia traskiana, as well as *P. californiensis*, is considered to be semi-terrestrial (Busath, 1979). It occurs along the west coast of N. America from Baja California to Unalaska, AK. Its habitat is variable, including rocky, sandy and muddy marine shores as well as salt marshes. It cannot survive prolonged submergence in either salt or fresh water (Busath, 1979). Busath (1979) found much variation in length of time populations of *T. traskiana* from different habitats in southern California could withstand submersion in sea water and in fresh water. In submersion experiments in the present study, *T. traskiana* was less able to tolerate long submersion in sea water than was *P. californiensis*, although all specimens of *T. traskiana* also lived while submerged, longer than intertidal organisms are faced with continual submersion in nature. *Traskorchestia traskiana* from Tomales Bay was considerably more tolerant of fresh water submersion than *P. californiensis*; one-half of the specimens immersed in fresh water lived 4.5 hours.

In feeding trials, a higher percent of *P. californiensis* specimens from Jetty Road fed than specimens from Elkhorn Slough Bridge, and a higher percent of Bridge specimens fed than did Tomales Bay specimens. Also, seven of the nine animals that fed in 50% or more feeding trials were from Jetty Road, whereas 18 of 23 individuals that did not feed in any trials were from Tomales Bay. Jetty Road individuals fed on the highest diversity of prey as well, although Elkhorn Slough Bridge and Tomales Bay specimens were not offered all species of isopods. Tomales Bay worms never fed on *Ligia*, in several trials.

Nemerteans from Tomales Bay were larger than those from Elkhorn Slough Bridge. Jetty Road individuals were smallest. Nemertean size was strongly correlated with major prey abundance, with largest animals found at Tomales Bay where *T. traskiana* was most abundant. Nemertean abundance was also correlated with prey abundance, with highest densities of *P. californiensis* and *T. traskiana* at Tomales Bay, intermediate densities of both at Elkhorn Slough Bridge and lowest densities of both at Jetty Road. These population parameters, along with the feeding patterns noted above for the three nemertean populations, suggest strongly that *P. californiensis* populations are regulated by distribution and abundance of *T. traskiana*. This study indicates that the distribution of *P. californiensis* in the high intertidal is a reflection of prey distribution. I suggest that *P. californiensis* lives as high as it can without drying out too much, or encountering too much fresh water from rain, or other terrestrial conditions, because its prey is most abundant there. This hypothesis is supported by the abundance pattern of *T. traskorchestia*, which extends higher in the intertidal than does the worm, and which becomes rare in lower parts of the worm zone and below worm distribution.

In addition to the morphological characteristics of terrestrial nemerteans listed by Moore & Gibson (1985), in reproduction, terrestrial nemerteans usually secrete a few to several hundred eggs into gelatinous cocoons and development is invariably direct (Moore & Gibson, 1985). *Pantinonemertes agricola*, related to *P. californiensis*, is unique among terrestrial nemerteans in being ovoviviparous (Moore & Gibson, 1985). *Pantinonemertes californiensis* is oviparous, and spawns hundreds to thousands of eggs at a time. Development is by way of a planktonic larva, morphologically similar to the larvae of other marine hoplonemerteans (pers. obs.). In its reproductive biology, *P. californiensis* is not at all similar to terrestrial species.

I was unable to determine the stimulus for spawning by *P. californiensis* in natural conditions. Since, however, I was able to induce spawning in the laboratory and in disturbed specimens in nature by covering them with sea water, I hypothesize that submersion during high tides might be the stimulus for spawning in nature. The problem with this hypothesis is that worms in nature often spawned when they were uncovered and exposed to air. Such disturbances should not be common under normal circumstances, and, with planktonic larvae, it would be adaptive to spawn when water covers worms.

I was unable to determine if individuals die in nature after spawning, although I know that they do not immediately die. Since individuals often spawn only part of their gametes at any one time, some individuals certainly do not die through several spawning sessions, and probably live for more than one reproductive season. I also do not know if spawning in late summer and fall is by individuals that were slow to mature or if individual worms can develop more than one complete set of gametes per reproductive season.

Larvae of *P. californiensis* are apparently planktotrophic. They lived longer when dissolved amino acids and sugar were added to culture water, but even with added nutrients, they did not continue to develop. They also became misshapen and accumulated brown occlusions with age. I was unable to determine on what they normally feed, nor was I able to determine the stimulus for settlement. I did not see individuals at any study site that I recognized as young, or recruits. I was unable to determine if the very small worms at Jetty Road were young, or small due to lack of food.

Pantinonemertes californiensis appears to occur as high in the intertidal as it can, because that is where its prey occurs. Its ecology is similar to other intertidal hoplonemerteans such as *Parane-*

mertes peregrina (Roe, 1976). It might be better to consider *P. californiensis* simply as a marine form that lives high in the intertidal, rather than as an important transitional or semiterrestrial species. Since the term semi-terrestrial does not appear to be well defined for nemerteans in general, I suggest that such terms as high intertidal, supralittoral or upper littoral replace semi-terrestrial as descriptors for those species previously described as semi-terrestrial. At least in the populations I studied, the life history characteristics of this species provide little evidence of adaptations for terrestrial life or similarities to terrestrial nemerteans. The life history parameters of *P. californiensis* are not very helpful in elucidating the role of semi-terrestrial nemerteans in the evolution of terrestrial species.

Acknowledgements

I wish to thank my students Mavis Carey-Williams for doing much of the laboratory feeding study of individual worms over time, Bruno Pernet for maintaining larvae while I was away for two weeks, and Kip Scott for doing respiration studies. These three, as well as Richard Anderson, Bill Broenkow, Glenn Callahan, Jay Christofferson, Joan Ferraris, Gary Greene, Nancy Jaureguy, Chris Johnson, Marvin Johnson, Susanne Laurenz-Miller, Dana Reneau, Peter Slattery, Guy Van Cleve, Tamara Voss and Dan Wickham helped with various other aspects of the study. I thank all of them for their many contributions.

References

Busath, A. L., 1979. Geographic Variation in the Semiterrestrial Amphipod *Orchestia traskiana*: Phenotypic Plasticity or Incipient Speciation. Ph.D. Thesis, University of California, Santa Barbara.

Coe, W. R., 1902. The nemertean parasites of crabs. Am. Nat. 36: 431–450.

Gibson, R., 1982a. Nemertea. In S. P. Parker (ed.), Synopsis and classification of living organisms, vol. 1. McGraw-Hill, N.Y.: 823–846.

Gibson, R., 1982b. Nemerteans of the Great Barrier Reef 5. Enopla Hoplonemertea (Monostilifera). Zool. J. linn. Soc. 75: 269–296.

Gibson, R., 1986. The macrobenthic nemertean fauna of Hong Kong. In B. Morton (ed.), Proceedings of the Second International Marine Biological Workshop: The Marine Flora and Fauna of Hong Kong and Southern China, Hong Kong, 1986. Hong Kong University Press, Hong Kong: 33–212.

Gibson, R., J. Moore & F. B. Crandall, 1982. A new semi-terrestrial nemertean from California. J. Zool., Lond. 196: 463–474.

Moore, J., 1985. The distribution and evolution of terrestrial nemertines. Am. Zool. 25: 15–21.

Moore, J. & R. Gibson, 1981. The *Geonemertes* problem (Nemertea). J. Zool., Lond. 194: 175–201.

Moore, J. & R. Gibson, 1985. The evolution and comparative physiology of terrestrial and freshwater nemerteans. Biol. Rev. 60: 257–312.

Riser, N. W., 1988. *Notogaeanemertes folzae* gen. n., sp. n., an additional ecologically restricted hoplonemertean from New Zealand. Hydrobiologia 156: 125–133.

Roe, P., 1976. Life history and predator-prey interactions of the nemertean *Paranemertes peregrina* Coe. Biol. Bull. 150: 80–106.

Sundberg, P., 1989. Physiology and cladistic classification of terrestrial nemerteans: the genera *Pantinonemertes* Moore & Gibson and *Geonemertes* Semper. Zool. J. linn. Soc. 95: 363–372.

U.S. Dept. Commerce, 1989. Tide Tables 1989 High and Low Water Predictions West Coast of North America and South America Including the Hawaiian Islands. U.S. Government Printing Office, 238 pp.

U.S. Dept. Commerce, 1990. Tide Tables 1990 High and Low Water Predictions West Coast of North America and South America Including the Hawaiian Islands. U.S. Government Printing Office, 238 pp.

U.S. Dept. Commerce, 1991. Tide Tables 1991 High and Low Water Predictions West Coast of North America and South America Including the Hawaiian Islands. U.S. Government Printing Office, 238 pp.

Hydrobiologia **266**: 45–56, 1993.
R. Gibson, J. Moore & P. Sundberg (eds), Advances in Nemertean Biology

The infestation and dispersion patterns of *Carcinonemertes* spp. (Nemertea) on their crab hosts

Jeffrey D. Shields
Department of Biological Sciences, The University of California, Santa Barbara, CA 93106, USA; Department of Parasitology, The University of Queensland, St. Lucia, Brisbane, QLD 4072, Australia; (present address: Chesapeake Bay National Estuarine Research Reserve, Virginia Institute of Marine Science, P.O. Box 1346, Gloucester Point, VA 23062, USA

Key words: Cancer anthonyi, Carcinonemertes epialti, C. mitsukurii, C. regicides, Hemigrapsus oregonensis, Paralithodes camtschaticus, Portunus pelagicus, aggregation, egg predation

Abstract

Seasonal changes in the infestation and dispersion patterns of egg predatory nemerteans on their crab hosts were analyzed. Marked differences in the seasonal patterns of infestation were noted between worm species, yet, common patterns in the aggregation of the worms were found. Worm aggregation increased at the onset of the reproductive seasons of the hosts, and at the nadirs of the reproductive seasons for those hosts with year round breeding. The aggregation patterns of two worm species fluctuated with the physical environment of their estuarine hosts. Salinity changes as a result of seasonal rains may have caused changes in the underlying dispersion patterns of *Carcinonemertes epialti* on *Hemiarapsus oregonensis*, and *C. mitsukurii* on *Portunus pelagicus*. Lastly, the embryogenic cycle of the host species was significant in shaping the infestation and aggregation patterns of *C. epialti* on *Cancer anthonyi* and *C. regicides* on *Paralithodes camtschaticus*. Worm immigration and emigration were linked to crab embryogenesis and directly influenced the dispersion patterns of the worms.

Introduction

The dispersion of parasite populations among their host populations is typically aggregated. Such dispersion patterns are characterized by many hosts possessing few parasites and few hosts possessing many parasites. Several different theoretical frequency distributions (*i.e.*, negative binomial, log normal, log series, Neyman type A, Polya-Aeppli) may describe aggregated populations (Anscombe, 1950; Evans, 1953; Williams, 1964), but the negative binomial frequency distribution has received the most attention in parasite-host systems (e.g., Crofton, 1971; Pennycuick, 1971; Anderson & May, 1978; Hirsch, 1980; Kitron, 1980; Scott, 1987).

The forces that generate aggregated dispersion patterns are varied, and are broadly classified as either demographic or environmental in nature (May,1975; Anderson & Gordon, 1982). Demographic factors such as immigration and death rates, or host and parasite genetic variability may have different effects on the dispersion of parasites (May, 1975; Anderson & Gordon, 1982; McCallum & Anderson, 1984). Demographic factors are, however, greatly affected by environmental factors, e.g., temporal or spatial changes in the environment.

The underlying mechanisms that lead to aggregation can be examined by an analysis of the infestation and dispersion patterns of parasites (Crofton, 1971; Iwao & Kuno, 1972; Kitron, 1980). For example, changes in parasite prevalence and mean intensity in the host population appear to reflect changes in the degree of aggregation of the parasite population (Kitron, 1980; Kitron & Higashi, 1985; Scott, 1987). Fluctuations in both parameters can, however, obscure their contributions to the degree of aggregation in the population.

Worms in the genus *Carcinonemertes* von Kölliker are specialized egg predators of decapod crustaceans. The distinctive patterns of prevalence and intensity of the different species suggest that their infestation dynamics may provide insight into the causal mechanisms that lead to their observed aggregation patterns. For example, *Carcinonemertes epialti* on the yellow crab, *Cancer anthonyi*, has a high prevalence (mode 100%) and low mean intensity (Shields *et al.*, 1990), whereas *Carcinonemertes epialti* on the shore crab, *Hemigrapsus oregonensis*, has a markedly variable prevalence and mean intensity (Kuris, 1978; Roe, 1979; Shields & Kuris, 1988). Different infestation patterns have been observed for species of *Carcinonemertes* on the Dungeness crab, *Cancer magister*, the red king crab, *Paralithodes camtschaticus*, and the sand crab, *Portunus pelagicus* (Wickham, 1986; Kuris *et al.*, 1991; Shields & Wood, 1993). The patterns of infestation appear well tuned to the seasonal production of host eggs and their embryogenesis.

The objectives of the present study were: (1) to compare the seasonal changes in abundance and aggregation of *Carcinonemertes epialti* on two of its hosts, *Cancer anthonyi*, and *H. oregonensis*, that of *Carcinonemertes regicides* on *Paralithodes camtschaticus*, and that of *C. mitsukurii* on *Portunus pelagicus*; (2) to investigate the influence of the infestation parameters, prevalence and intensity, on the underlying dispersion patterns; and (3) to determine what demographic processes lead to these patterns of aggregation.

Material and methods

The methods of collection and examination of the crabs and nemerteans have been previously described. *Cancer anthonyi* with *Carcinonemertes epialti* were collected from the Santa Barbara Channel, California (1981–1983) (Shields *et al.*, 1990); *Hemigrapsus oregonensis* with *C. epialti*, from Bodega Harbor, California (1982–1985) (Shields & Kuris, 1988); *Paralithodes camtschaticus* with *C. regicides*, from Kachemak Bay, Alaska (1983–1985) (Kuris *et al.*, 1991); and *Portunus pelagicus* with *C. mitsukurii* from Moreton Bay, Australia (1990) (Shields & Wood, 1993). Ovigerous crabs were used in the analysis, except in the case of *Portunus pelagicus*, where mature, non-ovigerous and ovigerous crabs were used. Only data that had seasonal or monthly samples of hosts and symbionts were used in the analysis.

The dispersion of the nemertean populations was determined to be either random or aggregated. The variance (s^2):mean abundance ($\bar{x}$) ratio (hereafter the variance:mean ratio, also known as the index of dispersion), Lloyd's mean crowding ($\overset{*}{x}$), and the related patchiness ($\overset{*}{x}/\bar{x}$) indices were used to measure overdispersion and for comparisons between aggregated distributions (Lloyd, 1967; Elliott, 1977). The variance:mean ratio, while calculated, has serious flaws which limit its usefulness as an index of dispersion (e.g., the ratio is not independent of mean abundance; distributions that are clearly not Poisson may be indicated as such by the ratio, etc. – see Hurlbert, 1990). The mean crowding index is defined as the mean number of individuals per individual on the same host. It is estimated using the equation $\overset{*}{x} = \bar{x} + (s^2/\bar{x}) - 1$, where $\overset{*}{x} = \bar{x}$ for a random distribution, $\overset{*}{x} < \bar{x}$ for a regular or uniform distribution, and $\overset{*}{x} > \bar{x}$ for an aggregated distribution, such as the negative binomial distribution. Patchiness represents the mean crowding of an individual if the population it is drawn from has a random distribution (Lloyd, 1967). The χ^2-test of David & Moore (1954) was employed to examine differences between variance:mean ratios; the usefulness of this test was, however, limited by unequal sample sizes.

Two other indices of dispersion were used to examine changes in aggregation patterns. The regression of mean crowding on mean abundance, $\overset{*}{x} = a + b\ \bar{x}$, provided information on dispersion and contagion (Iwao, 1968, 1970; Iwao & Kuno, 1971). The Y-intercept, *a*, represents the index of basic contagion, and the slope, *b*, represents the density-contagiousness coefficient. The index of basic contagion (*a*) estimates the numbers of parasites that arrive as a simultaneous unit on the host. It may consist of a single individual or a group of individuals. When $a = 0$, then single individuals compose the basic unit of contagion; when $a > 0$, then more than one individual represents the basic unit of contagion; $a < 0$ represents evidence of negative density dependence or repulsion. The density-contagiousness coefficient (*b*) presents the dispersion pattern of the symbionts. When $b = 1$, then a random distribution is indicated; when $b < 1$, a regular or uniform distribution; and when $b > 1$, an aggregated distribution is indicated.

In addition, *tau*, the mortality index of Iwao & Kuno (1971) was used to examine differences in the population dynamics during the course of host embryogenesis for *Carcinonemertes epialti* on *Cancer anthonyi*. Tau = $\text{Patchiness}_1/\text{Patchiness}_0$, where Patchiness_0 represents the initial patchiness, and Patchiness_1 represents patchiness after the operation of mortality. *Tau* represents recruitment if the mean intensity increases from Patchiness_0 to Patchiness_1. When $tau = 1$, mortality/recruitment is independent of parasite density; when $tau < 1$, mortality/recruitment is density-dependent; and when $tau > 1$, mortality/recruitment is either inversely density-dependent or it resembles an all-or-none type of response.

Standard statistical tests are from Sokal & Rohlf (1981). The terms prevalence, abundance, mean density, and mean intensity are defined in Margolis *et al.*(1982). However, mean abundance was used in lieu of mean density since density has been used to describe the number of worms per 1000 crab eggs (Wickham, 1979). The terms dispersion and distribution are defined in Pielou (1977) and Anderson & Gordon (1982).

Results

Seasonal patterns

Carcinonemertes epialti exhibited two seasonal peaks in abundance on *Cancer anthonyi* (Fig. 1). The late summer peaks in worm abundance were associated with a highly aggregated dispersion of worms among hosts, but the winter peaks in worm abundance were not as highly aggregated. Prevalence was 100% in every month but August when prevalence was over 80% (1982 & 1983). Late summer peaks in the variance:mean ratio were significantly greater than other periods (χ^2 of David & Moore, $P < 0.05$). Patchiness showed an increase in relative aggregation only in the late summer months (Fig. 1B). In every case, the distribution of worms on *C. anthonyi* was highly overdispersed. The variance:mean ratio was never consistent with the expected ratio of the Poisson distribution ($\chi^2 > 200$, d.f. = 9–27).

Carcinonemertes epialti on *Hemigrapsus oregonensis* experienced a seasonal peak in worm intensity in the fall season (Fig. 2). The peak in worm abundance was paralleled by an increase in the variance:mean ratio but not by the patchiness index. The variance:mean ratios did not agree with the expected ratios of the Poisson distribution ($\chi^2 > 100$, $P < 0.05$, d.f. = 8–54), with the exception of the winter sample of 1983, ($\chi^2 = 30$, $P > 0.05$, d.f = 29). In March, 1983, the dispersion pattern fit a Poisson distribution because both prevalence and intensity were extremely low. In other months, the dispersion of the worm was highly overdispersed.

Carcinonemertes regicides from *Paralithodes camtschaticus* experienced a long summer plateau in worm abundance (Fig. 3A). The summer increase in worm abundance coincided with an increase in patchiness (Fig. 3B), but the decline in abundance in the fall did not. The variance:mean ratios never agreed with the expected values of the Poisson distribution ($\chi^2 > 1000$, $P < 0.05$, d.f. = 8–29). Patchiness was highest during the nadir in crab reproduction and worm abundance (April).

Carcinonemertes mitsukurii from non-ovigerous

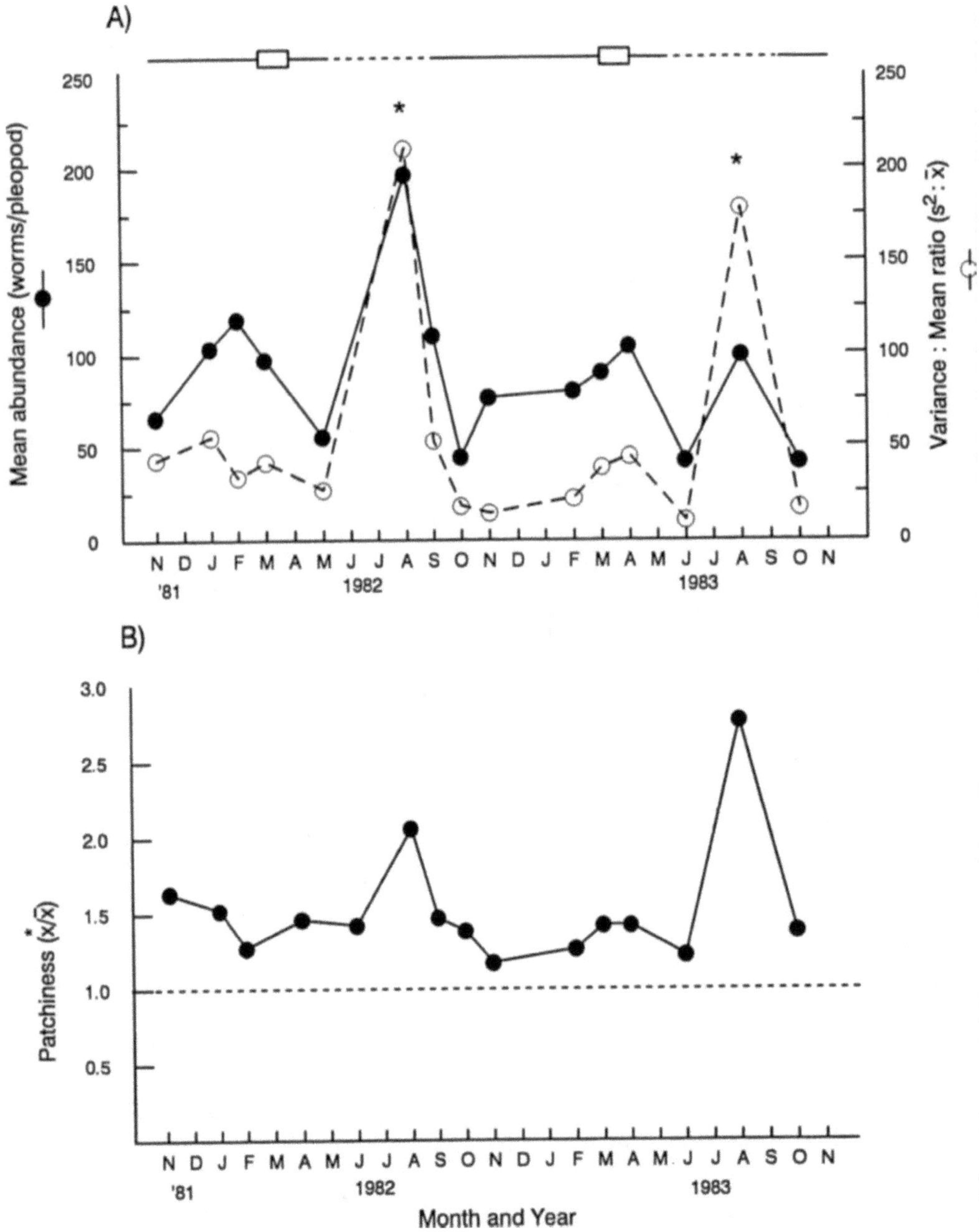

Fig. 1. Changes in abundance and relative aggregation of *Carcinonemertes epialti* on *Cancer anthonyi* from 1981–1983. A) Mean abundance and the variance:mean ratio. Data on crab reproduction are presented as a bar (peak in numbers of ovigerous females), line (moderate numbers of ovig. females), and dotted line (few ovig. females) [Data from Reilly (1987) and Shields *et al.* (1991)]. *Variance:mean ratio significantly greater than other ratios (David & Moore, 1954). B) Patchiness; the reference line indicates distributions approaching random dispersion ($\overset{*}{x} = \bar{x}$).

and ovigerous *Portunus pelagicus* exhibited a summer peak in abundance (Figs 4A & 4C). For worms on non-ovigerous hosts, the variance: mean ratio paralleled the fluctuations in worm abundance but patchiness showed a strong increase in aggregation in the late fall and early

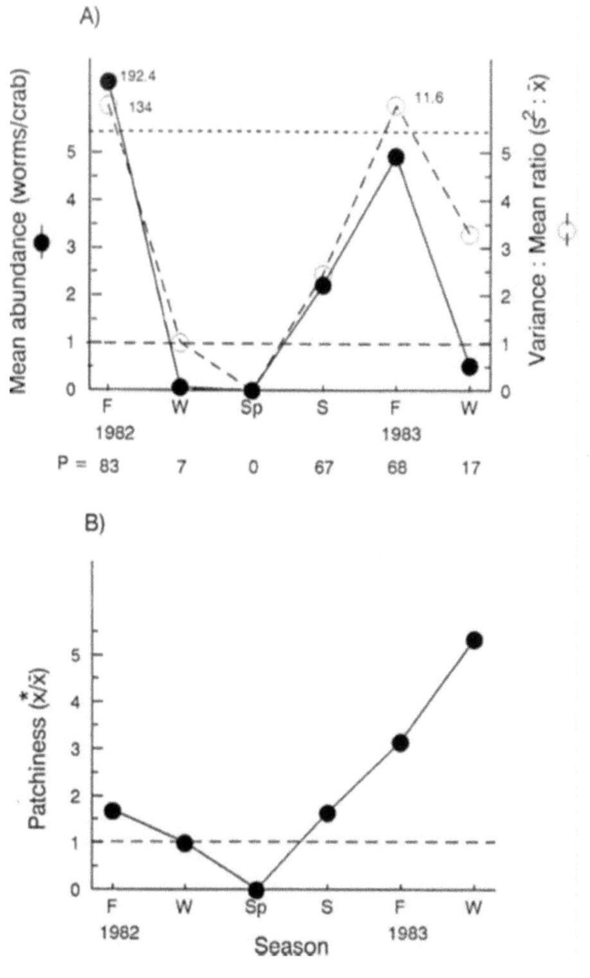

Fig. 2. Changes in abundance and relative aggregation of *Carcinonemertes epialti* on *Hemigrapsus oregonensis* over one year. A) Mean abundance and the variance:mean ratio; P = prevalence (%). B) Patchiness; the reference line indicates random dispersion ($\overset{*}{x} = \bar{x}$).

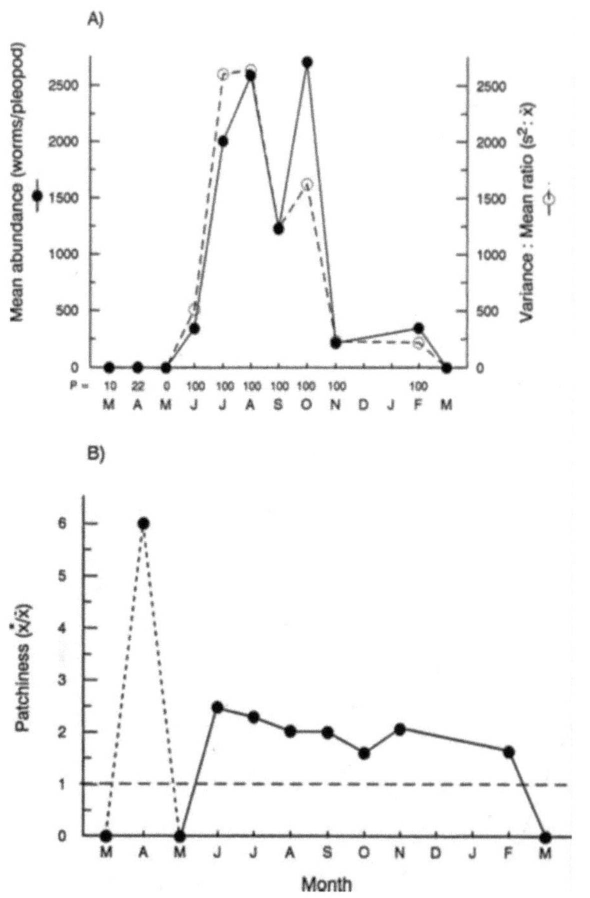

Fig. 3. Changes in abundance and degree of aggregation of *Carcinonemertes regicides* on *Paralithodes camtschaticus* with season (figure represents composite data of 3 years after Kuris *et al.*, 1991). A) Mean abundance and the variance:mean ratio; P = prevalence (%). B) Patchiness; the reference line indicates random dispersion ($\overset{*}{x} = \bar{x}$); dotted line indicates small size of host population.

winter months (Fig. 4B). The underlying aggregation pattern was different for worms from ovigerous crabs (Fig. 4D). In August, patchiness declined for worms on non-ovigerous hosts but increased sharply for worms on ovigerous hosts; these changes occurred at the onset of crab reproduction (Figs 4B & 4D).

The regression of mean crowding ($\overset{*}{x}$) on mean abundance ($\bar{x}$) was calculated for each worm species (Table 1). The correlation coefficient was high in every relationship ($R > 0.700$, $P < 0.05$). The regression coefficient (b), termed the density contagiousness coefficient, was significantly greater than one for the dispersion of each worm species ($t_s > 2.42$, $P < 0.05$) (Table 1). The Y-intercept, a, the index of basic contagion, was not significantly different from zero for any of the worm species ($t_s < 1.604$, $P > 0.05$).

Patterns with host embryogenesis

The dispersion pattern of *C. epialti* on *Cancer anthonyi* changed over the course of host embryogenesis. From early to late embryogenesis, the

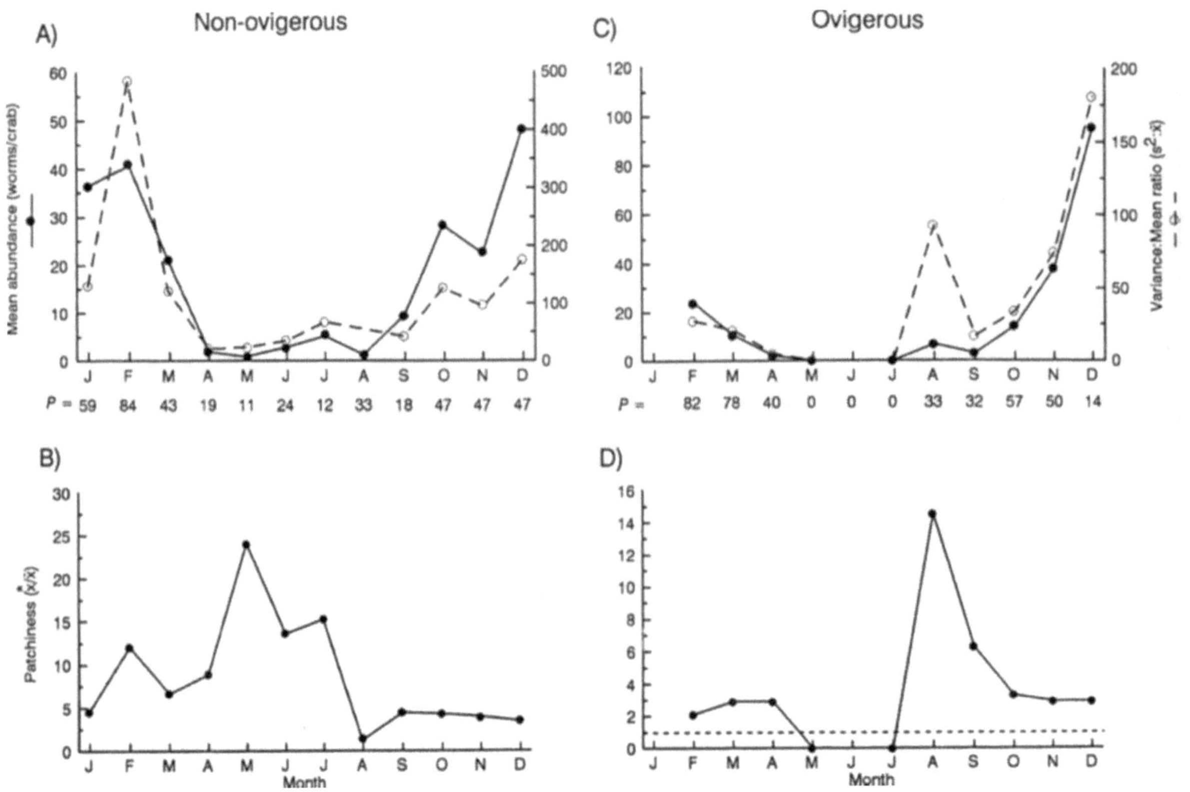

Fig. 4. Changes in abundance and degree of aggregation of *Carcinonemertes mitsukurii* on *Portunus pelagicus* with season; non-ovigerous and ovigerous hosts (A and B), ovigerous hosts (C and D). A) Mean abundance and the variance:mean ratio; P = prevalence (%). B) Patchiness; the reference line indicates random dispersion ($\overset{*}{x} = \bar{x}$). C) Mean abundance and the variance-:mean ratio; P = prevalence (%). D) Patchiness; the reference line indicates random dispersion ($\overset{*}{x} = \bar{x}$).

underlying distributions became more highly aggregated, possibly even polymodal, until host eclosion when the population was most highly overdispersed (Fig. 5, EDS VIII). The χ^2-test (for

Table 1. Regression statistics for mean crowding ($\overset{*}{x}$) on mean abundance ($\bar{x}$). All correlation coefficients were significantly greater than zero ($P < 0.05$). All regression coefficients were significantly greater than one. Data for *Cancer magister* are from Wickham (1986).

Host	Nemertean	*r*	a	b	*n*
Cancer anthonyi	*C. epialti*	0.908	−47.34	2.128	15
Hemigrapsus oregonensis	*C. epialti*	0.988	12.89	1.666	7
Cancer magister	*C. errans*	0.979	53.48	1.533	8
Paralithodes camtschaticus	*C. regicides*	0.982	67.03	1.887	10
Portunus pelagicus	*C. mitsukurii*[b]	0.700	40.20	4.486	12
Portunus pelagicus	*C. mitsukurii*[o]	0.947	15.59	2.637	8
Portunus pelagicus	*C. mitsukurii*[n]	0.749	9.21	6.081	12

[b] Worms from non-ovigerous and ovigerous hosts.
[o] Worms from ovigerous hosts only.
[n] Worms from non-ovigerous hosts only.

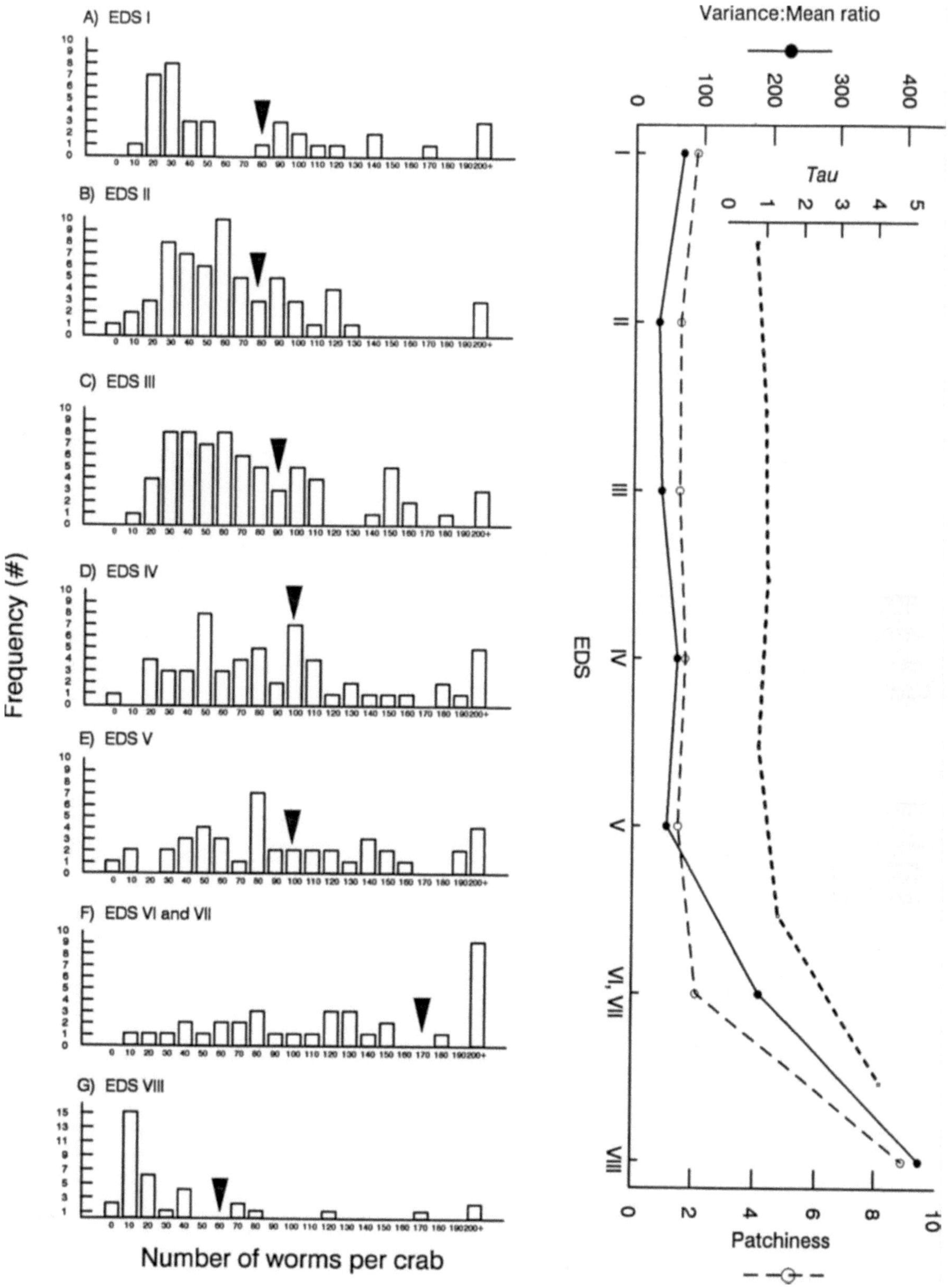

Fig. 5. Frequency distributions, mean abundance (arrow), variance:mean ratio, patchiness, and *tau* for *Carcinonemertes epialti* on *Cancer anthonyi* in relation to crab embryogenesis. Developmental stages of the crab eggs (EDS) are detailed in Shields *et al.* (1990).

the variance:mean ratio) was significantly different from the Poisson distribution for the dispersion of worms at every developmental stage of embryogenesis ($\chi^2 > 1591$, $P > 0.05$, d.f. = 8–61). In every case, the distributions of the worms were highly overdispersed.

The index of mortality/recruitment (*tau*) for *Carcinonemertes epialti* on *Cancer anthonyi* changed little during the early and middle periods of host embryogenesis (Fig. 5). Worm recruitment during this time was most likely independent of worm intensity. Later, near host eclosion, mortality operated in what may have been a many-or-few type of response (e.g., either most of the worms died on a host or relatively few died).

Discussion

Several models of infection dynamics may give rise to the aggregated dispersion patterns of parasites on their hosts (e.g., Bliss & Fisher, 1953; Crofton, 1971; Pennycuick, 1971). These are: (1) A heterogeneous host or parasite population (e.g., genetic variability); (2) constant birth-immigration-death rates of parasites through time; (3) infection by random clumps of individuals *versus* single individuals; (4) true contagion, wherein the presence of a parasite increases or decreases the probability of the host acquiring more parasites; and (5) superimposed or heterogeneous exposures of a parasite, in which the probability of infection is random (heterogeneous Poisson transmission) (see Elliott, 1977). Few tests have been devised to evaluate the mechanisms of aggregation, but the relationships between mean abundance, mean crowding, and patchiness may provide some insight (Iwao & Kuno, 1972; Kitron, 1980).

Four common patterns in the aggregation of the worms emerged and these were related to the timing and synchrony of reproduction of the host species (Table 2). (1) For worms on *Cancer anthonyi* and non-ovigerous *Portunus pelagicus*, aggregation increased at the nadir of the reproductive season of the host. For example, the patchiness of *C. epialti* on *Cancer anthonyi* increased with worm abundance only during the

Table 2. General patterns in the infestation and dispersion dynamics of *Carcinonemertes* spp., and related patterns in the reproductive ecology of their hosts.

Symbiont	*C. epialti*	*C. epialti*	*C. regicides*	*C. mitsukurii*
Prevalence	high	intermediate	high	intermediate
Intensity	low	low	high	intermediate
# seasonal peaks	2	1	1	1
Season of peak	summer, winter	fall	summer–fall	summer
Peak in $(\overset{*}{x})/\bar{x}$	summer	winter	spring–summer	summer, winter
Peak in $s^2/\bar{x}$	summer	fall	early summer	summer
Host	*Cancer anthonyi*	*H. oregonensis*	*P. camtschaticus*	*P. pelagicus*
Broods/year	multiple	multiple	single	multiple
Seasonal peak in reproduction	winter	winter	summer–fall	late winter, summer
Seasonal nadir in reproduction	summer	spring	spring	fall
Duration of embryogenesis	intermediate	intermediate	long	short
Aggregation pattern	1, 4	1, 3	1, 2, 4	1, 2, 3

Key to aggregation pattern – worm dispersion changes in relation to:
1 – Nadir in host reproduction.
2 – Onset of host reproduction.
3 – Physical/environmental change.
4 – Embryogenic cycle of host.

late summer. The worm population reached its maximum abundance and was most highly aggregated in association with the reproductive nadir of the crab population (Shields *et al.*, 1990). Worm aggregation (and abundance) is relatively lower in the winter than in the summer because more hosts are available for completion of the life cycle of the worm and because heavy infestations (outliers) have a greater impact on smaller populations (summer months).

(2) For worms on ovigerous *P. pelagicus* and *Paralithodes camtschaticus*, aggregation (as measured by patchiness) increased at the onset of the reproductive season of the host. For example, the abundance and aggregation of *Carcinonemertes regicides* on the red king crab, *Paralithodes camtschaticus*, followed a distinctive pattern that may be related to both season and crab embryogenesis. Female red king crabs oviposit in March–May, and embryos hatch in February–March (Marukawa, 1934; Kuris *et al*, 1991). Reproduction is synchronized in the crab population and this synchronization reduces the heterogeneity of the population to transmission by the worms. The degree of aggregation changed most notably with host oviposition and eclosion, *i.e.*, events which respectively increase opportunities for worm transmission and mortality (see #4 below). In general, as intensity increased, the degree of aggregation of *C. regicides* on *P. camtschaticus* decreased. This phenomenon was not observed in the *Cancer anthonyi-C. epialti* relationship, perhaps as a result of the seasonal fluctuations in the proportion of ovigerous hosts in the population.

(3) For worms on *P. pelagicus* and *H. oregonensis*, aggregation may have changed with fluctuations in the physical environment of the estuarine hosts. Changes in salinity are known to limit the abundance of *C. errans* on estuarine *Cancer magister* (McCabe *et al.*, 1987). For *C. mitsukurii*, abundance decreased while aggregation increased with decreased salinity during the rainy season (February–June; Shields & Wood, 1993). In addition, the aggregation pattern shifted in July and August (late winter) prior to and during the onset of crab reproduction. This was reflected in the differences between aggregation patterns of worms from ovigerous and non-ovigerous hosts. For *C. epialti* on *H. oregonensis*, worm dispersion was less aggregated at low densities (spring), than that at high densities (fall). The spring nadir in worm abundance and aggregation coincided with a sharp decline in the salinity of Bodega Bay (the El Niño storms of Spring 1983; Shields, pers. obs.) and with the nadir in crab reproduction (Kuris, 1978). Seasonality and the pattern of host reproduction could not be separated in the *C. epialti – H. oregonensis* association.

Changes in the dispersion patterns and related changes in parasite densities show the dynamic nature of the underlying factors that shape animal distributions. At low densities, many parasitic insects are distributed randomly among their host plants; while at higher densities, their underlying distribution is over-dispersed (Finney, 1941; Harcourt, 1961; Forsythe & Gyrisco, 1963). Further increases in density may shift the distribution back to an under-dispersed form (Waters, 1959; Dybas & Davis, 1962). Similarly, at low densities, worms on *H. oregonensis* were less aggregated than those at high densities, and occasionally were randomly distributed on the host population. Worms on *P. camtschaticus* were highly aggregated early in the reproductive season, but the degree of aggregation declined with increasing worm abundance and prevalence.

(4) Changes in the abundance and aggregation pattern of *Carcinonemertes epialti* over the developmental period of *Cancer anthonyi* documented two important forces acting on the dispersion of the worms. Firstly, the dramatic changes in the relationship between mean crowding and mean abundance with host embryogenesis provided evidence for heterogeneous waves of infestation (immigration, Fig. 6). Immigration into the egg mass had, however, two quite different effects on aggregation. After host eclosion, planktonic larvae colonized their new hosts. The effect of colonization was to reduce the degree of aggregation. Later in the course of embryogenesis, immigration caused an increase in the degree of aggregation. The sharp increase in abundance, mean crowding, and patchiness in the late stages

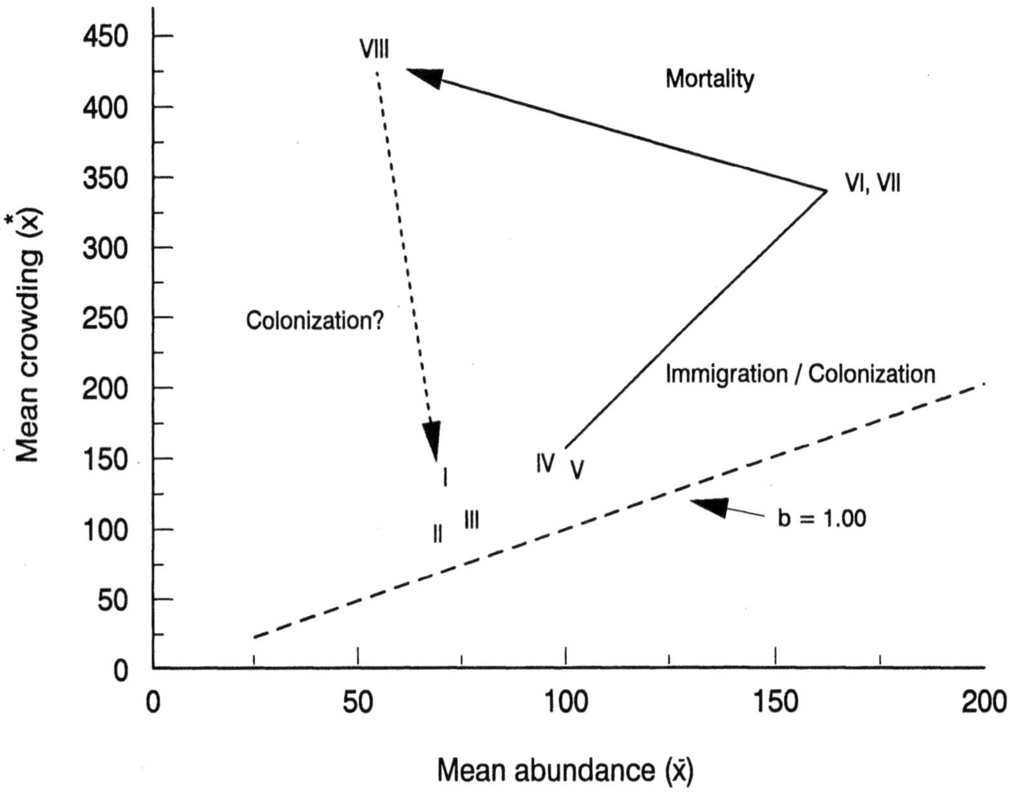

Fig. 6. Changes in the aggregation pattern of *Carcinonemertes epialti* on *Cancer anthonyi* in relation to embryogenesis of the host. Developmental stages of the crab eggs are detailed in Shields *et al.* (1990). The effect of colonization/immigration and worm mortality on aggregation are evident.

of host embryogenesis suggests that the immigration rates are not constant among hosts.

Secondly, death rates varied with the embryogenesis of the host. Species of *Carcinonemertes* either die or emigrate out of the egg masses of their hosts immediately prior to or during host eclosion (for review see Wickham & Kuris, 1985; Shields & Kuris, 1990; Kuris, 1993). Worm mortality/emigration increased just prior to host eclosion and acted to increase the degree of aggregation (Fig. 6). Worm immigration, emigration and mortality strongly shaped the observed aggregation patterns.

The life histories of the nemerteans revolve closely around the duration of the embryogenic cycle of their respective hosts (Shields & Kuris, 1990). In an analogous manner, the infestation and dispersion patterns of each worm species was tightly linked to the reproductive ecology of their hosts, even though details of the life histories of the host species and the worms vary considerably.

Acknowledgements

Drs Uriel Kitron, and Armand Kuris improved the manuscript. This work is a result of research sponsored in part by NOAA, National Sea Grant College Program, Department of Commerce, under grant number NA80AA-D-00120, project number R/F-75, through the California Sea Grant College Program, in part by the California State Resources Agency, and in part by an Australian Research Council grant (#AA8931718) to JDS and RJG Lester. The U.S. Government is authorized to reproduce and distribute for governmental purposes.

References

Anderson, R. M. & R. M. May, 1978. Regulation and stability of host-parasite population interactions: I. Regulatory processes. J. anim. Ecol. 47: 219–48.

Anderson, R. M. & D. M. Gordon, 1982. Processes influencing the distribution of parasite numbers within host populations with special emphasis on parasite-induced host mortalities. Parasitology 85: 373–398.

Anscombe, F. J., 1950. Sampling theory of the negative binomial and logarithmic series distributions. Biometrika 37: 340–432.

Bliss, C. I. & R. A. Fisher, 1953. Fitting the negative binomial distribution to biological data. Biometrics 9: 176–200.

Crofton, H. D., 1971. A model of host-parasite relationships. Parasitology 63: 343–364.

David, F. N. & P. G. Moore, 1954. Notes on contagious distributions in plant populations. Ann. Bot. 18: 47–53.

Dybas, H. S. & D. D. Davis, 1962. A population census of seventeen-year periodical cicadas (Homoptera: Cicadidae: Magicicada). Ecology 43: 432–444.

Elliott, J. M., 1977. Some methods for the statistical analysis of samples of benthic invertebrates, 2. Sci. Publ. Freshw. Biol. Assoc. 25: 1–148.

Evans, D. A., 1953. Experimental evidence concerning contagious distributions in ecology. Biometrika 40: 186–211.

Finney, D. J., 1941. Wireworm populations and their effects on crops. Ann. appl. Biol. 28: 282–295.

Forsythe, H. Y. & G. G. Gyrisco, 1963. The spatial pattern of the pea aphid in alfalfa fields. J. econ. Ent. 56: 104–107.

Harcourt, D. G., 1961. Spatial pattern of the imported cabbageworm, *Pieris rapae* (L.) (Lepidoptera: Pieridae), on cultivated Cruciferae. Can. Ent. 93: 849–859.

Hirsch, R. P., 1980. Distribution of *Polymorphus minutus* among its intermediate hosts. Int. J. Parasitol. 10: 243–248.

Hurlbert, S. H., 1990. Spatial distribution of the montane unicorn. Oikos 58: 257–271.

Iwao, S., 1968. A new regression method for analyzing the aggregation pattern of animal populations. Res. Pop. Ecol. 10: 1–20.

Iwao, S., 1970. Problems of spatial distribution in animal population ecology. In G. P. Patil (ed.), Random counts in scientific work. University Park & London: Pennsylvania State Univ. Press, 2: 117–149.

Iwao, S. & E. Kuno, 1971. An approach to the analysis of aggregation pattern in biological populations. In G. P. Patil, E. C. Pielou & W. E. Waters (eds), Statistical Ecology. University Park and London: Pennsylvania State Univ. Press 1: 461–513.

Kitron, U. D., 1980. The pattern of infestation of the beach-hopper amphipod *Orchestoidea corniculata*, by a parasitic mite. Parasitology 81: 235–249.

Kitron, U. D. & G. I. Higashi, 1985. *Schistosoma haematobium* in upper Egypt: analysis of dispersion patterns. Am. J. Trop. Med. Hyg. 34: 331–340.

Kuris, A. M., 1978. Life cycle, distribution and abundance of *Carcinonemertes epialti*, a nemertean egg predator of the shore crab, *Hemigrapsus oregonensis*, in relation to host size, reproduction and molt cycle. Biol. Bull. 154: 121–137.

Kuris, A. M., 1993. Life cycles of nemerteans that are symbiotic egg predators of decapod Crustacea: adaptations to host life histories. Hydrobiologia 266: 1–14.

Kuris, A. M., S. F. Blau, A. J. Paul, J. D. Shields & D. E. Wickham, 1991. Infestation by brood symbionts and their impact on egg mortality of the red king crab, *Paralithodes camtschatica*, in Alaska: geographic and temporal variation. Can. J. Fish. aquat. Sci. 48: 559–568.

Lloyd, M., 1967. Mean crowding. J. anim. Ecol. 36: 1–30.

Margolis, L., G. W. Esch, J. C. Holmes, A. M. Kuris, G. A. Schad, 1982. The use of ecological terms in parasitology (report of an ad hoc committee of the American Society of Parasitologists). J. Parasitol. 68: 131–133.

May, R. M., 1975. Stability and complexity in model ecosystems, 2nd Ed. Princeton Univ. Press, Princeton.

Marukawa, H., 1933. Biological and fishery research on Japanese king crab *Paralithodes camtschatica* (Tilesius). J. imper. Fish. Exp. St., Tokyo 4: 1–152.

McCabe, G. T., R. L. Emmett, T. C. Coley & R. J. McConnell, 1987. Effect of river-dominated estuary on the prevalence of *Carcinonemertes errans*, an egg predator of the Dungeness crab, *Cancer magister*. Fish. Bull. 85: 140–142.

McCallum, H. I. & R. M. Anderson, 1984. Systematic changes in host susceptibility to infection: demographic mechanisms. Parasitology 89: 195–208.

Pennycuick, L., 1971. Frequency distributions of parasites in a population of three-spined sticklebacks, *Gasterosteus aculeatus* L., with particular reference to the negative binomial distribution. Parasitology 63: 389–406.

Pielou, E. C., 1977. An introduction to mathematical ecology. 2nd edition. Wiley-Interscience, New York.

Reilly, P. N., 1987. Population studies of rock crab, *Cancer antennarius*, yellow crab, *C. anthonyi*, and Kellet's whelk, *Kelletia kelletii*, in the vicinity of Little Cojo Bay, Santa Barbara County, California. Calif. Fish Game 73: 88–98.

Roe, P., 1979. Aspects of development and occurrence of *Carcinonemertes epialti* (Nemertea) from shore crabs in Monterey Bay, California. Biol. Bull. 156: 130–140.

Scott, M. E., 1987. Temporal changes in aggregation: a laboratory study. Parasitology 94: 583–595.

Shields, J. D. & A. M. Kuris, 1988. Temporal variation in abundance of the egg predator *Carcinonemertes epialti* (Nemertea) and its effect on egg mortality of its host, the shore crab, *Hemigrapsus oregonensis*. Hydrobiologia 156: 31–38.

Shields, J. D. & A. M. Kuris, 1990. *Carcinonemertes wickhami* n. sp. (Nemertea), an egg predator on the California lobster, *Panulirus interruptus*, with notes on symbiont-host adaptations. Fish. Bull. 88: 279–287.

Shields, J. D., R. K. Okazaki & A. M. Kuris, 1990. Brood mortality and egg predation by the nemertean, *Carcinon-*

emertes epialti, on the yellow rock crab, *Cancer anthonyi*, in southern California. Can. J. Fish. aquat. Sci. 47: 1275–1281.

Shields, J. D., R. K. Okazaki & A. M. Kuris, 1991. Fecundity and the reproductive potential of *Cancer anthonyi* in southern California. Fish. Bull. 89: 299–305.

Shields, J. D. & F. E. I. Wood, 1993. Impact of parasites on the reproduction and fecundity of the blue sand crab *Portunus pelagicus* from Moreton Bay, Australia. Mar. Ecol. Prog. Ser. 92: 159–170.

Sokal, R. R. & F. J. Rohlf, 1981. Biometry, 2nd edn. W. H. Freeman & Co., San Francisco, California, 859 pp.

Waters, W. E., 1959. A quantitative measure of aggregation in insects. J. econ. Ent. 52: 1180–1184.

Wickham, D. E., 1979. Predation by *Carcinonemertes errans* on eggs of the Dungeness crab, *Cancer magister*. Mar. Biol. 55: 45–53.

Wickham, D. E., 1986. Epizootic infestations by nemertean brood parasites on commercially important crustaceans. Can. J. Fish. aquat. Sci. 43: 2295–2302.

Wickham, D. E. & A. M. Kuris, 1985. The comparative ecology of nemertean egg predators. Am. Zool. 25: 127–134.

Williams, C. B., 1964. Patterns in the balance of nature. Academic Press, London, 324 pp.

Hydrobiologia **266**: 57–80, 1993.
R. Gibson, J. Moore & P. Sundberg (eds), Advances in Nemertean Biology

Carcinonemertes pinnotheridophila sp. nov. (Nemertea, Enopla, Carcinonemertidae) from the branchial chambers of *Pinnixa chaetopterana* (Crustacea, Decapoda, Pinnotheridae): description, incidence and biological relationships with the host

John J. McDermott[1] & Ray Gibson[2]
[1]*Department of Biology, Franklin and Marshall College, Lancaster, Pensylvania 17604, USA;* [2]*School of Natural Sciences, Liverpool Polytechnic, Byrom Street, Liverpool L3 3AF, UK*

Key words: host-symbiont relationships

Abstract

A new species of *Carcinonemertes*, *C. pinnotheridophila*, is described and illustrated. The worms were found in the pinnotherid crab *Pinnixa chaetopterana* collected from the coasts of New Jersey, North Carolina and Florida. The anatomy of the new species is compared and contrasted with that of other members of the genus *Carcinonemertes* and an emended generic diagnosis is provided. The nemerteans only inhabit female hosts, crabs being infested with one or two mature female worms but with no more than one ensheathed in each branchial chamber; one or more smaller male nemerteans may be associated with each female. Sheaths are attached to the medial portion of the host branchial exoskeleton, and project through an opening in the floor of the chamber to exit via another aperture in the sternum; the anterior part of the sheath opens in the excurrent canal of the branchial chamber. Female worms cement their oval egg sacs on the pleopods to which the crab's eggs are also attached. Attachment, development and hatching of both host and symbiont eggs are synchronous. The incidence of infestation, reproductive potential of the nemertean, damage to its host and tolerance of the crab's growth cycles are described.

Introduction

Since Kölliker (1845: 93) first briefly described a worm that he called *Nemertes cartinophilos* 'in dem Eierklumpen einer kleinen Krabbe', there have been many reports of enoplan nemerteans found living on the gill filaments, among the egg masses or attached to the carapace of a wide variety of decapod crustaceans. One genus in particular, *Carcinonemertes*, has a world-wide distribution with nine described species, one of which is divided into two subspecies; these are *C. australiensis* Campbell, Gibson & Evans, 1989, *C. carcinophila* (Kölliker, 1845) (with the subspecies *C. c. carcinophila* [Kölliker, 1845] and *C. c. imminuta* Humes, 1942), *C. coei* Humes, 1942, *C. epialti* Coe, 1902, *C. errans* Wickham, 1978, *C. humesi* Gibson & Jones, 1990, *C. mitsukurii* Takakura, 1910, *C. regicides* Shields, Wickham & Kuris, 1989, and *C. wickhami* Shields & Kuris, 1990. Members of this genus are all regarded variously as egg predators or brood parasites (Wickham, 1986) and are typically found on brachyuran decapods (Campbell *et al.*, 1989). Two other genera, both monospecific, have also been reported from the egg masses of decapod Crustacea; these are *Alaxinus oclairi* Gibson, Wickham & Kuris, 1990, from an anomuran crab, and *Pseudocarcinonemertes homari* Fleming & Gibson, 1981, from a lobster; a third genus, *Coenemertes*, is known to occur on the thoracic limbs of *Callianassa* (Corrêa, 1966), but a definite relationship between the nemerteans and crab has not been established.

Identified species of *Carcinonemertes* have been recorded on some 47 species of Brachyura, representative of nine families (Campbell *et al.*, 1989), but to date none has been described from pinnotherid crabs, all of which are found as commensals or parasites in bivalve molluscs, worm tubes or echinoderms (Schmitt *et al.*, 1973; Abele & Felgenhauer, 1982). McDermott (1966, 1967) was the first to report nemerteans associated with pinnotherids when he found worms of apparently the same species in the branchial chambers of two crab host species, *Pinnixa chaetopterana* Stimpson, 1860, and *Pinnotheres ostreum* Say, 1817, although subsequently (McDermott, 1981) he could find no evidence of nemerteans in *Pinnixa cylindrica* (Say, 1818). The present studies have shown that the nemerteans from *Pinnixa chaetopterana*, a well-known symbiotic crab living with a variety of tubicolous polychaetes (particularly *Amphitrite ornata* [Leidy, 1855] and *Chaetopterus variopedatus* [Renier, 1804]), belong in the genus *Carcinonemertes*, but that they possess anatomical features which exclude them from any of the known species; the name *Carcinonemertes pinnotheridophila* sp. nov. is accordingly proposed for them. The purpose of the present study, apart from describing the new species, is to provide details of the intimate host-symbiont relationship, and to compare it with those involving other species of the Carcinonemertidae and their decapod hosts.

Materials and methods

In New Jersey, crabs were collected during the period May 1963 to October 1964 from the fragile, U-shaped tubes of *Amphitrite ornata* located on a muddy-sand tidal flat inside the mouth of Hereford Inlet (39° 02′ N, 74° 48′ W). This site has been under study for more than 30 years and has a salinity range of about 28–32‰ and water temperature range of 0.5–27.3 °C. Crabs from North Carolina were all obtained from the flexible, U-shaped tubes of *Chaetopterus variopedatus* dug from muddy-sand tidal flats mostly in the vicinity of Beaufort (34° 43′ N, 76° 40′ W) (Pivers Island at the Duke University Marine Laboratory, Bird Shoal and Cape Lookout). Collections were made from June to August 1965 and 1967, and June to October 1966; hydrographic data were not routinely recorded but water temperatures over the flats were often > 25 °C and salinities always exceeded 30‰. In Florida some crabs came from *Chaetopterus* tubes, but others were found in the J-shaped burrows of *Arenicola cristata* Stimpson, 1856, the siphonal burrows of *Tagelus plebeius* (Lightfoot, 1786), or from unknown hosts. All the hosts were located on tidal, calcareous-sand flats in the vicinity of the University of Miami's School of Marine and Atmospheric Science on Virginia Key (25° 40′ N, 80° 10′ W). Collections were made from December 1966 to April 1967, water temperatures ranging from 19–25 °C and salinities always being above 30‰.

Crabs from individual *Amphitrite* tubes were isolated in jars in the field, whilst those from *Chaetopterus* were brought to the laboratory within worm tubes secured at both ends with rubber bands. All were isolated in 9 cm diameter fingerbowls in the laboratory, fed green algae and flakes of an infant cereal (Pabalum), given fresh water every other day, and observed for varying lengths of time. They were sexed, measured with vernier calipers (carapace width to the nearest 0.1 mm) and eventually sacrificed to be examined for the presence of nemerteans. Ovigerous crabs liberating the biocular, ciliated nemertean larvae were maintained for longer periods of time, some for more than one year. A total of 1234 *Pinnixa chaetopterana* (633 females) from all locations was examined. With the exception of a single male crab that had an immature worm under its abdomen, all infested crabs were female.

The length of living, naturally relaxed and moving worms was measured with an ocular micrometer in a dissecting microscope. The dimensions of egg sacs produced by mature females were measured in a similar manner; the numbers of eggs in individual egg sacs were counted. Living worms were gently flattened to detect internal structures, but particularly to observe and measure (ocular micrometer in compound microscope) the stylet apparatus.

Some worms removed from crabs were preserved in 10% seawater formalin, 70% ethyl alcohol or Bouin's original fluid. Variously preserved whole mounts, usually gently flattened, were stained in Semichon's Carmine. A set of serial cross-sections of a female in a sheath with two males was preserved in the 10% formalin and stained with Alum Haematoxylin-Eosin. Other individuals were fixed in Bouin's fluid and stained with the same dyes. Histological sections were cut at 8 μm in paraffin wax.

Systematics

Family Carcinonemertidae Sumner, Osburn & Cole, 1913

Sumner *et al.* (1913) were the first to use the family name Carcinonemertidae, although they neither indicated that it was a new taxon nor provided a diagnosis. Definitions of the family used subsequently have been based on that given by Humes (1942), later modified by Wickham (1978), but Shields *et al.* (1989) have recently amended this diagnosis. Fleming & Gibson (1981) included the genus *Pseudocarcinonemertes* in the Carcinonemertidae, but Uhazy *et al.* (1985) transferred this genus to the Tetrastemmatidae; Campbell *et al.* (1989: 261) comment that 'since no secure diagnosis can be made for [the Tetrastemmatidae], and disagreement exists concerning the systematic significance of different features of *Pseudocarcinonemertes*, its removal from the Carcinonemertidae cannot at present be either substantiated or justified'. Shields *et al.* (1989: 927) remove from their familial diagnosis features which had previously been included, namely the absence of accessory stylet pouches and stylets from the proboscis, the small non-glandular form of the anterior proboscis chamber and the absence of an excretory system; they comment, however, that 'These characters may ... be useful for generic diagnosis'.

Genus Carcinonemertes *Coe, 1902*

Diagnosis of the genus: The following diagnosis of the genus is amended from that given by Gibson & Jones (1990: 196) to accord with comments made by Shields *et al.* (1989):

Monostiliferous enoplan nemerteans living as ectohabitants on the gills, under the abdomen, in apodemes and axillae of the limbs, or in and on the egg masses of anomuran, brachyuran or palinuran Decapoda; proboscis with single central stylet but neither accessory stylets nor pouches, very short, anterior chamber usually extremely short and non-glandular; rhynchocoel restricted to anteriormost portion of body, with wall lacking distinct musculature; lateral nerve cords lying internal to well developed layer of submuscular glands; cephalic glands well developed; cerebral sensory organs missing; with two eyes; cephalic muscle fibres present; excretory system indistinguishable histologically; blood system a simple vascular loop, without mid-dorsal vessel; Takakura's duct present in males; internal fertilisation and ovoviviparity commonly occur; extensive development of spermatozoa and ova; most species secrete mucous sheath attached to endopodite hairs of ovigerous hosts, less commonly to host gill filaments.

Carcinonemertes pinnotheridophila *sp. nov.*

Etymology

The specific epithet is derived by combining the familial name of the host with the Greek *philos* (having an affinity for) and is indicative of the nemerteans' association with pinnotherid crabs.

Type specimens
All type specimens are deposited with the National Museum of Natural History, Smithsonian Institution, Washington, D.C.; they comprise the holotype, USNM 146149, complete set of transverse sections of an immature female, paratypes, set of sections of female with two males in sheath, USNM 146150, stained whole mount of immature female, USNM 146151, and stained whole mounts of five mature males, USNM 146152.

Type locality
In the branchial chambers of the pinnotherid crab, *Pinnixa chaetopterana*, originally collected during 1963 in New Jersey and 1965 in North Carolina (McDermott, 1966).

Description

External features
In life the nemerteans have a typical carcinonemertean appearance, with a slender body bluntly rounded at either end. Dimensions of living worms, measured when relaxed and moving, were: males, 0.9-3.3 mm long (mean, 2.27 ± 0.78 mm, $n = 19$), 0.05-0.17 mm wide; females, 4.7-15.0 mm long (mean, 8.40 ± 2.91 mm, $n = 15$), 0.2-0.5 mm wide; small unsexed individuals, 0.6-2.3 mm long (mean, 1.15 ± 0.47 mm, $n = 26$), 0.07-0.21 mm wide. Mature females when contracted in their mucous sheaths within the crab gill chamber are about 3.4 mm long (mean, 3.39 ± 0.89 mm, $n = 7$), *i.e.*, they are about as long as the medial part of the chamber. Figure 1 shows the normal appearance of a mature male, but one larger worm, an immature female, possessed a distinct narrowing of the body just in front of the head (Fig. 2), a similar feature being reported in male *C. wickhami* by Shields & Kuris (1990). Males and small sexually undifferentiated worms are off-white to tan in colour, but mature females are a deep orange-red; immature females are paler but never pink as, for example, seen in *C. carcinophila*.

Body wall, musculature and parenchyma
The epidermis is about 8–10 μm in maximum thickness. Its construction resembles that described for other carcinonemerteans by Humes (1942), Norenburg (1985) and Campbell *et al.* (1989). Internally the epidermal basement membrane is exceedingly delicate and virtually indistinguishable except under high power magnification.

The body wall muscles possess a typical enoplan arrangement; in most parts of the body the outer circular layer is 4–6 μm thick, the inner longitudinal muscles 10–15 μm across, but in the mid-body region the latter muscle layer is in places up to 40–45 μm deep. Both muscle layers reach the tip of the head but are pre-cerebrally gradually reduced to only one or two fibres' thickness. The anteriorly undivided nature of the longitudinal muscles and absence of a diagonal layer are both characteristic of other members of the genus. Dorsoventral muscle fibres, which have been reported from the intestinal region of some species, were not found in the present form.

Below the body wall muscle layers the submuscular glands (referred to as hypodermal glands by Humes, 1942) form a distinct zone some 20–30 μm in maximum depth (Fig. 3); their attenuated distal portions extend peripherally between the fibres of the muscle layers towards the body surface. The glands are most strongly developed in the foregut and anterior intestinal regions of the body, whilst in front of the brain their density decreases anteriorly and they are missing from the cephalic tip. In males the submuscular glands are particularly pronounced along the lateral body margins. The extent of these glands throughout the body was regarded as a generic feature by Coe (1902a, b), but similar glands are also found in *Pseudocarcinonemertes* (Fleming & Gibson, 1981).

Parenchymatous connective tissues are nowhere abundantly developed.

Proboscis apparatus
The proboscis pore opens into a thin-walled rhynchodaeum whose delicate epithelium is neither ciliated nor glandular.

The rhynchocoel, as in other *Carcinonemertes*,

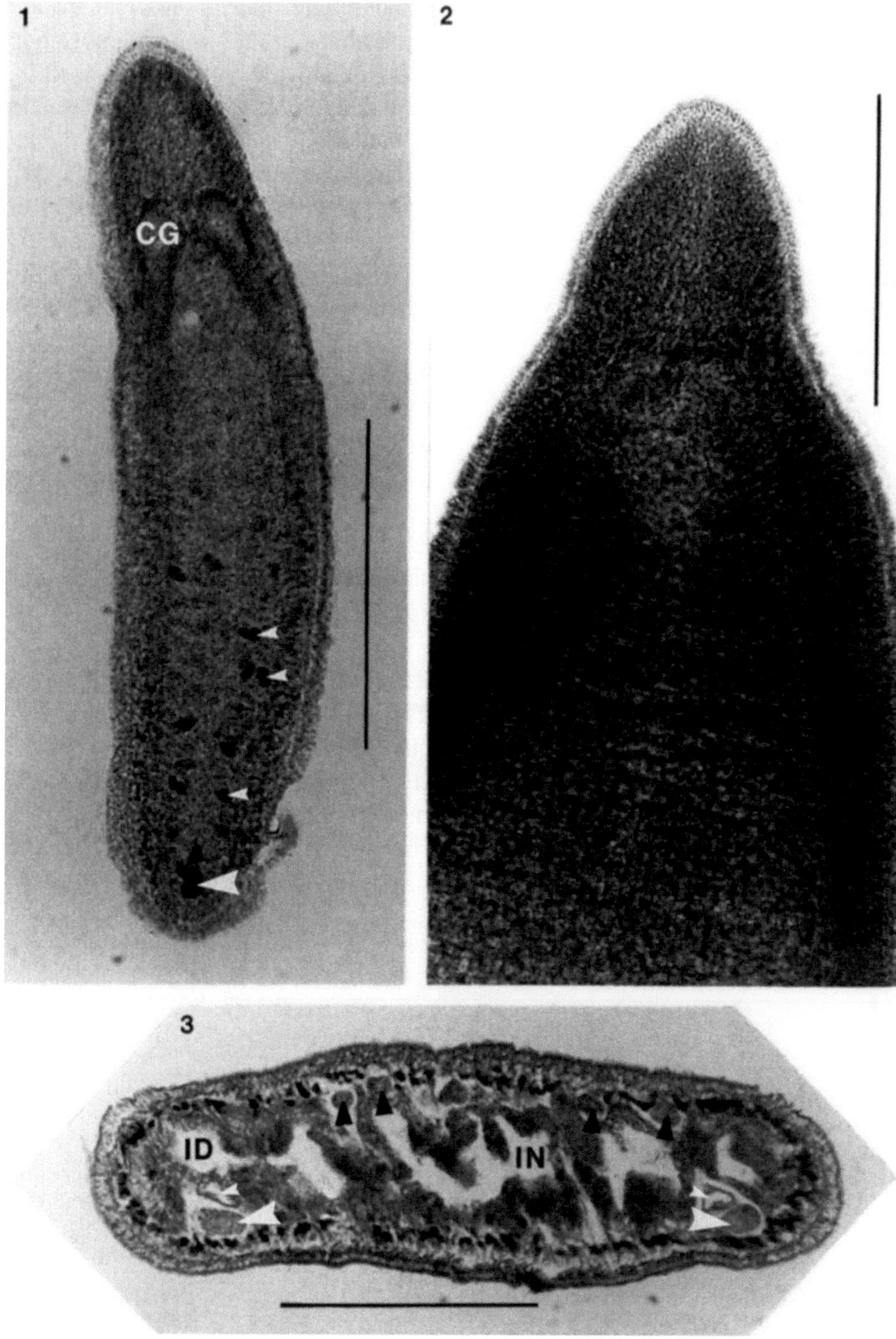

Figs 1–3. Carcinonemertes pinnotheridophila sp. nov. *1*. Whole mount preparation of mature male; note the testes and seminal vesicle, indicated by small and large arrowheads respectively. CG, cerebral ganglia. Scale bar = 250 μm. *2*. The anterior portion of an immature female showing the distinct narrowing of the head just in front of the brain. Scale bar = 250 μm. *3*. Transverse section through the intestinal region of an immature female. Subepidermal glands show black just below the body wall muscle layers. The lateral nerve cords and blood vessels are indicated by large and small white arrowheads respectively, the four rows of ovaries are indicated by black arrowheads. ID, branch of intestinal diverticulum; IN, intestine. Scale bar = 200 μm.

is short; in an individual 3.75 mm long it reaches to about 285 μm behind the tip of the head, *i.e.*, approximately 7.5% of the body length. In other examples with body lengths of 0.7 mm and 2.1 mm the rhynchocoel extends back between 110 μm and 320 μm from the anterior tip, *i.e.*, 15–16% of the total length. This suggests that, even allowing for differential contraction during fixation, rhynchocoel length becomes progressively shorter relative to body length as the worms increase their overall size. The wall of the rhynchocoel is thin, indistinct and appears to lack muscles; in these respects it resembles that of other carcinonemerteans.

The proboscis is small but retains the characteristic enoplan division into three regions. The anterior portion is the smallest, lined by a non-papillate epithelium in which no gland cells could be distinguished, the posterior chamber the largest and with a glandular epithelium. In a worm 3.75 mm long dimensions measured for the three proboscis regions were 20 μm long and 25 μm diameter (anterior chamber), 25 μm long by 30 μm diameter (stylet bulb) and 70 μm long and 35 μm maximum diameter (posterior chamber). Comparison with figures given by Shields & Kuris (1990: Table 1) for other species of *Carcinonemertes* shows that the proboscis of the present taxon is significantly smaller than that of any other form, although the size of the posterior chamber approximates to that reported for *C. c. carcinophila* and *C. epialti*. No retractor muscle could be distinguished.

The stylet armature comprises a single central stylet (Fig. 4) with no accessory stylets or pouches; the absence of reserve stylets and pouches is characteristic of *Carcinonemertes* adults, although one of us (R.G., unpublished observations) has seen these structures in two juvenile specimens identified as *C. epialti* from the Californian coast, and Shields *et al.* (1989: 927) note that accessory stylets and pouches are present in 'several carcinonemertid forms from *Paralithodes camtschatica* ... [which] appear related to the type genus, *Carcinonemertes*'. Dimensions of the central stylet and basis of the present species are given in Table 1.

Alimentary canal

The short oesophagus emerges from the rhynchodaeum a short distance in front of the brain. Its epithelium is unciliated but contains small numbers of gland cells. In one specimen the oesophagus is partially everted (Figs 5, 6) and ap-

Table 1. Dimensions of the stylet apparatus recorded for eleven specimens of the carcinonemertean found in *Pinnixa chaetopterana*; all measurements, made under oil immersion, are given in μm. ND = small worms, sex not determined.

Specimen number	Sex	Stylet length	Basis length	Basis width	Stylet:basis ratio
1	M	7.0	16.0		0.438
2	M	7.0	17.0	6.5	0.412
3	M	5.5	18.0	5.5	0.306
4	M	6.6	17.6	5.5	0.375
5	M	7.5	20.5	6.0	0.366
6	M	7.0	20.0		0.350
7	M	5.5	17.6		0.313
8	ND	7.7	14.3	6.6	0.538
9	ND	7.5	17.5	7.0	0.429
10	ND	6.6	19.8	7.7	0.333
11	F	8.0	16.0	5.5	0.500
Mean values					
all specimens		6.9	17.7	6.3	0.390
males		6.6	18.1	5.9	0.365

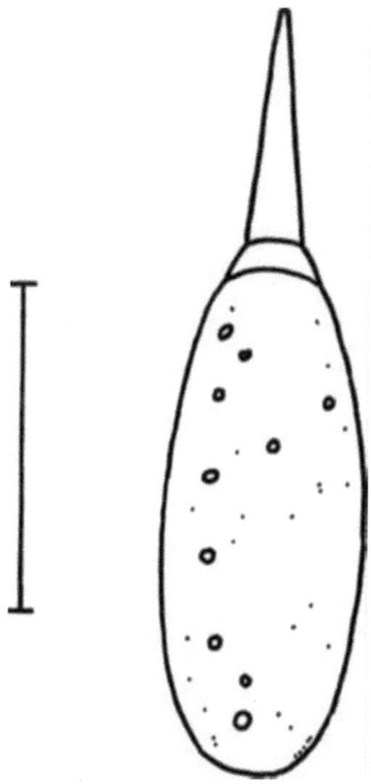

Fig. 4. Carcinonemertes pinnotheridophila sp. nov. Drawing of central stylet and basis of a male obtained from a female *Pinnixa chaetopterana* collected 11 July 1965 at Beaufort, North Carolina. Scale bar = 10 μm; total length of proboscis armature was 24 μm.

pears similar to that illustrated by Humes (1942: pl. IV, Figs 50–52) for *C. c. imminuta.*

The glandular and ciliated stomach, whose epithelium is mostly 10–15 μm thick, is not folded as in many enoplans. At its rear it opens directly into an intestine which, even in immature individuals, bears deeply branched lateral diverticula (Fig. 3); Humes (1942: 34) noted for *C. c. imminuta* that 'In sexually mature females the intestine is deeply lobed on both sides of the body [but that] ... In young worms and those [with] ovaries [which] are undeveloped, there is only a very slight and somewhat irregular crenation of the intestinal wall'. Other authors (e.g., Shields & Kuris, 1990; Gibson & Jones, 1990) have shown deep but unbranched lateral pouches extending along either side of the intestine. A pair of short lateral pouches (Shields & Kuris, 1990: Fig. 1, refer to these as the lateral horns of the intestine in *C. wickhami*) projects on either side of the stomach-intestine junction. The anus opens at the posterior tip of the body. The general organisation of the alimentary tract thus closely resembles that described for other members of the genus.

Blood system

The blood system comprises a pair of longitudinal vessels (Fig. 3) which communicate with each other only via a short pre-cerebral loop anteriorly and a supra-intestinal connective posteriorly. There is no mid-dorsal vessel and pseudometameric transverse links in the intestinal region are missing. Except for the position of the posterior connective, which in some *Carcinonemertes* species is sub-intestinal, the organisation of the blood system is typical of the genus. The blood vessels throughout the body possess distinct walls.

Nervous system

The brain (Fig. 1) is small, with similar sized dorsal and ventral lobes enclosed by an extremely thin outer neurilemma; no inner neurilemma between the fibrous and ganglionic cerebral components could be distinguished. The dorsal cerebral commissure, 10 μm in diameter, is slightly anterior to and smaller than the ventral commissure, which has a width of about 15 μm.

The lateral nerve cords possess neither accessory nerves nor neurochords, but in some sections evidence of neuromuscular fibres (myofibrillae), similar to those described from *C. australiensis* (Campbell *et al.*, 1989), could be discerned at the interface between fibrous and ganglionic tissues. The lateral nerves for most of the body length run ventrolaterally just internal to the submuscular glands (Fig. 3); the nerves meet posteriorly by a supra-intestinal commissure just in front of the anus.

No part of the peripheral nervous system could be distinguished in any of the sections examined.

Sense organs

The present species resembles all other *Carcinonemertes* in lacking cerebral sensory organs. It differs, however, in that whereas all previously described species as adults normally possess two

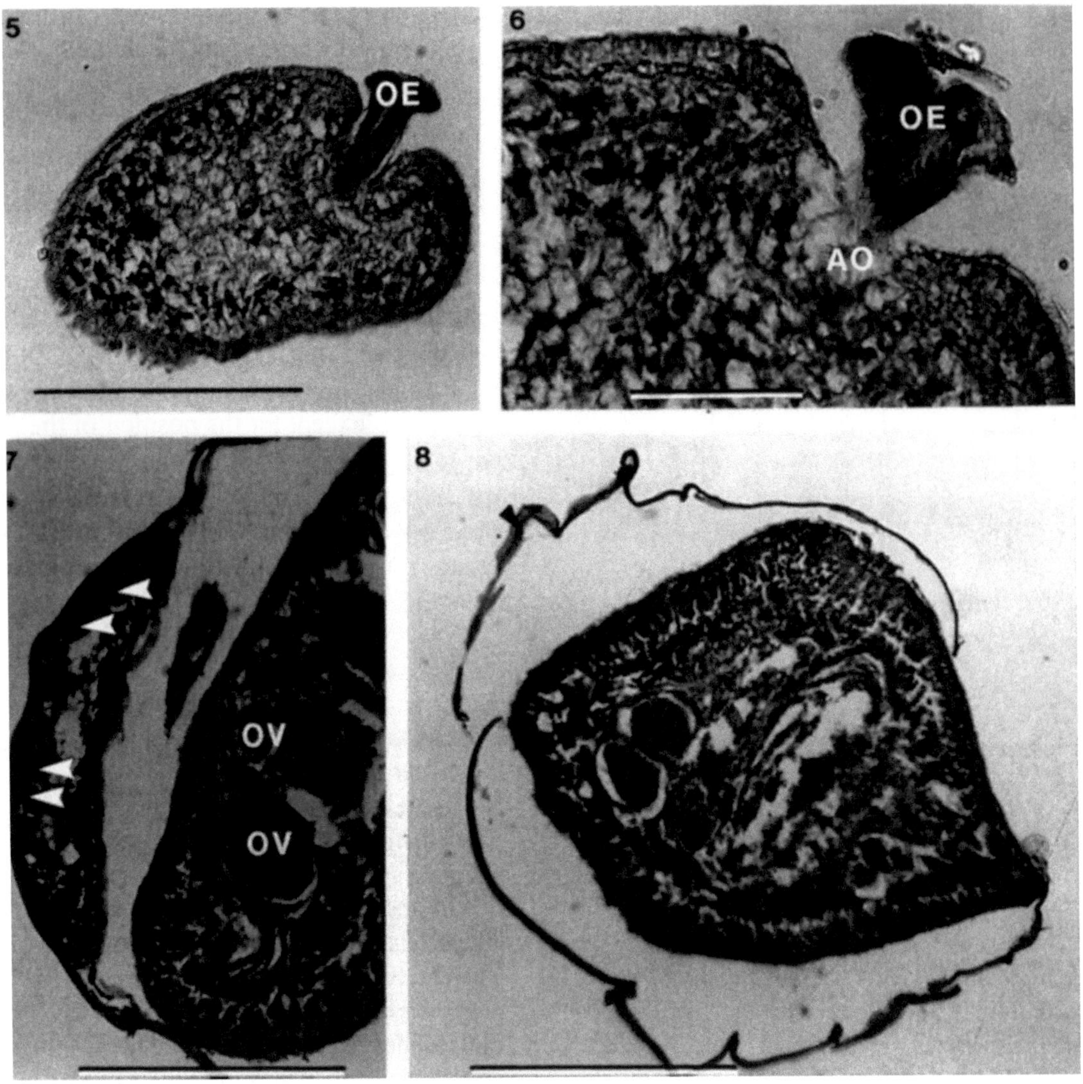

Figs 5–8. Carcinonemertes pinnotheridophila sp. nov. *5.* Oblique section through the tip of the head to show the partially everted oesophagus (OE). Note the absence of submuscular glands from the extreme anterior portion of the body. Scale bar = 200 μm. *6.* Oblique section through the tip of the head to show the apical organ (AO) and part of the everted oesophagus (OE). Scale bar = 50 μm. *7.* Section through a sheath to show the presence of male (to the left) and female (to the right) worms. Testes containing ripe spermatozoa in the male are indicated by arrowheads. OV, ovary. Scale bar = 200 μm. *8.* Transverse section through a female specimen in her sheath; one of the surface lapillae on the sheath is indicated by the arrowhead upper left. Scale bar = 200 μm.

distinct eyes (occasional irregular eye fragmentation is known to occur [Coe, 1902a, b; Gibson & Jones, 1990] and Humes [1942] cites an example of *C. c. imminuta* with only a single eye), the worms from pinnotherid crabs are totally eyeless as adults (Figs 1, 2), though newly hatched larvae possess two black, irregularly rounded eyespots. The loss of eyes with growth marks a major distinction between the present species and all other known members of the genus.

Apical organ and cephalic glands

A single, pit-like apical organ, 25–30 μm in diameter, opens at the tip of the head immediately above the proboscis pore (Fig. 6). A similar structure has been recorded in only one other species, *C. australiensis* (Campbell *et al.*, 1989). The cephalic glands of the present form are, as in all carcinonemerteans, well developed and extend back to the cerebral region; no evidence of the numerous improvised pores distributed over the cephalic surface through which the glands discharge in *C. c. imminuta*, *C. coei* and *C. humesi* (Humes, 1942; Gibson & Jones, 1990) could be found and their method of secretion is unknown.

Excretory system

No trace of structures which could be interpreted as components of an excretory system was seen in any of the sections studied. Shields *et al.* (1989) were the first to find an excretory system in carcinonemerteans (*C. epialti*, *C. regicides*), most evident in living juvenile worms but occasionally distinguishable in adults. Campbell *et al.* (1989) found a pair of short dorsolateral ducts close behind the proboscis apparatus in *C. australiensis* but were unable to confirm that these were excretory. Other authors have been unable to demonstrate excretory tubules in histological sections and whether or not the present species possesses such a system remains unknown.

Reproductive system

The sexes are separate, mature females being larger than mature males; worms less than 1 mm long have been found with testes filled with ripe spermatozoa, whereas no worms less than 4–7 mm long during normal movement were found with large ova.

Testes are irregularly scattered throughout the posterior half of the body (Fig. 1). Individual testes containing mature spermatozoa are up to 15–18 μm in diameter and connect with a Takakura's duct near the hind end of the body. This duct is initially 6–8 μm wide but posteriorly expands to form a distinct seminal vesicle (Fig. 1), 20 μm or more in diameter. In whole mount preparations the seminal vesicle appears somewhat similar in shape to the S-form described by Shields & Kuris (1990) for *C. wickhami*.

The ovaries in females are both much more numerous than testes and extend throughout most of the intestinal region of the body between the branches of the lateral diverticula (Figs 7, 8). They are arranged in four loosely disposed longitudinal rows, most evident in sections of juveniles (Fig. 3); in individuals with well developed ova the distribution of the ovaries is much more irregular. Developing eggs are up to 90 μm or more in diameter and contain a nucleus 15 μm across in which a single, often excentric, nucleolus 6–7 μm wide is distinguishable. Several eggs in any one ovary show a similar stage in development, with smaller immature eggs often being appressed against the walls of the ovarian chamber. No evidence of gonopores was seen in any of the specimens which were examined histologically.

Mucous sheaths

On their hosts the nemerteans secrete sheaths which they attach to the medial portions of the branchial chambers; one or more males may share the same sheath, with (Fig. 7) or without a female. The walls of the sheaths are 3–4 μm thick with a striated or laminated appearance in histological sections. Irregularly distributed and variably sized surface lapillae, up to about 30 μm thick and with a homogeneous or amorphous appearance, are scattered over the surface of the sheaths (Fig. 8). Wickham & Kuris (1985, 1988) have suggested that the ornamentation of the sheaths due to the size, shape and pattern of the lapillae may be of specific significance.

Biology of the host-symbiont relationships

Incidence of nemerteans

The numbers of *Pinnixa chaetopterana* collected along the east coast of the United States from New Jersey to Florida which were examined for nemerteans are given in Table 2. About half of the 1234 crabs examined were males and, with the exception of a single male on which an immature worm was found under its abdomen, only females were infested. Of the 633 females from all three locations, 140 harboured nemerteans (Fig. 9). Mature female worms involved in synchronous progeny development with their hosts occurred at

Table 2. The numbers, sex, maturity and female reproductive state of *Pinnixa chaetopterana* collected during the 1960's from New Jersey, North Carolina and Florida, which were examined for nemerteans.

Location	Males			Females			Total both sexes
	Total	Mature	Juvenile*	Total	Mature (ovig.**)	Juvenile*	
New Jersey	151	132	19	174	155 (83)	19	325
North Carolina	400	360	40	412	372 (262)	40	812
Florida	50	8	42	42	20 (8)	27	97
Totals	601	500	101	633	547 (353)	86	1234

* Juveniles somewhat arbitrarily defined as crabs with a carapace width of less than 7 mm.
** Ovigerous females, numbers in parentheses.

all three locations. In North Carolina, where most of the sampling was carried out, 116 of 412 females (28.2%) were infested, 30 of which harboured mature female worms (*i.e.*, those capable of producing egg sacs). Nemerteans were found only in mature female crabs, *i.e.*, those able to produce egg sponges, with a minimum carapace width of 7 mm. All 18 infested crabs from New Jersey contained mature female worms, but because these crabs were held for several months in the laboratory, for other studies, we are uncertain of the number infested with mature female worms at the time of collection. It is likely that at least some of the nemerteans were immature when collected and that they attained maturity in the laboratory prior to detection. Furthermore, because the observations of the crab-worm relationship in New Jersey were seminal, it is possible that some infestations with immature nemerteans were missed. Note, however, that worms were only detected in mature hosts.

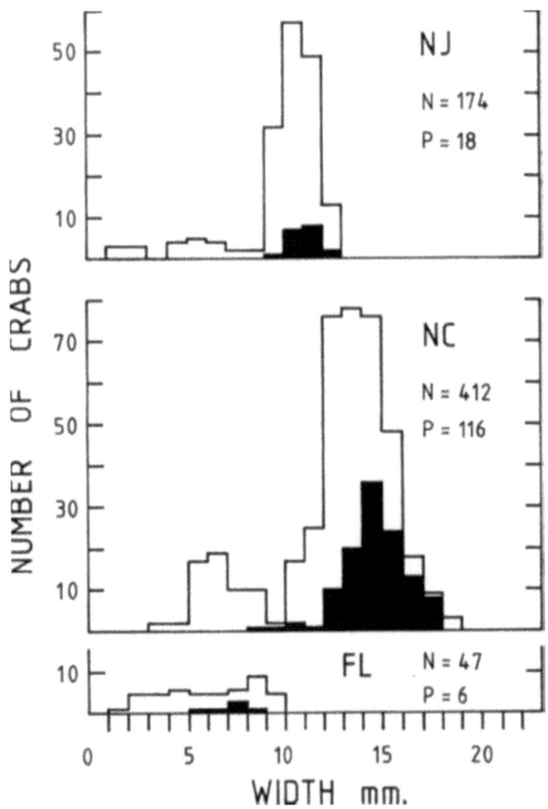

Fig. 9. Incidence of *Carcinonemertes pinnotheridophila* infestations (shaded) in carapace width size classes of female *Pinnixa chaetopterana* (unshaded) examined from New Jersey (NJ), North Carolina (NC) and Florida (F). N = number of crabs, P = number infested.

The distinctly smaller size range of New Jersey crabs, compared with those from North Carolina, may reflect the differences in the size of the worm host species and their tubes; *Chaetopterus* is the larger. The Florida collections yielded mostly relatively small crabs; in most cases their hosts were unknown. *Chaetopterus*, as well as *Arenicola* specimens, were relatively small compared with those of northern populations, and the space available to crabs in the bifurcated siphonal burrows of *Tagelus* is very restricted.

Figure 10 shows the incidence of nemerteans in female *Pinnixa* collected during the summers of 1965–1967 in North Carolina. The host populations sampled in 1965 and 1966 in number, size

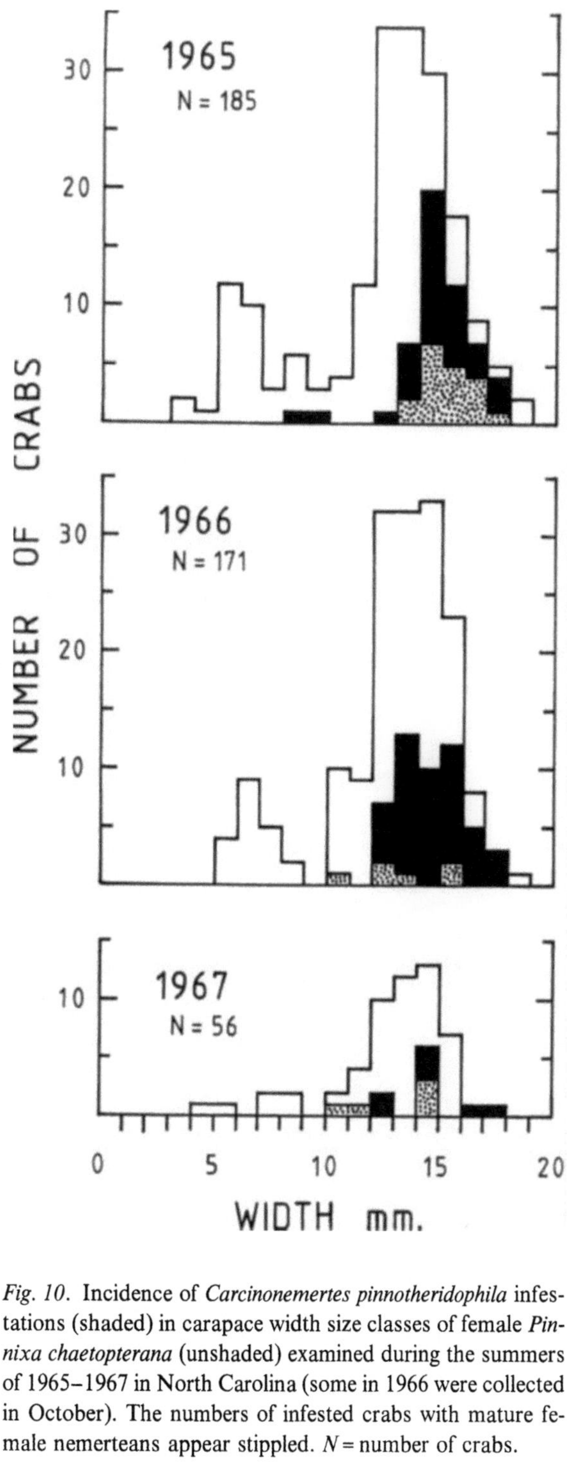

Fig. 10. Incidence of *Carcinonemertes pinnotheridophila* infestations (shaded) in carapace width size classes of female *Pinnixa chaetopterana* (unshaded) examined during the summers of 1965–1967 in North Carolina (some in 1966 were collected in October). The numbers of infested crabs with mature female nemerteans appear stippled. N = number of crabs.

distribution and proportion infested with nemerteans (26.8% and 29.8% respectively) were similar; note, however, that the 1966 data include October. A difference is evident in the percentage of crabs harbouring mature female nemerteans for the two years, with 19 of 53 (35.8%) for 1965 and 6 of 51 (11.8%) for 1966. Results for 1967, based on only one-third of the number of hosts examined during the previous two years, yielded a 21.4% infestation rate with 5 out of 12 (41.6%) containing mature female worms. In all years, it was only the mature crabs (7 mm or more in carapace width) that were infested.

Limited information on the seasonal incidence of nemerteans in *Pinnixa chaetopterana* may be derived from the more extensive collections made in North Carolina where, for three consecutive years, crabs were examined throughout the summer months. Figure 11 shows that for 1965 and 1966 the total incidence of worms increased

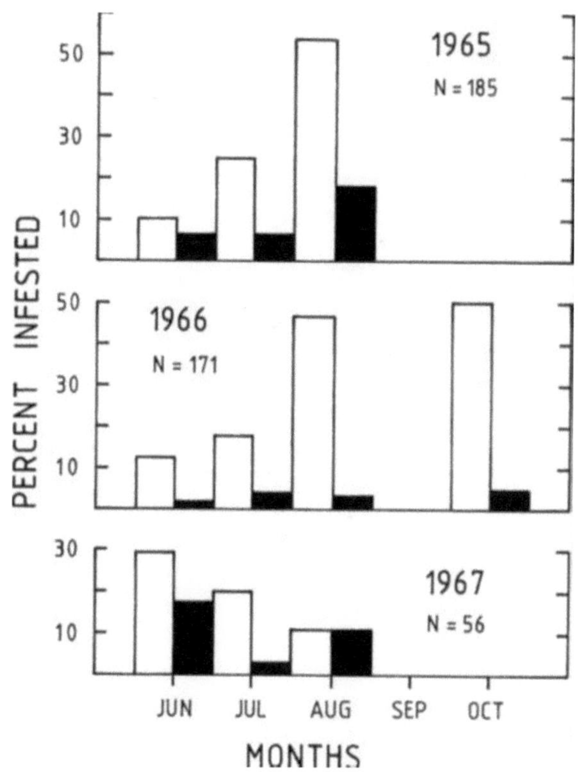

Fig. 11. The monthly percent incidence of *Carcinonemertes pinnotheridophila* (unshaded for female *Pinnixa chaetopterana* examined in North Carolina during the summers of 1965–1967 – extended into October in 1966). Shaded bars are the percent of crabs with mature female nemerteans. N = number of crabs.

gradually from June to August (and in 1966 also remained high in October). The numbers of crabs harbouring mature female worms showed no distinct trend, although we had anticipated that their numbers might tend to increase towards autumn following a decline in reproduction by the host populations. In both New Jersey and North Carolina *Pinnixa chaetopterana* produces egg sponges from at least May to October, peaking during the summer months (J.J.M., unpublished observations). The fact that mature, reproducing female worms occur in June suggests that some of these may have overwintered after reproducing the previous year; data from New Jersey on the longevity of crabs and worms and their reproduction under laboratory conditions support this suggestion.

Location and activities of mature worms

Mature female nemerteans live in secreted mucous sheaths cemented to the medial floor and wall of the host branchial chambers (Fig. 12). There is no significant difference between habitation in the left or right chamber. The anterior part of the sheath extends through the excurrent canal of the chamber, and it and its contained worm may sometimes be observed extending out of the chamber when the third maxillipeds are opened during gill ventilation. This portion of the sheath may be cemented to the base of one of the crab's mouthparts.

The extension of the posterior end of the sheath to eventually make subabdominal (sternal region) contact among the pleopods of the host is more variable. It is in the subabdominal region that the worms, without moving completely out of their sheaths, cement their egg sacs among the crabs' eggs. Localization of the sheath in this critical area is accomplished basically by one of two methods: 1) the sheath extends out of the postero-medial corner of the branchial chamber and curves under the abdomen; 2) the sheath

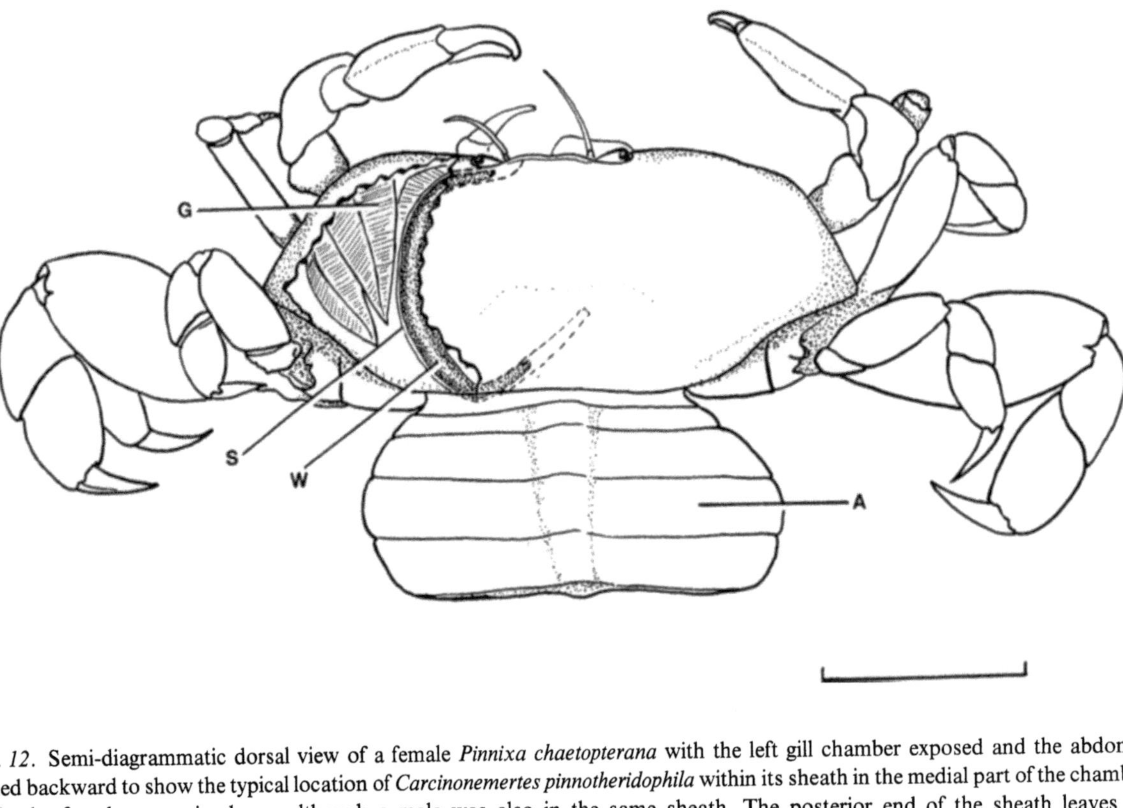

Fig. 12. Semi-diagrammatic dorsal view of a female *Pinnixa chaetopterana* with the left gill chamber exposed and the abdomen flexed backward to show the typical location of *Carcinonemertes pinnotheridophila* within its sheath in the medial part of the chamber. Only the female worm is shown although a male was also in the same sheath. The posterior end of the sheath leaves the postero-medial end of the chamber and extends under the sternum (dashed lines). A, abdomen; G, gill; S, sheath; W, female worm. Scale bar = 4 mm.

penetrates through one of the segmental sutures in the floor of the chamber, and extends to the sternum either via the same or a different suture (Fig. 13). In both cases the sheath ends up projected among the pleopods. Most commonly the sheath both penetrates the suture between the seventh and eighth thoracic segments (equivalent to the suture between the third and fourth walking legs) and exits to the sternum via the same aperture, but there are variations where the sheath wanders through the endophragmal skeleton of the host before making subabdominal contact. Such variations include the posterior portion of the sheath leaving the gill chamber by the natural

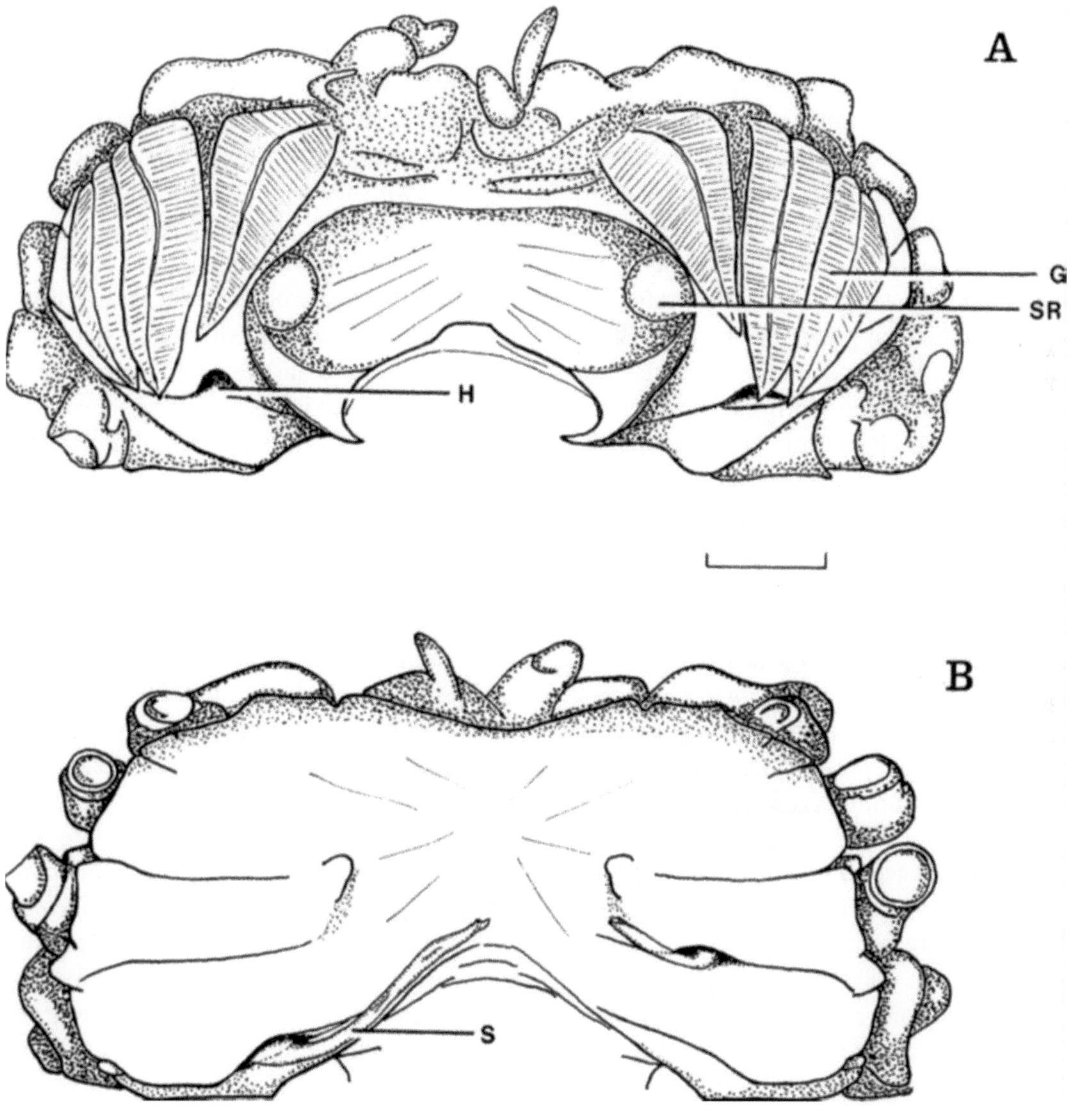

Fig. 13. A. Semi-diagrammatic dorsal view of a female *Pinnixa chaetopterana* with the carapace, abdomen and legs removed to show the holes produced by mature female nemerteans (one formerly in each chamber) in the floor of the branchial chambers between the 7th and 8th thoracic segments. *B.* Ventral view of the same female with the abdomen removed to show the sheaths of *Carcinonemertes pinnotheridophila* exiting via the sternum; the sheath on the left side of the sternum (right side in the drawing) exited via the suture between the 6th and 7th segments, whereas that on the right exited via the 7th and 8th segments. G, gill; H, hole; S, sheath; SR, seminal receptacle. Scale bar = 2 mm.

opening at the posterior corner, or exiting between the sixth and seventh thoracic segments of the sternum (Fig. 13); the sheath may also leave the branchial chamber via the suture between the fourth and fifth segments. In instances of double infestation (one mature female worm in each branchial chamber) the sheaths may follow identical or different pathways to the sternal region. There was one case of an unusual orientation of a mature female with its three males; its sheath did not run forward to the excurrent canal of the gill chamber, but projected laterally over the gills. This sheath passed through a suture in the anterior floor of the right chamber and exited via the suture between the seventh and eighth segments of the sternum. Careful dissection of this sheath showed that it followed suture lines having no apparent contact with the crab's tissues. Even in instances where the worms leave the branchial chamber via the natural posterior opening, there is sometimes a shallow groove in the exoskeleton at the junction where the worms turn anteriad under the abdomen. Penetration of the sheath from the branchial chamber to the sternum via the sutures can only occur some time during the crab's ecdysis when the female worm presumably wedges herself into the suture between the softened segments.

There were two exceptions to the rule that only one female worm inhabits a branchial chamber, both occurring in crabs from North Carolina which were preserved prior to examination. In one instance a 15.8 mm ovigerous crab had two apparently immature worms in its right chamber, with no egg sacs present on the host egg sponge; both worms were relatively large, their size indicating that they were female. Their sheaths, however, penetrated the gill chamber floor, one between the seventh and eighth segments and the same at the sternum, the other penetrating the same suture but exiting between the fifth and sixth sternal segments. In the other example an ovigerous 15.3 mm crab, without attached egg sacs, had two mature female nemerteans in the right chamber. Both had the anterior parts of their sheaths in the excurrent canal of the chamber. One sheath used the natural posterior opening to make subabdominal contact, the other penetrated the suture between the fifth and sixth segments in the floor of the chamber and exited between the sixth and seventh sternal segments; the sheath of the latter worm was covering the outer (third) gill. One female nemertean had four presumed males around it and was partially coiled over the first and second gills, the other female had two presumed males around it. This crab thus harboured a total of eight nemerteans, the maximum number recorded for any single host.

A female nemertean can reverse itself in its sheath *in vitro*, and no doubt also does so *in vivo* because worms may be found facing either anteriorly or posteriorly. One or more males, residing in their own sheaths, may be found inside or outside the female's sheath (Figs 7, 14). Mature worms have lived for more than a year in the laboratory, some producing successive broods of progeny in conjunction with successive broods of crab eggs. Male and female worms have been easily maintained in dishes within their sheaths in the laboratory for at least a month prior to their being sacrificed.

Shortly after the host ovulates eggs on to its pleopods, the nemerteans produce their own egg sacs. Under laboratory conditions this occurred within hours of the crab's egg deposition to no more than two days later. The first two pleopods* of *Pinnixa* are biramous, and it is to the endopod of each that both crab eggs and worm egg sacs are secured; the third and fourth pleopods are uniramous, represented only by the endopods, to which the crab and nemertean eggs are attached (Fig. 15). The developmental time from ovulation to the emergence of first zoeae in *P. chaetopterana* is dependent on temperature, but is approximately two weeks as determined in laboratory observations (random sample of some of the reproducing crabs held in the laboratory: mean 14.6 ± 1.4 days, range 11–18 days, $n = 50$). Thus, due to the synchronized development between host and sym-

* There are no pleopods on the first abdominal segment, thus the first two pairs are in reality pleopods 2 and 3, followed by 4 and 5.

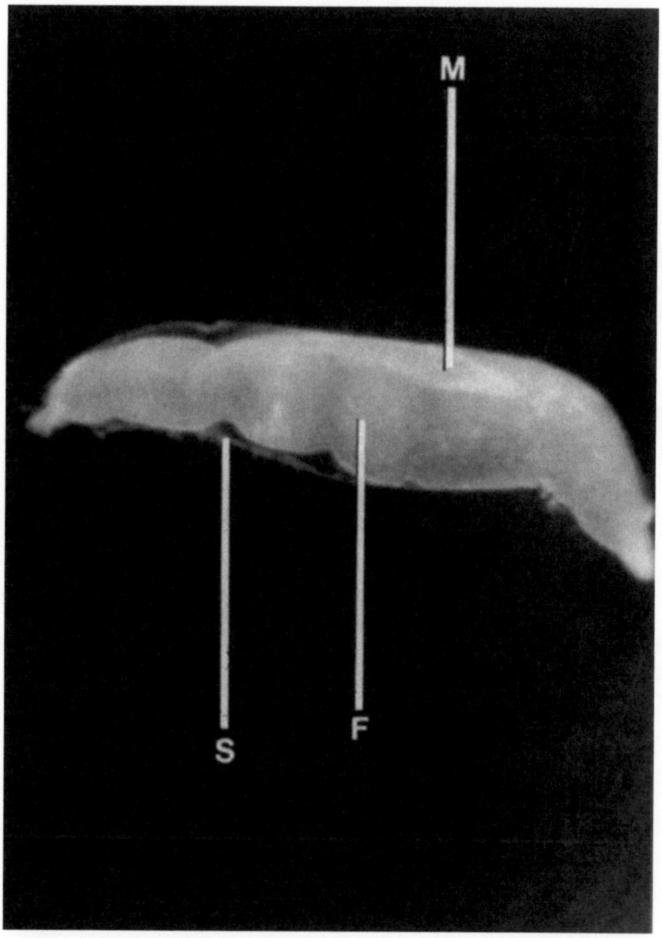

Fig. 14. Photograph of a living ensheathed female and male *Carcinonemertes pinnotheridophila* just as they were removed from a 10.6 mm female crab host. The female worm is greatly contracted within the remains of its sheath (approximately 3.5 mm long) whereas the male is not. F, female; M, male; S, sheath.

biont, developmental time for the nemertean embryos is slightly shorter.

Nemertean egg sacs and reproductive potential

The nemertean egg sacs consist of a mucous sheath surrounding an oval to lemon-shaped mass of eggs, which in some cases gives them a mulberry-like appearance (Fig. 15). The sheath is clear and the eggs are white, so the gross development of the larvae can easily be observed. Seventy-seven egg sacs, taken from an 11.5 mm crab collected 17 June 1964 in New Jersey, were 522 ± 98 μm long by 341 ± 46 μm wide; the mean number of eggs/sac was 77.5 ± 22.7, range 6–120. Individual undeveloped nemertean eggs are round and measured 84.3 ± 1.7 μm in diameter ($n = 10$). The female worm that produced the sacs was 9.8 mm long, with an accompanying male 1.3 mm long and 0.1 mm wide. Another female worm, 8.0 mm long and inhabiting the gill chamber of a 7.4 mm crab collected 12 July 1966 in Florida, produced 37 egg sacs in the laboratory (Table 3). The crab itself produced only 210 eggs, this being the second sponge. The worm egg sacs were 425 ± 60 μm long by 259 ± 31 μm wide ($n = 31$), the mean number of eggs/sac 18.9 ± 5.5, range 12–20 ($n = 21$); fully developed larvae were in these sacs when they were removed from the host.

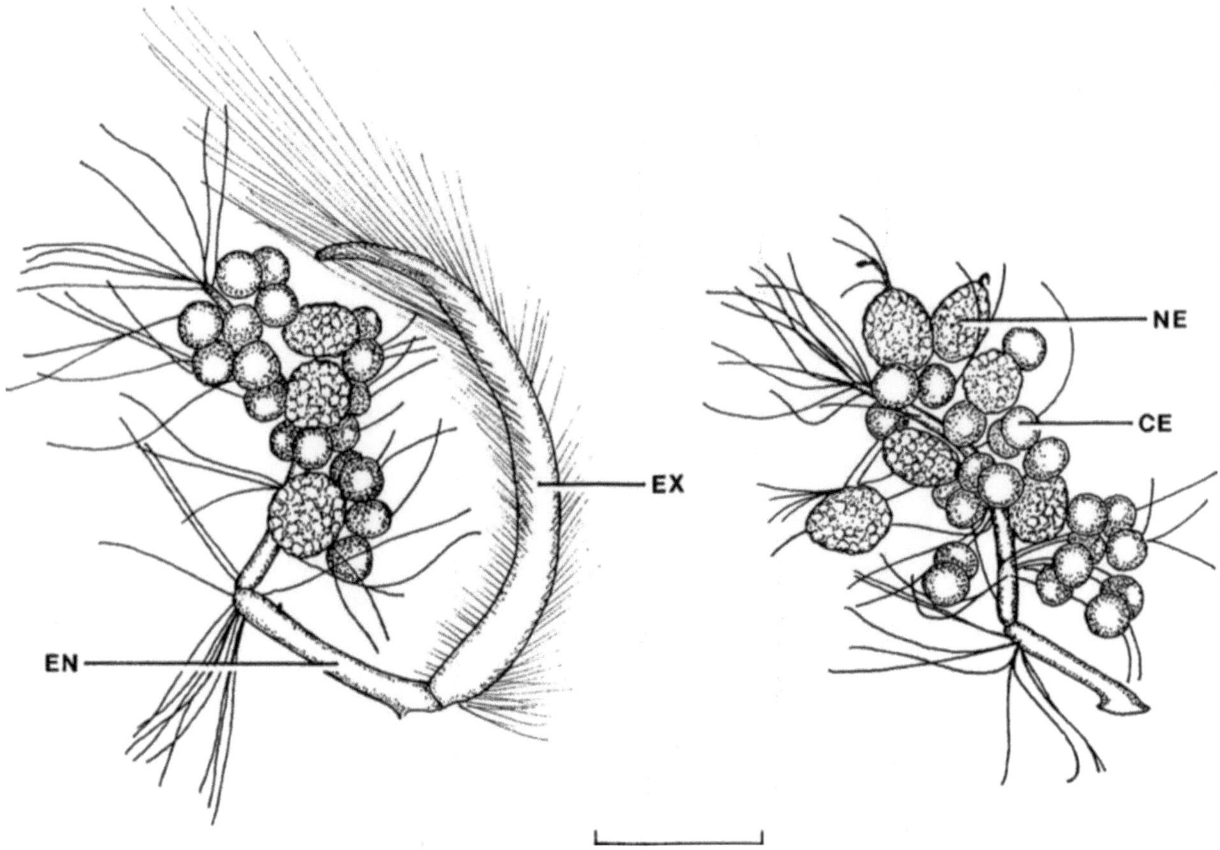

Fig. 15. Second pleopod (left) and third pleopod (right) of a female Pinnixa *chaetopterana* infested with *Carcinonemertes pinnotheridophila,* showing the oval, mulberry-like, nemertean egg sacs which are attached to the hairs of the endopods along with the crab eggs. The third pleopod is uniramous (endopod only). CE, crab egg; EN, endopod; EX, exopod; NE, nemertean egg sac. Scale bar = 1 mm.

The number of egg sacs on ovigerous *P. chaetopterana*, together with the numbers of crab embryos and the other data from all three locations are given in Table 3. More egg sacs were found in crabs with double infestations. Table 4 compares crab egg and nemertean egg sac distribution on the individual pleopods of crabs infested with one or two mature female worms. In the former, the numbers of nemertean embryos from each pleopod are also given. In the doubly infested crab the numbers of sacs were more than three times as large as in single infestations. Here, as in other cases, there seems to be a trend for greater numbers of sacs to be cemented to the first and fourth pleopods, which tend to harbour fewer crab eggs.

Synchrony of development

Because *C. pinnotheridophila* deposits its egg sacs on the host pleopods shortly after the crabs' eggs are attached, and because hatching of host and symbiont larvae is synchronous, the mean developmental time for the nemertean is slightly less than 14 days. Observations of individual crabs when first collected and then examined daily in the laboratory have shown that the nemerteans produce at least two broods. *Pinnixa chaetopterana* produces several broods during the summer months, and if a crab is infested with a mature female worm at the beginning of the reproductive period, it is likely that the nemertean will also produce a comparable number of broods.

Table 3. Distribution and numbers of crab eggs and *Carcinonemertes pinnotheridophila* egg sacs on the left and right pleopods of *Pinnixa chaetopterana* collected in New Jersey, North Carolina and Florida, and the gill chamber locations of mature female worms. Crabs listed according to increasing carapace widths (mm).

Crab width, location and date of collection	Pleopods						Number of female worms in left (L) or right (R) chamber
	Left		Right		Total		
	Crab eggs	Worm egg sacs	Crab eggs	Worm egg sacs	Crab eggs	Worm egg sacs	
7.4 Florida 7 December 1966	117	3	93	34	210	37	1 L*
9.0 Florida 27 January 1967	538	35	473	77	1011	112	2 L & R*
10.6 New Jersey 14 August 1963	–	12	–	57	–	69	1 L*
10.7 New Jersey 14 August 1963	631	1	291	21	922	23**	1 R*
11.8 New Jersey 28 May 1964	864	9	597	72	1461	81	1 L
13.7 North Carolina 18 June 1966	–	1	–	32	–	33	1 L
14.0 North Carolina 25 July 1965	4120	74	3967	101	8087	175	2 L & R
14.8 North Carolina 23 June 1967	4616	0	4415	69	9031	69	1 L
14.9 North Carolina 11 July 1965	4033	164	4826	91	8859	255	2 L & R
16.4 North Carolina 11 July 1965	3355	9	3676	0	7296***	9	1 L*

* This was the second egg sponge produced by the crab.
** One egg sac was not designated either left or right, therefore a total of 23 is indicated.
*** 265 crab eggs dislodged, giving a total of 7296; the first brood released 10611 zoeae.

Newly deposited eggs of *C. pinnotheridophila* are round and measure approximately 84 μm in diameter, whereas fully developed larvae still within their egg membranes are more oval, measuring 100 μm by 92 μm (mean 99.9 $\pm$ 6.6 μm by 92.1 $\pm$ 2.9 μm, $n = 11$). Hatching of the larvae involves their liberation from their outer developmental membranes as well as from the mucous sheath making up the wall of the egg sacs. The mechanism for hatching is not known.

Freshly hatched larvae are oval in outline, dorsoventrally flattened, completely ciliated, with anterior and posterior cirral tufts and possess two black, irregularly rounded eyespots. The animals are positively phototactic and rotate on their antero-posterior axis as they move forwards. During steady cruising movements, they may suddenly undergo increased bursts of speed, and can also turn rapidly by differential longitudinal contraction of their muscles. The newly-hatched and naturally relaxed larvae (minimal ciliary action) are about 100–140 μm in length (mean 132 $\pm$ 12.6 μm, $n = 6$), and they are approximately half as wide. The anterior cirral tuft is nearly twice as long as the posterior.

Table 4. The distribution of crab eggs, *Carcinonemertes pinnotheridophila* egg sacs and embryos on individual pleopods of two female *Pinnixa chaetopterana* infested with one and two females worms.

	Pleopod number	Numbers					
		Left side			Right side		
		Crab eggs	Worm		Crab eggs	Worm	
			egg sacs	embryos		egg sacs	embryos
1 Worm*							
	1	152	2	135	127	5	407
	2	257	0	0	242	33	2627
	3	313	1	33	176	22	1514
	4	142	6	516	52	12	1062
	Totals	864	9	685	597	72	5610
2 Worms**							
	1	868	29	–	895	23	–
	2	1436	34	–	1693	7	–
	3	1134	41	–	1498	17	–
	4	595	60	–	740	44	–
	Totals	4033	164	–	4826	91	–

* 11.8 mm crab, collected 28 May 1964 in New Jersey; worm in left chamber.
** 14.9 mm crab, collected 11 July 1965 in North Carolina; one worm in each chamber.

Synchrony with host ecdysis

Since *C. pinnotheridophila* is essentially an attached ectosymbiont (*i.e.*, it is attached by means of its cemented sheath to the exoskeleton of the crab's gill chamber), it would, like anything else adhering to the exoskeleton, be at risk of being shed during the host's ecdysis. Laboratory observations made it possible to demonstrate that the worms in some unknown way anticipate pending ecdysis, move out of their sheaths and relocate in the new gill chamber where a new sheath is then secreted. Eighteen singly infested crabs harbouring mature female worms were observed for several months, during which time all but three moulted; five crabs moulted once, five twice and five three times. Following all moults the nemerteans were relocated in the new gill chambers of the soft crabs. One of these crabs died in the last phase of its third ecdysis (the old exoskeleton was completely ruptured), but the female nemertean was already in a partially secreted sheath within the new gill chamber. The worms do not rupture their old sheaths before migrating to the new chamber, but rather move out from the opening at one end or the other.

Discussion

Both the habits and anatomy of the present nemerteans accord closely with the familial definition given by Shields *et al.* (1989) and the modified generic diagnosis of Gibson & Jones (1990). Shields *et al.* (1989: 926–927) concluded that 'The family [Carcinonemertidae] appears to be too narrowly defined ... We view the following as diagnostic familial characters: (i) symbiotic relationship with a decapod crustacean, (ii) short proboscis, (iii) absence of cerebral organs, (iv) presence of Takakura's duct [in mature males], and (v) a rhabdocoel-like hoplonemertean larvae'. The present nemerteans exhibit all these characteristics defined for the family. The only feature the worms show which is at variance with the generic diagnosis given by Gibson & Jones (1990) is in lacking eyes as adults, yet the newly-hatched larvae possess a pair of irregular black eyespots; the evident loss of eyes which accompanies growth and development of the worms must therefore be regarded as a characteristic unique to this species which may possibly be a consequence of their host itself living as a symbiont within polychaete tubes. There are no grounds for excluding the nemertean from the genus *Carcinonemertes* and the worms are accordingly placed in this taxon.

Gibson & Jones (1990: 201) comment that the genus 'is unusual in that the constituent species are remarkably alike in structure ... [and] ... species identification depends upon points of detail.' From a strictly anatomical viewpoint this remains true but, as rightly suggested by Shields & Kuris (1990), life-history patterns may also be useful for distinguishing between the morphologically similar species of *Carcinonemertes*. Because we are dealing with a symbiotic relationship, such features as host specificity, habitat within the host, synchrony of development, relationship to host moulting, mating (Roe, 1984, 1986), etc., should be considered. Shields & Kuris have detected three patterns which relate to embryogenesis in the host (*i.e.*, oviposition to eclosion [or hatching] of the host's first larval stage). The first involves a *short* embryogenesis of 13–16 days as seen in the portunid crabs, and as characterised by the life cycle of *Carcinonemertes carcinophila* in *Callinectes sapidus* (Humes, 1942; Hopkins, 1947). Here the development of both host and symbiont are rapid. The invasive stages of *C. carcinophila* show a specificity for mature female crabs, and become mature themselves as the crabs reach the oviposition period. They then migrate out of the gill chambers to the pleopods, on to which they cement their egg strings. Relatively rapid development of their larvae as well as the host's zoeae is followed by synchronous hatching, after which the worms migrate back to the branchial chambers. The second is a *moderate* embryogenetic period of 40–120 days, typical of cancrid and grapsid crabs. Here both male and female crabs are invaded, some juveniles are transferred from males to females during copulation (Wickham *et al.*, 1984), and the worms mature on the host sponge where egg strings are deposited. Synchrony of development is not involved. *Carcinonemertes errans* and *C. epialti* are characteristic of this pattern. In the third case, there is a *long* embryogenesis (120–300 + days) as in spiny lobster (*Panulirus)* and anomuran king crab (*Paralithodes*) hosts. Worm larvae settle and mature only on ovigerous crabs. Egg strings are deposited on the host sponge, there is no apparent synchrony of development, and adult worms supposedly die or leave their host after the decapods' larvae hatch.

Without doubt, *C. pinnotheridophila* is morphologically distinct from all of the previously described species in the genus (Table 5). However, its life-history characteristics ally it to the first group of carcinonemerteans with a short embryogenesis and synchrony of development. In the *Pinnixa – C. pinnotheridophila* relationship several differences can be distinguished in comparison with portunid – *C. carcinophila* relationships; these are:

1) there is usually only one mature female worm per gill chamber versus hundreds of potential mature females in the *C. carcinophila* cycle.
2) worms in the gill chambers do not ensheath themselves on the leaflets of the gills.
3) the mature female probably remains in its branchial sheath as it deposits its egg sacs.
4) distortion of the crab's exoskeleton is the usual result of the worm in the branchial chamber making contact with the pleopodal area.
5) oval or lemon-shaped egg sacs are distinct from the long egg strings of *C. carcinophila*.
6) mature *Pinnixa* females may moult several times following the pubertal moult (they may also moult between successive broods); worms in the infested crabs relocate in the new gill chamber following each moult rather than being cast off with the old exoskeleton – the pubertal moult of *Callinectes*, in contrast, is the terminal moult and death ensues following a reproductive period, so that the nemerteans apparently die with their host.

The morphological distinction of the present species from all other known carcinonemerteans is thus further justified by differences in their life-

Table 5. Summary of some of the features which can be used to distinguish between the known species of *Carcinonemertes* compared with those of the nemerteans from *Pinnixa chaetopterana*. Data from Campbell *et al.* (1989), Shields *et al.* (1989), Gibson & Jones (1990), Shields & Kuris (1990) and the present paper.

Species	Sex	Length (mm)	I	II	III	IV	V	VI	VII	VIII	IX	X
C. australiensis	M & F	7.0	40.0	15–18	0.413	75	90	A	?	2	1	?
C. carcinophila	M	20.0										
carcinophila	F	70.0	25.0	9.0	0.360		63	B	+	2	1	S
C. carcinophila	M	8.7										
imminuta	F	16.5	21.0	7.3	0.348	50	139	B	+	2	1	S
C. coei	M	4.2										
	F	6.6	22.7	8.7	0.383	32 +	78	?	+	2	2	?
C. epialti	M	2.2										
	F	4.3	31.2	14.5	0.465	75	63	?	+	2	1	M
C. errans	M & F	4.0–6.0	35.2	11.0	0.313	46 +	100	?	0	2	1	M
C. humesi	F	15.0	31.0	7.5	0.242	35	60	?	+	2	1	?
C. mitsukurii	M	10.0										
	F	30.0	27.0	8.0	0.296	48	86	A	+	2	1	?
C. regicides	M	1.6										
	F	2.1	40.0	17.0	0.425	76	82	?	+	2	1	L
C. wickhami	M	10.0				79	149					
	F	30.0	40.0	20.0	0.500	98	125	B?	+	2	2	
Present species	M	2.3	18.1	6.6	0.365							
	F	8.4	16.0	8.0	0.500	20	70	A	+	0	2	S

I Average length of stylet basis (μm)
II Average length of central stylet (μm)
III Average value of stylet:basis ratio
IV Average length of anterior chamber of proboscis (μm)
V Average length of posterior chamber of proboscis (μm)
VI Posterior blood vessel connective supra- (A) or subintestinal (B)
VII Adults do (+) or do not (0) secrete mucous sheats
VIII Normal number of eyes in adults
IX Ovaries in single (1) or double (2) row on either side of intestine
X Embryogenesis short (S), moderate (M) or long (L)

Lengths indicated for the sexes represent a mixture of living and preserved measurements; no male specimens of *C. humesi* have yet been found.

cycle and host-symbiont relationships; these differences, which are summarised in Table 5 and Fig. 16, warrant the establishment of the new species, *Carcinonemertes pinnotheridophila*, for them.

The highly evolved ectosymbiotic relationship demonstrated between *Pinnixa chaetopterana* and *Carcinonemertes pinnotheridophila* most probably involves a variety of chemical messengers, released by the host, which are received and interpreted by the nemerteans. If this is true, these messengers must be external metabolites, because the worms have no contact with the crab's living tissues. Presumably there are differences in the variety and/or concentrations of crab metabolites released – some related to reproduction, others to the growth cycle. The nemertean's invasive stage, still unknown but succeeding metamorphosis of the free-living larvae, is apparently attracted only

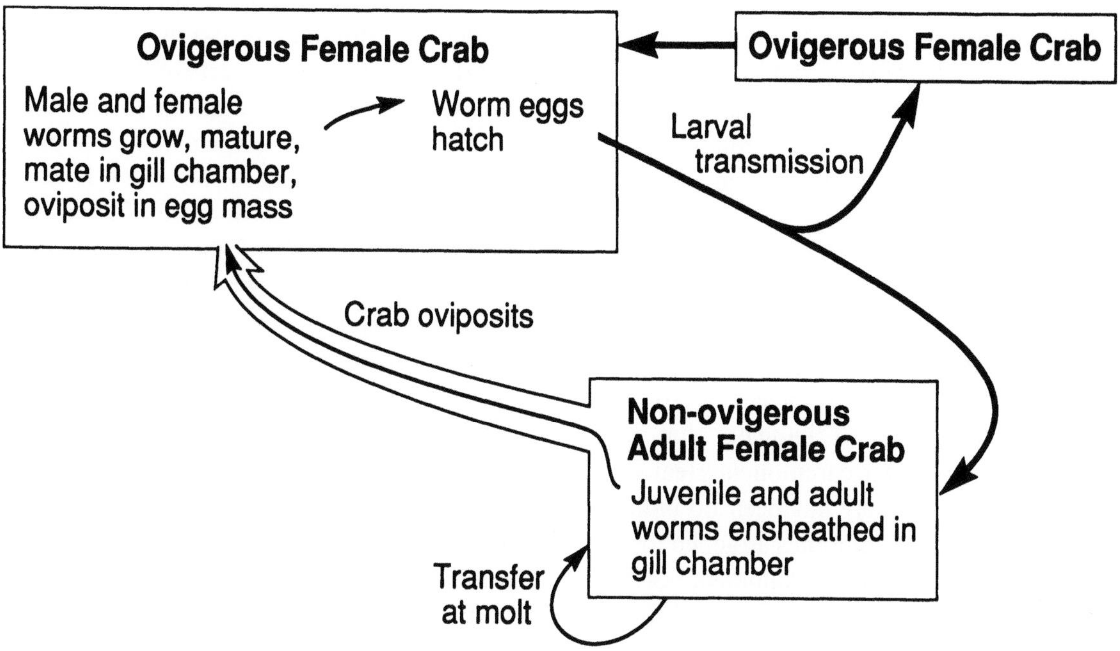

Fig. 16. Diagrammatic representation of the life cycle of *Carcinonemertes pinnotheridophila* on the symbiotic crab *Pinnixa chaetopterana*. Heavy arrows represent worm movement *between* crabs, light arrows movement *on* crabs, and the open arrow represents a *change* in host condition, *i.e.*, oviposition. Drawn by A. M. Kuris in the format consistent with other carcinonemertean life cycles illustrated by him (see this volume).

to female crabs and mainly to those that have reached maturity. This suggests that female external metabolites released by non-ovigerous or ovigerous crabs may be effective over short distances in inducing invasion by immature worms. Perhaps metabolites released from a preovulating or ovulating crab induces female nemerteans residing in the gill chamber to mobilise their own reproductive resources, relating to oocyte development or ovulation, to coincide with its host activities. It is possible that external crab metabolites differ during different phases in their reproductive cycle, such as the period of oocyte maturation, preovulation and postovulation. Whatever these messages might be, and whenever they may be released, *C. pinnotheridophila* becomes mobilised to reproduce synchronously with the hatching of its host's zoeae. This synchrony is highly adaptive for the nemertean, because immediately after the zoeae are released, *Pinnixa* vigorously cleans the egg membranes from its pleopods with its chelipeds. If the developmental period for the nemertean was longer than that of the crab, the worm egg sacs would be torn off with the host egg membranes and likewise be discarded into the sediments. Furthermore, the characteristic abdominal undulations of the ovigerous crab, which aerate its eggs, cease after zoeal release. This in itself would probably be unfavourable for further development of the nemertean larvae (Roe, 1988).

External metabolites related to the intermoult cycle of the crab may be sensed by the ensheathed worm living in the branchial chamber, or the stimulus may be some combination between metabolite release and mechanical activity related to incipient ecdysis; either way, the nemertean mobilises itself during its host's ecdysis to exit its attached sheath and relocate in the new gill chamber, as observed so many times during the present studies. If this were not the case and the worm remained in its sheath, or responded but did not act in time, the old exoskeleton containing the nemertean would be cast off into the sediments, with a consequent reduction in the likelihood of survival. *Carcinonemertes pinnotheridophila* is similar to *C. carcinophila* in much of what has been said about the intimacy of their relationships with their hosts, but since *Pinnixa* continues to moult after maturity (unlike *Callinectes*), *C. pinnotheridophila* may have had to evolve a receptivity to the moult cycle metabolites of its host in order to survive.

The trophic aspect of the *C. pinnotheridophila – P. chaetopterana* relationship remains a mystery. What kind of nutrition supplies the energy necessary to produce mature female nemerteans capable of ovulating thousands of eggs? No evidence has been found to indicate that egg predation, typical of other carcinonemerteans, is the nutritional source; there has been no indication of empty host egg membranes attached to the pleopods of the many sponges examined, nor did limited examinations of faecal material from worms removed from ovigerous crabs reveal prezoeal parts or the dark body pigment of developing zoeae. The numbers of crab eggs listed for the first sponges of infested crabs held in the laboratory (Table 3), and other infested crabs not listed, are within the normal range found in uninfested crabs (J.J.M., unpublished data). Male nemertean nutrition is also unknown. Uptake of dissolved organic matter, demonstrated for some nemertean species (Fisher & Oaks, 1978; Roe *et al.*, 1981; Crowe *et al.*, 1982), may be a partial option for *C. pinnotheridophila*, but it is difficult to believe that this could be entirely responsible for the species' nutritional needs, particularly for female worms.

In recent years symbiotic relationships involving other nemertean genera and crustaceans have been described. One of these involves *Pseudocarcinonemertes homari*, an egg predator of *Homarus americanus* and *H. gammarus* (Aiken *et al.*, 1985; Brattey *et al.*, 1985); questions concerning the familial placing of *Pseudocarcinonemertes* have been briefly addressed elsewhere in the present paper. This species may occur in thousands on lobster egg masses, producing brood sacs on the abdomen or host egg sponge. The ciliated larvae, which lack anterior and posterior cirral tufts, hatch within the brood sacs and eventually break out of these as immature worms to reinfest the same host's egg mass if this is still present. Following destruction of the host brood by the worms, the latter disperse to various parts of the exoskeleton or to the gill chambers. In this life-history there is a distinct predilection for female hosts, particularly if they are ovigerous, but there is no synchrony of development. Because of their general shape, the brood sacs of *Pseudocarcinonemertes* are somewhat reminiscent of those produced by *C. pinnotheridophila*. The subspherical eggs of *P. homari* are, however, much larger and less numerous than those in the egg sacs of the nemertean from *Pinnixa*.

Shields *et al.* (1989) described a new symbiotic species, *Carcinonemertes regicides*, found on ovigerous red king crabs, *Paralithodes camtschatica*. This egg predator has not been found in the host branchial chambers. Gibson *et al.* (1990) described a new genus and species of monostiliferous nemertean, *Alaxinus oclairi*, from the egg mass of the same king crab species. As yet there is little published information on this relationship except that the single crab host found harboured a 'large' number of worms which caused host brood mortality, and that the nemerteans produced 'big' egg sacs.

Only one other pinnixid has been examined for nemerteans. In New Jersey, *Pinnixa cylindrica*, which lives in the burrows of the lugworm *Arenicola cristata*, is sympatric (on the same intertidal flats) with the *P. chaetopterana* used in the present study. Forty-one females and seven males, collected in 1960, 1962 and 1963, were examined in the same manner; all were negative in this relatively small sample (McDermott, 1981).

Pinnotheres ostreum, an ectoparasite of the American oyster *Crassostrea virginica* (Gmelin, 1791) (Christensen & McDermott, 1958) is infested with what appears to be the same nemertean species found in *Pinnixa chaetopterana* (McDermott, 1967). *Pinnotheres* belongs to the subfamily Pinnotherinae. Oysters living sympatrically with *Chaetopterus variopedatus* in the vicinity of Beaufort, North Carolina, were collected and opened to find crabs which live in the mantle cavity; it is there they feed on plankton brought in by the host, mature and reproduce. The crabs were maintained and examined as with *Pinnixa*. Nemerteans were found almost exclusively in the mature (stage V) female crabs, and their biology was similar to that demonstrated for the *Pinnixa* relationship. Beach (1969), working with *Pinnotheres ostreum* in the same location, confirmed these findings. The major differences in the *Pinnotheres* versus the *Pinnixa* relationships are that more than one female worm matures in each gill chamber, both chambers are more routinely infested, and the worms tend to ensheath themselves over the gills. Thus the nemerteans may be potentially more damaging to this host than to *Pinnixa*. A report on the incidence and biology of this nemertean from the oyster crab, and a clarification of its specific identity, is in progress.

The results of the present investigations indicate that an emended diagnosis of the genus *Carcinonemertes* is now necessary; the genus is accordingly defined as comprising monostiliferous enoplan nemerteans living as ectohabitants on the gills, in the gill chambers, under the abdomen, or in apodemes and axillae of the limbs, or on and in the egg masses of anomuran, brachyuran or palinuran Decapoda; proboscis with single central stylet but, at least as adults, with neither accessory stylets nor pouches, very short, anterior chamber usually extremely short and non-glandular; rhynchocoel restricted to anteriormost portion of body, with wall lacking distinct musculature; lateral nerve cords lying internal to well developed layer of submuscular glands; cephalic glands well developed; cerebral sensory organs missing; adults with no or two eyes; cephalic muscle fibres present; excretory system distinguishable in living specimens; blood system a simple vascular loop, without mid-dorsal vessel; Takakura's duct present in males; internal fertilisation and ovoviviparity may occur; extensive development of spermatozoa and ova; most species secrete mucous sheaths as adults, typically attached to endopodite hairs of ovigerous hosts but sometimes cemented to gill chamber wall or gill filaments; embryogenesis short with hatching synchronized with that of host, moderate without synchrony, or long without synchrony.

Acknowledgements

J.J.M. appreciates the helpfullness of many individuals during several years of field studies along the Atlantic coast of the United States; these are, for providing research facilities in New Jersey, H. H. Haskin of Rutgers University, in North Carolina C. G. Bookhout, J. D. Costlow and F. J. Vernberg of the Duke University Marine Laboratory, in Florida A. A. Myrberg and the late F. G. W. Smith of the University of Miami; for histological preparations, R. L. Snyder; and for assistance in the field, W. R. Belzer, J. J. McDermott III, T. M. McDermott and M. H. Taylor. We thank A. M. Kuris of the University of California, Santa Barbara, for providing us with the life cycle diagram of *C. pinnotheridophila*. J.J.M. is indebted for the financial assistance provided by Franklin and Marshall College, National Science Foundation Postdoctoral Awards from the Duke Marine Laboratory, and N.S.F. Research Grants G11448, G22068 and B959. R.G. gratefully acknowledges the support of Natural Environmental Research Council Grant Number GR3/7168, which provided certain consumable items.

References

Abele, L. G. & B. E. Felgenhauer, 1982. Decapoda. In S. P. Parker (ed.), Synopsis and classification of living organisms. McGraw-Hill, New York 2: 296–326.

Aiken, D. E., S. L. Waddy & L. S. Uhazy, 1985. Aspects of the biology of *Pseudocarcinonemertes homari* and its association with the American lobster, *Homarus americanus*. Can. J. Fish. aquat. Sci. 42: 351–356.

Beach, N. W., 1969. The oyster crab, *Pinnotheres ostreum* Say, in the vicinity of Beaufort, North Carolina. Crustaceana 17: 187–199.

Brattey, J., A. Campbell, A. E. Bagnall & L. S. Uhazy, 1985. Geographic distribution and seasonal occurrence of the nemertean *Pseudocarcinonemertes homari* on the American lobster, *Homarus americanus*. Can. J. Fish. aquat. Sci. 42: 360–367.

Campbell, A., R. Gibson & L. H. Evans, 1989. A new species of *Carcinonemertes* (Nemertea: Carcinonemertidae) ectohabitant on *Panulirus cygnus* (Crustacea: Palinuridae) from Western Australia. Zool. J. linn. Soc. 95: 257–268.

Christensen, A. M. & J. J. McDermott, 1958. Life-history and biology of the oyster crab, *Pinnotheres ostreum* Say. Biol. Bull. mar. biol. Lab., Woods Hole 114: 146–179.

Coe, W. R., 1902a. The genus *Carcinonemertes*. Zool. Anz. 25: 409–414.

Coe, W. R., 1902b. The nemertean parasites of crabs. Am. Nat. 36: 431–450.

Corrêa, D. D., 1966. A new hermaphroditic nemertean. Anais Acad. bras. Cienc. 38: 365–369.

Crowe, J. H., L. M. Crowe, P. Roe & D. Wickham, 1982. Uptake of DOM by nemertean worms: association of worms with arthrodial membranes. Am. Zool. 22: 671–682.

Fisher, F. M. & J. A. Oaks, 1978. Evidence for a nonintestinal nutritional mechanism in the rhynchocoelan, *Lineus ruber*. Biol. Bull. mar. biol. Lab., Woods Hole 154: 213–225.

Fleming, L. C. & R. Gibson, 1981. A new genus and species of monostiliferous hoplonemerteans, ectohabitant on lobsters. J. exp. mar. Biol. Ecol. 52: 79–93.

Gibson, R. & D. S. Jones, 1990. A new species of *Carcinonemertes* (Nemertea: Enopla: Carcinonemertidae) from the egg masses of *Naxia aurita* (Latreille) (Decapoda: Brachyura: Majidae) collected in the Albany region of Western Australia. In F. E. Wells, D. I. Walker, H. Kirkman & R. Lethbridge (eds), Proceedings of the third international marine biological workshop: the marine flora and fauna of Albany, Western Australia. Western Australian Museum, Perth 1: 195–202.

Gibson, R., D. E. Wickham & A. M. Kuris, 1990. A new genus and species of monostiliferoidean nemertean (Nemertea: Enopla) found on an egg mass of the anomuran decapod *Paralithodes camtschatica*. Zool. J. linn. Soc. 98: 185–198.

Hopkins, S. H., 1947. The nemertean *Carcinonemertes* as an indicator of the spawning history of the host, *Callinectes sapidus*. J. Parasit. 33: 146-150.

Humes, A. G., 1942. The morphology, taxonomy, and bionomics of the nemertean genus *Carcinonemertes*. Illinois biol. Monogr. 18: 1–105.

Kölliker, A., 1845. Drei neue Gattungen von Würmern. Verh. schweiz. naturf. Ges. 29: 86–98.

McDermott, J. J., 1966. The biology of a nemertean parasite of pinnotherid crabs. Am. Zool. 6: 331.

McDermott, J. J., 1967. The oyster crab, *Pinnotheres ostreum*, as a host for a parasitic nemertean. Am. Zool. 7: 772–773.

McDermott, J. J., 1981. Biology of the symbiotic crab *Pinnixa cylindrica* (Say) (Decapoda, Pinnotheridae). Proc. Pa acad. Sci. 55: 23–27.

Norenburg, J. L., 1985. Structure of the nemertine integument with consideration of its ecological and phylogenetic significance. Am. Zool. 25: 37–51.

Roe, P., 1984. Laboratory studies of feeding and mating in species of *Carcinonemertes* (Nemertea: Hoplonemertea). Biol. Bull. mar. biol. Lab., Woods Hole 167: 426–436.

Roe, P., 1986. Parthenogenesis in *Carcinonemertes* spp. (Nemertea: Hoplonemertea) Biol. Bull. mar. biol. Lab., Woods Hole 171: 640–646.

Roe, P., 1988. Ecological implications of the reproductive biology of symbiotic nemerteans. Hydrobiologia 156: 13–22.

Roe, P., J. H. Crowe, L. M. Crowe & D. E. Wickham, 1981. Uptake of amino acids by juveniles of *Carcinonemertes errans* (Nemertea). Comp. Biochem. Physiol. 69A: 423–427.

Schmitt, W. L., J. C. McCain & E. S. Davidson, 1973. Decapoda I, Brachyura I, Fam. Pinnotheridae. In H.-E. Gruner & L. B. Holthuis (eds), Crustaceorum Catalogus 3. Dr W. Junk Publishers, The Hague: 1–160.

Shields, J. D. & A. M. Kuris, 1990. *Carcinonemertes wickhami* n. sp. (Nemertea), a symbiotic egg predator from the spiny lobster *Panulirus interruptus* in Southern California, with remarks on symbiont-host adaptations. Fish. Bull. 88: 279–287.

Shields, J. D., D. E. Wickham & A. M. Kuris, 1989. *Carcinonemertes regicides* n. sp. (Nemertea), a symbiotic egg predator from the red king crab, *Paralithodes camtschatica* (Decapoda: Anomura), in Alaska. Can. J. Zool. 67: 923–930.

Sumner, F. B., R. C. Osburn & L. J. Cole, 1913. A biological survey of the waters of Woods Hole and vicinity. Section III. – A catalogue of the marine fauna of Woods Hole and vicinity. Bull. Bur. Fish., Wash. 31: 549–794.

Uhazy, L. S., D. E. Aiken & A. Campbell, 1985. Morphology and systematics of the nemertean *Pseudocarcinonemertes homari* (Hoplonemertea: Monostilifera) from the American lobster, *Homarus americanus*. Can. J. Fish. aquat. Sci. 42: 342–350.

Wickham, D. E., 1978. A new species of *Carcinonemertes* (Nemertea: Carcinonemertidae) with notes on the genus from the Pacific coast. Proc. biol. Soc. Wash. 91: 197–202.

Wickham, D. E., 1986. Epizootic infestations by nemertean brood parasites on commercially important crustaceans. Can. J. Fish. aquat. Sci. 43: 2295–2302.

Wickham, D. E. & A. M. Kuris, 1985. The comparative ecology of nemertean egg predators. Am. Zool. 25: 127–134.

Wickham, D. E. & A. M. Kuris, 1988. Diversity among nemertean egg predators of decapod crustaceans. Hydrobiologia 156: 23–30.

Wickham, D. E., P. Roe & A. M. Kuris, 1984. Transfer of nemertean egg predators during host molting and copulation. Biol. Bull. mar. biol. Lab., Woods Hole 167: 331–338.

Hydrobiologia **266**: 81–87, 1993.
R. Gibson, J. Moore & P. Sundberg (eds), Advances in Nemertean Biology

Ecological studies of the nemertean fauna in an estuarine system of the northwestern Gulf of Mexico

Judy O. Wern
Texas A & M University at Galveston, Department of Marine Biology, P.O. Box 1675, Galveston, TX 77553, USA

Key words: Nemertea, abundance, feeding, predation, estuary, Texas

Abstract

The distribution and abundance of nemerteans in the brackish-water lakes of Sea Rim State Park, Texas, near the Louisiana border, were studied and compared with other macrobenthos during one year. Six of 93 macrobenthic species collected were nemerteans (0.9% of the total number of specimens). Only one species of nemertean, *Carinoma* sp., was consistently present. This species is the most ubiquitous and, probably, the most abundant nemertean in the estuarine systems of the Texas coast. *Carinoma* sp. was collected at Sea Rim from a salinity range of 0–21 ppt and at other Texas estuaries from 2–26 ppt. Preliminary experiments with *Carinoma* sp. as predator and as prey indicated that it feeds on polychaete worms and in turn is fed upon by white (*Penaeus setiferus*) and brown (*P. aztecus*) shrimp.

Introduction

The taxonomy and ecology of nemerteans (phylum Nemertea) of the northwestern Gulf of Mexico have been inadequately studied. The only survey was reported by Coe (1951a, b; 1954), who identified sixteen species. Coe believed more species were present but not collected because of sporadic efforts in surveying. This paper reports on a study of nemerteans collected over one year in the brackish-water lakes of Sea Rim State Park, Jefferson County, Texas, near the Louisiana border. Water exchange into the 1214 ha lake system occurs at two points: (1) the North Weir along the Intracoastal Waterway and (2) a pass at the Port Arthur Canal which leads to the Gulf of Mexico. Freshwater influence is from rainfall, local runoff and discharge from the Sabine and Neches Rivers that enters the system mixed with salt water at the passes.

Material and methods

Monthly benthic samples were taken in triplicate at 12 stations of Sea Rim State Park from September 1978 through August 1979 using an Ekman grab ($232\ cm^2$) (Fig. 1). Samples were washed on a 0.5 mm sieve, fixed in 5% seawater formalin and stored in 70% ethanol. The macrobenthos was sorted and counted. Nemertean specimens were embedded in 56 °C melting point paraffin, sectioned at 6–7 μm, stained by Mallory trichrome and examined.

Salinity, temperature, dissolved oxygen, and depth were recorded at each site. Sediment samples were analyzed quarterly for graphic mean (M), sorting and percentages of sand, silt and clay.

Data from stations 11 and 12 were analyzed separately because the number of individuals of nemerteans and the macrobenthos in general were

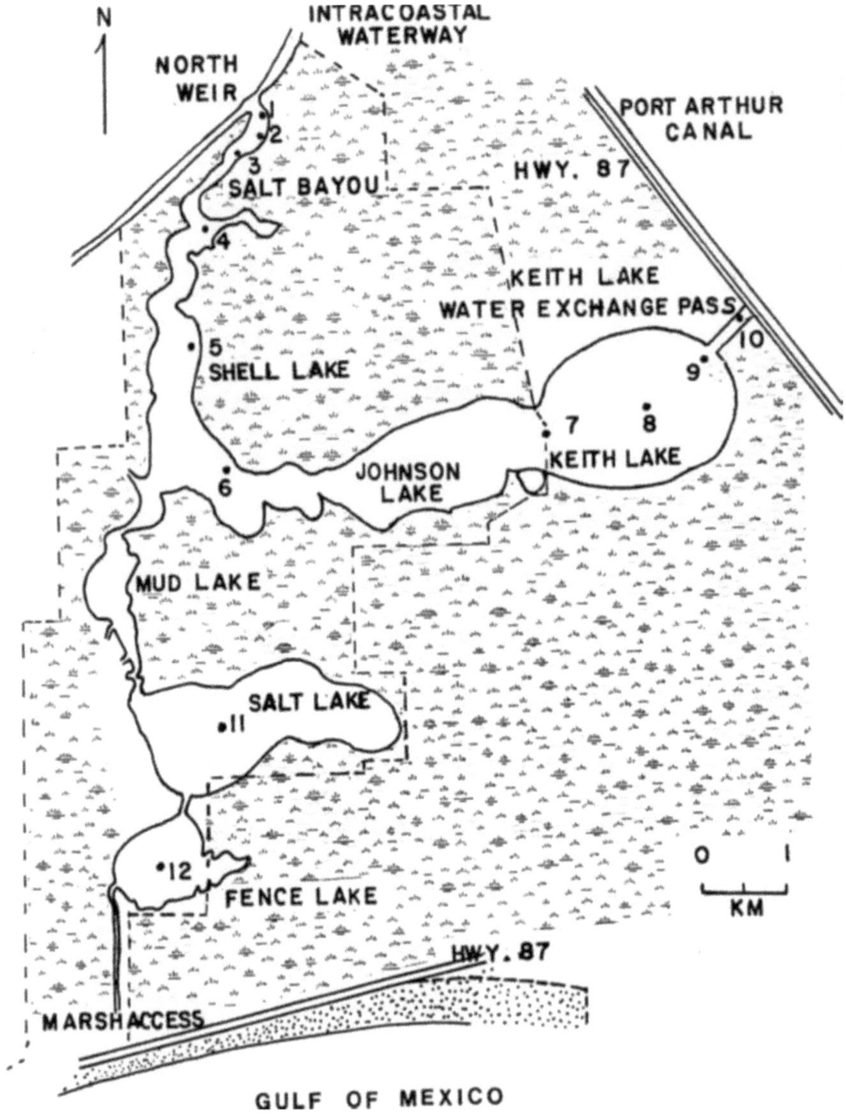

Fig. 1. Location of study sites in Sea Rim State Park, Texas.

much larger than at the other stations. Moreover, these stations had similarities in salinity that separated them from the other ten stations.

A feeding study with *Carinoma* sp. as a predator was done with 27 specimens of the worm collected from six Texas estuarine areas. Ten specimens collected from East Matagorda Bay were offered possible prey items: a bivalve, polychaetes (including *Streblospio benedicti* and Capitellidae spp.), amphipods (*Ampelisca abdita* and an unidentified species), a chironomid, an oligochaete, and an unidentified live nemertean fragment; all of these were collected simultaneously with the nemerteans. The polychaetes *Streblospio benedicti* (a spionid) and *Mediomastus ambiseta* (a capitellid) were always collected with *Carinoma* sp. at Sea Rim and hence considered possible prey. Seventeen specimens of *Carinoma* sp., collected over five locations west of Galveston to the central Texas coast, were also offered similar prey.

Another experiment was performed to determine if nemerteans are prey of white (*Penaeus setiferus*) and brown (*P. aztecus*) shrimp of size 55–110 mm. Both nemerteans and shrimp were collected from the Galveston Bay system. Three white and two brown shrimp were tested separately by placement in 26 ppt sea water in a 3 l glass beaker that set on a square of clear plexiglass mounted over a platform with a circular hole cut on the stage. A mirror was mounted under the platform for observation of activity in the beaker after an individual *Carinoma* sp. was introduced.

Results

Physical data

Table 1 summarizes abiotic data. For a complete description of the data, see Wern (1980). The principal component of the sediment was clay at most of the sampling stations. The graphic mean size of the sediment at station 12 remained basically unchanged, but most of the stations exhibited some variations.

Average depths at the stations ranged from 0.2–1.3 m. Because of shallowness, the system was mostly isothermal not only from surface to bottom but also among the stations (Table 1). In January the lowest temperature was 5.0 °C; while the highest in September was 30 °C (Wern, 1980). Dissolved oxygen was also nearly the same among the stations. The lowest average concentration of approximately 4 mg l^{-1} occurred when temperatures were the highest (September 1978 and August 1979) (Wern, 1980).

Between September–December 1978, average bottom salinity at stations 1–10 increased from 12.8 ppt to 19.1 ppt and then decreased to 12.7 ppt. In contrast, only a slight variation, between 9.5 and 11.0 ppt, occurred at stations 11 and 12 (Fig. 2). Salinities were less than 5 ppt in the entire system between January and June. Stations 1–10 experienced an increase in salinity to 7.6 ppt in July, though stations 11–12 continued to have low salinity. Tropical storm Claudette brought rains in late July, lowering the salinity in the entire system to an average of 2 ppt.

Nemertean data

Nemertean species comprised 6% of the taxa (6 of 93 species) but only 0.9% of the abundance (441 specimens of 47834) (Table 2). The two overwhelmingly dominant macrobenthic species were the polychaetes *Mediomastus ambiseta* (Capitellidae) and *Streblospio benedicti* (Spion-

Table 1. Biotic and abiotic characteristics of the 12 sampling sites.

Station	Average depth (m)	Average temperature (°C)	Average dissolved oxygen (mg l^{-1})	Average salinity (ppt)	Average sediment graphic mean (M)	Number of *Carinoma* sp.	Number of other nemerteans	Number of other macrobenthic sp.
1	0.89	21.4	6.3	5.9	9.2	12	0	2086
2	0.23	22.9	6.6	6.2	7.0	21	0	1278
3	0.56	21.5	6.3	6.0	9.4	22	0	1297
4	0.55	22.0	6.0	5.6	9.6	15	0	1342
5	0.77	22.3	6.3	5.6	9.1	26	0	2545
6	0.77	19.9	6.0	5.3	9.0	19	0	2736
7	0.93	20.2	6.7	6.0	9.2	13	0	1232
8	0.92	19.8	7.1	6.3	7.5	8	1	1064
9	1.30	20.1	6.4	6.6	8.0	14	1	1117
10	1.32	20.6	6.6	7.1	6.4	34	4	666
11	0.88	19.8	7.0	4.6	9.4	100	1	4023
12	0.75	20.0	7.2	4.6	9.6	150	0	28007

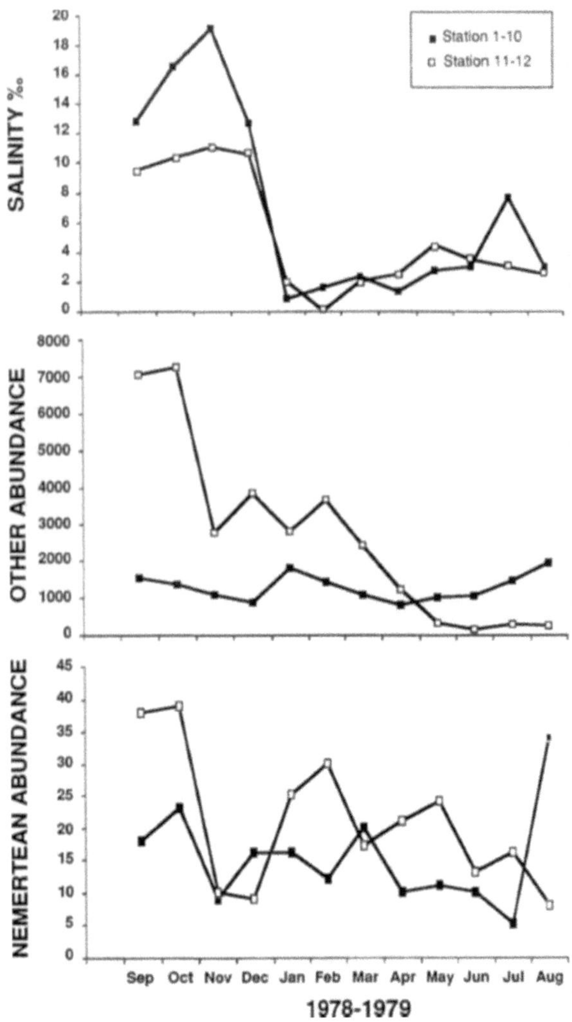

Fig. 2. Salinity and abundance of *Carinoma* sp. and other benthos at stations 1–10 and 11–12.

idae). Five transient nemertean species were collected only once from stations near the Gulf or station 11 (Table 2). *Carinoma* sp. (434 individuals) was the only nemertean consistently present, and over fifty percent of these were collected at two (stations 11 and 12) of the 12 stations (Fig. 1). Though *Carinoma* sp. occurred at most salinities between 0 and 21 ppt, 21% were collected at 10 ppt (Fig. 3). Subsequent to this study, I collected live *Carinoma* sp. in salinities ranging from 2–26 ppt at twenty-one locations from Fence Lake of Sea Rim State Park to Port Isabel near the Mexican border.

Table 2. Nemerteans collected at Sea Rim State Park.

	Abundance
Anopla	
Paleonemertea	
Carinomidae	
Carinoma sp.	434
Heteronemertea	
Sp. A	1
Sp. B	1
Valencinnidae	
Zygeupolia rubens (Coe, 1895)	1
Unknown higher taxon	
Sp. 1	1
Sp. 2	3

The abundance trend of *Carinoma* sp. at stations 11–12 paralleled that of other macrobenthic species with three exceptions (Fig. 2). This was not always the case for the abundances of *Carinoma* sp. and other benthos at stations 1–10. Also the trends of nemertean abundances between stations 11–12 and 1–10 were often in opposite directions.

There was a significant difference in abundance among the twelve stations (ANOVA, $P<0.001$). Duncan's multiple-range test ($\alpha = 0.005$) revealed no significant differences in abundance among stations 1–10; stations 11 and 12, however, differed significantly from each other and from the other stations.

Laboratory observations

The two largest intact *Carinoma* sp. measured 55 mm long × 1 mm wide and 50 mm long × 2 mm wide in preserved state but most of the collected specimens were much smaller. No observations were recorded on live worms collected during this study; however, I subsequently collected and observed specimens from other areas of the Texas coast. The worm burrowed shallowly head first in a seawater agar medium at an oblique angle and then reoriented its head to the surface. A specimen without a proboscis appeared to burrow just

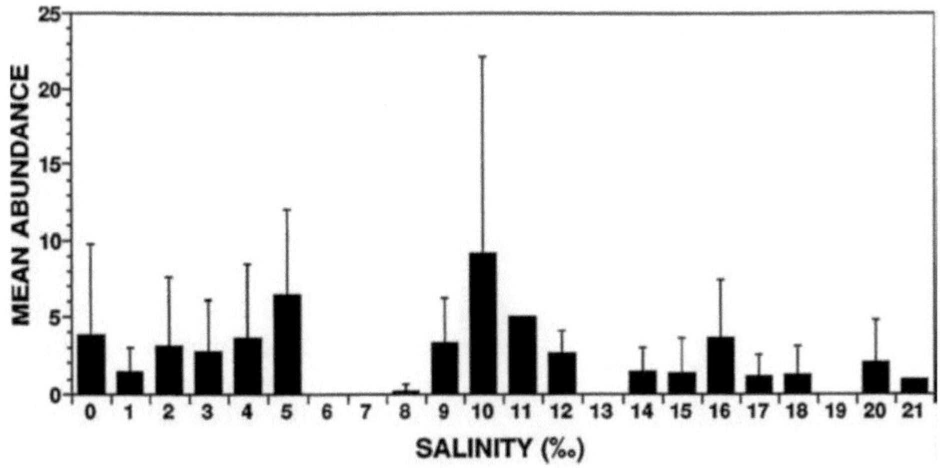

Fig. 3. Mean abundance of *Carinoma* sp. at salinity range of 0–21 ppt.

as efficiently as intact worms. The worms were not coiled when moving undisturbed on a substrate; however, as McCaul (1963) noted for *C. tremaphoros*, coiling occurred when distressed. The coiled posterior end often remained still as the anterior end probed about. To the naked eye, *Carinoma* sp. appeared to be orange with white anterior and posterior ends. Under a dissecting microscope the anterior was pale yellow or pale orange, the foregut cream, the intestinal area pale to deep orange and the posterior cream. Undulations of the thin, highly coiled proboscis were observed through the body wall of the foregut region.

Feeding experiments

Results of the feeding experiments are given in Table 3. Of the ten *Carinoma* sp. specimens from East Matagorda Bay, six rejected prey items, while two ate one capitellid polychaete each and one fully ate and another partially ate the spionid polychaete *Streblospio benedicti*. One Colorado River March specimen also ate a capitellid. The proboscis was not used directly in feeding. Occasionally the proboscis was everted when prey seemed to be detected. However, in all but one instance of proboscis eversion, the proboscis did not make contact with the prey. Upon retraction of an everted proboscis, the nemertean passed over the prey with its anterioventral surface, sometimes making several passes before accepting or rejecting it. One specimen, after spending about two minutes ingesting a worm, everted the proboscis at the time when the prey passed into the intestine. Sixteen specimens, which were collected over four locations from just west of Galveston to the far south Texas coast, did not respond to offerings of polychaetes and other potential prey.

All eight specimens of *Carinoma* sp. offered as prey to shrimp were accepted. Two white shrimp and one brown shrimp ate two nemerteans each, while one white shrimp and one brown shrimp ate one each.

Discussion

Carinoma sp. is the most ubiquitous and, probably, the most abundant nemertean in estuarine systems of the Texas coast. The genus *Carinoma* has six species two of which have been reported in North America. *C. tremaphoros* Thompson (1900) was collected in bays, harbors and estuaries from Cape Cod to northern Florida and in the Gulf from Florida to Louisiana (Thompson, 1900; Coe, 1943, 1951a, b, 1954; McCaul, 1963) and *C. mutabilis* Griffin (1898) = *C. griffini* Coe

Table 3. Feeding experiments with *Carinoma* sp. as predator.

Collection site data/specimen no.		Prey accepted	Rejected prey offerings
East Matagorda 3/6–3/10/85			
Specimen no.	1	none	A, C, E, F(2), J, M, P
	2	none	F, M
	3	none	F, I, M
	4	F	I, J, K, M, M
	5	none	A, B, F, G, I, M, N
	6	N	A, B, F, I
	7	F	A, B, J, K, M
	8	none	A, B, F, I
	9	none	A, F, I
	10	N	E, I
Colorado River 3/8/85			
Specimen no.	1	F	
Port Mansfield (Laguna Madre) 3/28/85			
Specimen nos	1–6	none	F, N
Chocolate Bayou 3/31–4/4/85			
Specimen nos	1–2	none	F, G
	3	none	F, N
	4–6	none	F
	7–8	none	E, F
Rockport Marsh (Copano Bay) 4/15/85			
Specimen no.	1	none	D, E, F, H, J, L, N, O
Indianola (Matagorda Bay) 4/16/85			
Specimen no.	1	none	F

A = *Ampelisca* sp., B = *Ampelisca* sp. (dead), C = amphipod, D = tanaid, E = bivalve, F = Capitellidae sp., G = Capitellidae sp. (dead), H = Orbiniidae sp., I = Cirratulidae sp., J = Nereidae sp., K = nemertean piece, L = polychaete piece (dead), M = oligochaete, N = *Streblospio benedicti*, O = debris, P = chironomid.

(1901) from the Northwest coast. *Carinoma* sp. did not appear to belong to either of the two species.

The habitat of stations 11 and 12 was conducive to the highest abundance not only of nemerteans but also of the total macrobenthos (67%). Monthly trawl catch data for Sea Rim State Park (Fish, unpublished) and abiotic data were compared with benthic abundance to determine causes of population fluctuation. Of the abiotic characteristics measured, no substantial differences existed among the stations with regard to temperature and dissolved oxygen (Table 1). Salinity, though, was more stable at these two isolated stations and the sediment at station 12 differed from the other sampling sites by being firmer.

A possible factor explaining the greater abundance of nemerteans at stations 11–12 was the large potential food supply of capitellid and spionid polychaetes; McDermott & Roe (1985) reported that paleonemerteans appeared to feed on annelids of comparable size. This study supported that contention.

Declines in general benthic populations at Sea Rim State Park were concomitant with peak or high abundances of either white shrimp (*Penaeus setiferus*) in November at stations 11–12 and in October at stations 1–10 or croakers (*Micropogonias undulatus*) in April at stations 11–12 and March at stations 1–10. At peak abundance, croakers were 40–90 mm long, the size range at which they are benthic feeders (Darnell, 1958). A period of low salinity (< 5 ppt) from January to August probably also had a deleterious effect on reproduction and/or survival of offspring. In comparing abundance data of nemerteans at stations 11–12 with general macrobenthic data at those stations, the direction of the curves (Fig. 2) was the same except in December, April and May. The differences in April and May at these stations suggested that nemerteans were not prey items for croakers. This agreed with the limited experimental studies cited by McDermott & Roe (1985) reporting that nemerteans were rejected by various species of predatory fish and decapod crustaceans. In this study, however, shrimp fed on nemerteans in the laboratory, and the decline of nemerteans in the fall (Fig. 2) suggested that nemerteans were prey of white shrimp.

Abundance trends of nemerteans at stations 1–10 (Fig. 2) were not usually similar to the abundance trends of nemerteans at stations 11–12 or other macrobenthos. This may have been a reflection of having a low number of nemerteans spread over ten stations.

Nemerteans, because of their relative low numbers, evidently do not make a great impact on ecological interactions as either predators or prey in this brackish-water system or in a number of other systems of the northwestern Gulf coast. Nemerteans (five distinct forms) represented 0.4% of the macrobenthos in Eckert's Bayou, Galveston (Potts, 1978), and Nance (1984) found that nemerteans made up 0.2% of the total number of macrobenthic specimens in New Bayou, a small gradient estuary in Brazoria County west of Galveston. Nemertean abundance was also low in two south Texas bay systems: approximately 0.6% in Aransas Bay (Harper, 1976) and 1.9% in San Antonio Bay (Harper & Hopkins, 1976). The low abundance of nemerteans is not peculiar to Texas since similar reports exist for the Atlantic coast also. Nemerteans comprised 0.04% of macrobenthic abundance at Pocasset River stations of a small estuary north of Woods Hole, Massachusetts (Sanders *et al.*, 1965) and 0.18% in Buzzards Bay (Sanders, 1960).

Acknowledgements

Macrobenthic collections in Sea Rim State Park were funded by Texas Parks and Wildlife Department Inter-Agency Cooperation Contract No. IAC (78–79) 1500. Collections in 1985 along the entire Texas coast were supported by Lerner-Gray Fund for Marine Research (American Museum of Natural History). I appreciate the assistance of Drs Melissa Tucker, Merrill H. Sweet III, Donald E. Harper, Jr., Mark Benfield, Darrell Smith and Merrill H. Sweet II.

References

Coe, W. R., 1901. Proceedings from the Harriman Alaska Expedition XX. The nemerteans. Proc. Wash. Acad. Sci. 3: 129–328.

Coe, W. R., 1943. Biology of the nemerteans of the Atlantic coast of North America. Trans. Conn. Acad. Arts Sci. 35: 129–328.

Coe, W. R., 1951a. The nemertean faunas of the Gulf of Mexico and of southern Florida. Bull. mar. Sci. Gulf Carib. 1: 149–186.

Coe, W. R., 1951b. Geographical distribution of the nemerteans of the northern coast of the Gulf of Mexico as compared with those of the southern coast of Florida, with descriptions of three new species. J. Wash. Acad. Sci. 41: 328–331.

Coe, W. R., 1954. The nemertean fauna of the Gulf of Mexico. Fish. Bull. Fish. Wildl. Ser. 55: 303–309.

Darnell, R. M., 1958. Food habits of fishes and larger invertebrates of Lake Ponchartrain, Louisiana, an estuarine community. Inst. Mar. Sci. 5: 353–416.

Harper, Jr D. E., 1976. The effects of oyster shell dredging on the biota of the Aransas National Wildlife Refuge. In A. H. Bouma (ed.), Shell Dredging and Its Influence on Gulf Coast Environments. Gulf Publishing Company, Houston: 221–231.

Harper, Jr D. E. & S. H. Hopkins, 1976. The effects of oyster shell dredging on macrobenthic and nektonic organisms in San Antonio Bay. In A. H. Bouma (ed.), Shell Dredging and Its Influence on Gulf Coast Environments. Gulf Publishing Company, Houston: 232–279.

McCaul, W. E., 1963. Rhynchocoela: nemerteans from marine and estuarine waters of Virginia. J. Mitch. Soc. 49: 111–123.

McDermott, J. J. & P. Roe, 1985. Food, feeding behavior and feeding ecology of nemerteans. Amer. Zool., 25: 113–125.

Nance, J. M., 1984. The seasonal distribution of macrobenthos and sediments chronically impacted by oil/gas field produced water. Ph.D. Dissertation. Texas A&M University, 267 pp.

Potts, D. L., 1978. A study of the distribution and ecology of macrobenthic communities in Eckerts Bayou on Galveston Island, Texas. Masters Thesis. Texas A&M University, 205 pp.

Sanders, H. L., 1960. Benthic studies in Buzzards Bay III. The structure of the soft-bottom community. Limnol. Oceanogr. 5: 138–152.

Sanders, H. L., P. C. Mangelsdorf, Jr & G. R. Hampson, 1965. Salinity and faunal distribution in the Pocasset River, Massachusetts. Limnol. Oceanogr. Supp. 10: R216–R229.

Thompson, C. B., 1900. *Carinoma tremaphoros*, a new mesonemertean species. Zool. Anz., 23: 627–630.

Wern, J. O., 1980. A study of the macrobenthos of the brackish lakes in Sea Rim State Park, Texas and contiguous Keith Lake. Masters Thesis. Texas A&M University, 215 pp.

Hydrobiologia **266**: 89–101, 1993.
R. Gibson, J. Moore & P. Sundberg (eds), Advances in Nemertean Biology

Methods of classifying nemerteans: an assessment

Janet Moore[1] & Ray Gibson[2]
[1] *Department of Zoology, Downing Street, Cambridge CB2 3EJ, United Kingdom;* [2] *School of Natural Sciences, Liverpool John Moores University, Byrom Street, Liverpool L3 3AF, United Kingdom*

Key words: classification, phenetics, cladistics, phyletics

Abstract

Phenetic, cladistic and phyletic methods of classifying animals are discussed with particular reference to nemerteans. It is concluded that phenetic (numerical) taxonomy is particularly inapplicable to any group of invertebrates for which well defined character differences are relatively few, whilst both the phenetic and cladistic methods fail through their fundamental assumption that convergent evolution is a rare occurrence. Terrestrial and freshwater nemerteans especially demonstrate convergent evolution in many ways; cladistic classifications proposed for these animals are therefore untenable. Convergence is shown to be a common occurrence in other nemerteans also. It is concluded that because the traditional phyletic approach does not implicitly assume that resemblances between organisms are more likely to be due to common ancestry than to convergence, it is far more likely to reveal true evolutionary relationships between taxa.

Introduction

The classification of animals fundamentally aims to reveal their evolutionary relationships. There are several methods which can be used, all of which are based on the same data, and no one method is intrinsically correct. There can be, however, only one correct evolutionary sequence: the problem is to choose the method of classification most likely to reveal this. The chief difficulty is that animal resemblances may be a consequence either of common ancestry or of convergent evolution. Studies of land and freshwater nemerteans in particular demonstrate this difficulty, since convergent evolution can be shown to be both frequent and widespread in these animals (Moore & Gibson, 1981, 1985, 1988a, b). In formulating the present paper we are greatly indebted to Sundberg's (1987) emphasis on the need to be explicit about taxonomic method, and to his application of a method different from our own to the classification of the land and freshwater nemerteans (Sundberg, 1989a, b). The particular questions we consider in our appraisal of classificatory methods are: (1) which methods of classification are least likely to obscure the occurrence of convergence; (2) which methods will accordingly be profitable for the classification of land and freshwater nemerteans in particular; and, (3) whether there are wider implications for the classification of nemerteans in general.

Possible methods of classification

There are three principal methods in current use:

1. PHENETIC or numerical taxonomy; the term is used in the restricted sense. Any classification based on perceived characters could be

called phenetic, but the term is here restricted to be synonymous with Numerical Taxonomy (Sneath & Sokal, 1973).
2. CLADISTIC ANALYSIS, the 'phylogenetic' method originating with Hennig (1966) (the modification known as 'transformed cladism' [Platnick, 1980; Patterson, 1980] is excluded from the present discussion, as being insufficiently aimed to reveal evolutionary relationships) (Charig, 1982).
3. PHYLETIC or 'traditional' taxonomy, as formulated by Simpson (1961) and Mayr (1982).

Phenetics

The procedure is to define as many characters as possible, record the character states for each species and use a computer. Those species with the greatest numbers of characters in common are then classified together. Characters are assumed to be independently assorted, and resemblance due to convergence is not distinguished from resemblance due to common ancestry. This method claims objectivity: there are however concealed subjective judgements underlying the whole procedure, firstly in the selection (and sometimes weighting) of characters and secondly in the assumption that convergence is a rare event which will be insignificant if sufficiently large numbers of characters are used.

Despite reservations, Sundberg (1989c) experimentally applied the phenetic method to the classification of one group of land nemerteans, using as his source of data serial sections of specimens in the Pantin collection (which contains some 20 good specimens for each of six of the Australian and New Zealand species). It proved difficult to find a sufficient number of qualitative character differences, and quantitative characters showed great intraspecific variation. The method appears to be particularly unsuitable where identifiable characters are relatively few and where convergence is known to be common. The phenetic method has been rejected by other workers as being insufficiently designed to reveal the course of evolution, a counsel of despair founded on the belief that to trace evolutionary affinities is impossible (Disney, 1983).

Cladistic analysis

This is the method currently favoured by a large number of taxonomists (Hennig, 1966; Eldredge & Cracraft, 1980; Wiley, 1981). The focus here is on differences between species, a step by step separation drawn as the familiar branching cladograms. First the primitive condition must be identified. This is often (but not always) possible by widespread comparison with 'outgroups'. Where two species have departed from the primitive condition in the same way, this is taken as proof that they are closely related: a primitive (plesiomorphic) character state has been replaced by a specialisation held in common by both species, *i.e.*, a shared derived character state or synapomorphy. Such steps are the units on which cladograms are constructed. The aim of the exercise is to identify monophyletic groups, which are defined as groups of species that include known or hypothesised ancestral species and all of their descendants. Degree of phylogenetic relationship is defined solely by recency of common ancestry: neither the passage of time nor the degree of further differentiation along a given line is taken into account.

Cladistic analysis can teach us extremely important lessons: for example, common primitive characters cannot unite species at the level where branching occurs. Emphasis on shared derived characters is something for which we must all be grateful to the cladists. Some questions, however, need to be asked:

(a) Is this procedure an objective test? As with phenetics, there is surely a hidden subjective step when characters are selected (Halstead, 1978; Hull, 1979). Cladistic analyses depend upon the definition of the primitive character state, so that changes from this state can be identified. Since this definition can only be possible for a limited number of characters, subjective selection is inevitable. Panchen (1982) compared this to predicting the result of a General Election by asking the voting intentions of the first dozen voters encountered. Objectivity may not be a prime requirement, but its absence needs to be admitted.

(b) Are cladograms too rigid? Did evolution really proceed in single separate steps which altered a finite number of discrete characters? Such a claim is simple-minded gradualism, in other contexts firmly dismissed: punctuated equilibria are seen as entirely compatible with traditional Darwinism, and this view is explicit in later editions of The Origin (Darwin, 1899: 85). These separate steps may be a requirement of method rather than theory, but method is then liable to dictate theory, unnoticed.

(c) How do we solve the problem that very different cladograms can be drawn from the same data? We are urged to be 'parsimonious'. This means the choice of the cladogram which minimises the assumption of convergence and reversal of character states. Hennig's parsimony only allows non-divergent evolution to be considered when there is strong independent evidence in its favour. Might not the sharing of a large number of characters be just as likely to result from convergence as from common ancestry?

Cladistics, like phenetics, rests on the assumption that convergence is a rare event – 'false synapomorphy' it is pejoratively called. Is this really a correct assessment of the evolutionary process? Friday (1987: 61, 63) gives a very helpful analysis of the overwhelming effect on our thinking of Darwin's Principle of Divergence which gave rise to 'the prevailing idealistic view of evolutionary change as a simple process of divergence that sometimes misbehaves ... Such is our familiarity with the ideal of evolutionary divergence that we have ceased to recognise it for what it is: an *a priori* model of evolutionary change' where 'best adapted' becomes equated with 'most divergent'. Again, Joysey (1988: 38) emphasises that phenetic and cladistic techniques 'are based on the assumption that divergent evolution is the norm and parallel [or convergent] evolution is the exception'.

Phyletic or traditional taxonomy

Objections to this method are not confined to the sense that it is 'old fashioned':

(a) It is subjective. Arguably, a controlled and disciplined subjectivity is helpful, since our brains can act as usefully sensitive computers. Certainly there is no substitute for thorough study and knowledge of the animals being classified (for example, to assess the degree of intraspecific variation) and certainly an open subjectivity must be preferred to a spurious objectivity with a concealed subjective step.

(b) There is a danger of emphasising similarities rather than differences. This is a real distinction: similarity is not merely the obverse of difference, similarity can be much less precisely defined. It is dangerously easy to assemble similarities and to be so much impressed by them that differences are ignored. The lessons of cladistic analysis, with its emphasis on shared derived characters, must be incorporated. It is not however essential to take the further, separate, step of erecting cladistic analysis into cladistic classification (Ashlock, 1974).

The cladistic and the traditional taxonomic approaches are based on the same raw data (Gibson, 1990a). What are the advantages of the traditional method? Firstly, we are driven to seek as many characters as we can to support each grouping which we make, and to evaluate the taxonomic importance of these characters. We are able to consider present characters, the 'autapomorphies' at the branch tips of the cladist's diagram, not merely the nodes of an assumed series of dichotomies. This brings us closer to evolution in action: as Mayr (1969) insisted, genealogical and genetic distance do not necessarily coincide. Autapomorphies can contribute to rank and we are allowed to recognise 'grades' defined by major differences (Mayr, 1982). This process involves the recognition of paraphyletic groups, anathema to a cladist. The cladist's definition of a monophyletic group must fail where the ancestral condition cannot be identified, and this difficulty is particularly acute in nemerteans. The traditional approach allows the recognition of important characters which are not necessarily governed by divergence. In general, the advantage is that we have fewer set rules to constrain us. Without the strait-jacket of assumption that divergence is normal, do we not have a better chance of recognising convergence when it occurs? Above all, we cannot forget that there is no substitute for study of the animals themselves: we cannot get away from subjective assessment based on knowledge of the species concerned.

Classification of terrestrial nemerteans

Marine nemerteans have colonised the land many times, from different ancestral stocks on different oceanic islands. The geographical distribution of species in general accords with morphological differences: islands close together have species much alike, while islands far apart may have species very different. Many characters clearly demonstrate parallel (more precisely, convergent) evolution (Pantin, 1969;Moore, 1985). Two morphologically distinct groups of land nemerteans, which can be traced back to two different hoplonemertean families, were established by Moore & Gibson (1981). Group I, the genera *Pantinonemertes* and *Geonemertes*, was placed in the family Prosorhochmidae (Moore & Gibson, 1988a). Group II, consisting of all other known terrestrial nemerteans and the freshwater genera *Potamonemertes* and *Campbellonemertes,* was placed in the Plectonemertidae (Moore & Gibson, 1988b).

Cladistic methods were then applied to the classification of land nemerteans, both in Group I (Sundberg, 1989a) and in Group II (Sundberg, 1989b) with results considerably different from those achieved by Moore and Gibson who used the traditional method. One point over which the cladists and traditionalists are in complete agreement is that the overall polarity within each group is not in doubt: no one supposes that nemerteans evolved on land and secondarily colonised the sea. The disadvantage for the cladist is the degree of convergent evolution shown by these animals: the separate origins of Groups I and II are beyond doubt, yet there is a pronounced degree of resemblance in characteristics which we can relate to the requirements of fully terrestrial life. All land nemerteans, from both groups, share the following characteristics:

1. Small size.
2. Ventral opening of cerebral organs (as a consequence of lateral furrows being lost).
3. Extensive cephalic glands and copious mucus production.
4. Large proboscis, often used for rapid locomotion.
5. Excretory system extending throughout the body and opening by a very large number of pores.

These characters are clearly of no help as indicators of close relationship. Difficulties begin when considering characters which cannot obviously be related to the common habitat: are resemblances due to common ancestry or to convergent evolution?

Sundberg's cladistic analysis of Group I, the genera *Pantinonemertes* and *Geonemertes,* is shown in Fig. 1 – we are grateful to him for allowing us to reproduce it with modification. It is rooted in the Prosorhochmidae, using the marine genera *Prosorhochmus* and *Prosadenoporus* as outgroups. In the succeeding sections, in which we re-examine the steps used by Sundberg to develop his cladistic classification, the various character states are indicated in bold numbers and relate to those given in Fig. 1.

What unites the two Group I genera?

23 Flame cells are fused in pairs and reinforced with cuticular bars.
22 Excretory ducts and pores are multiplied.
16 Lengthened oesophagus.

Comment: Flame cell structure (**23**) is a good shared derived character of *Pantinonemertes* and *Geonemertes*. Examination by light microscopy suggested that the patterns of reinforcement were so unalike that the two genera must have evolved separately. However we now have the benefit of an electron microscopy study by Jespersen & Lützen (1988a) which shows the reinforcement to

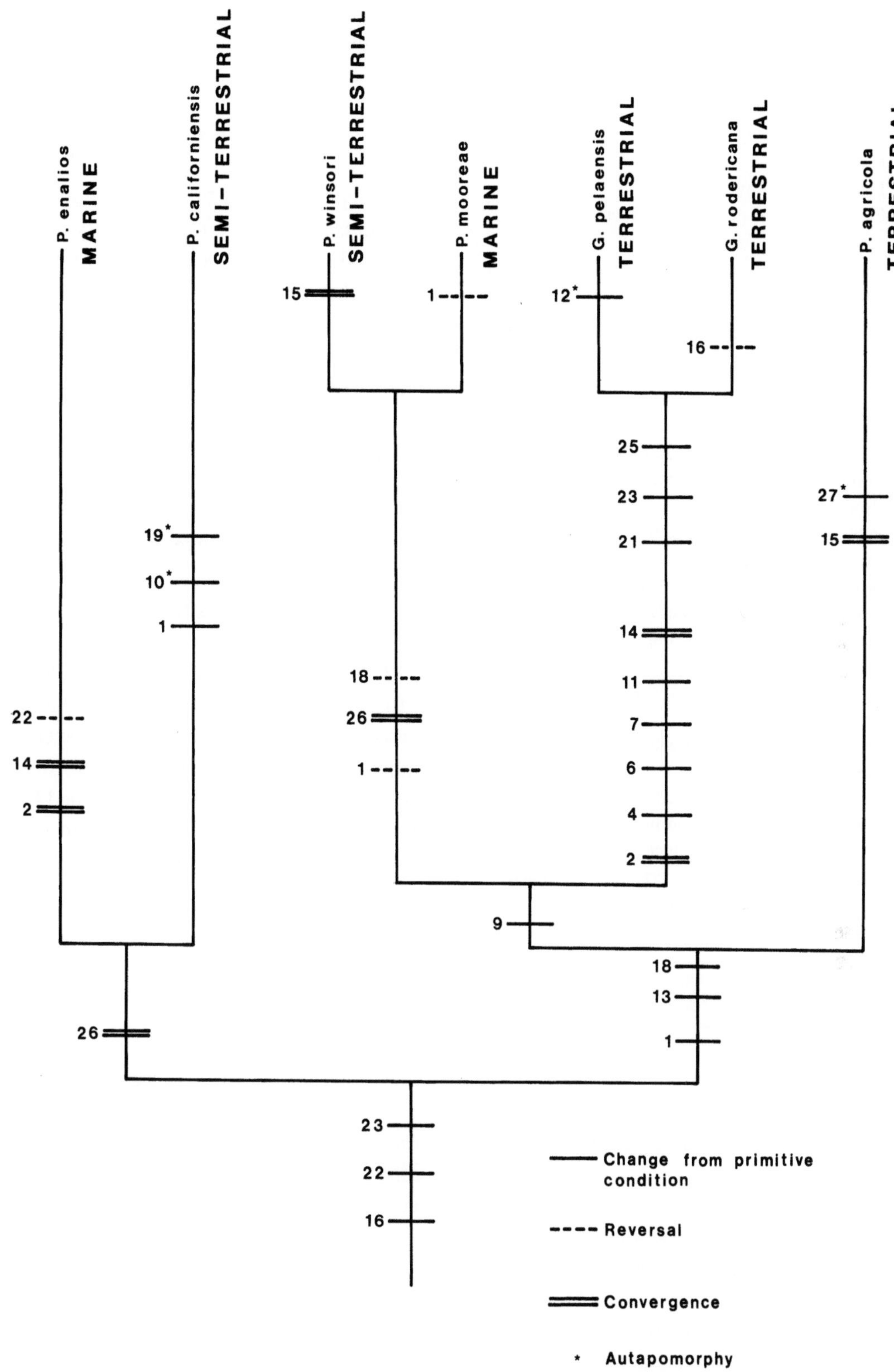

Fig. 1. Sundberg's cladogram showing the evolutionary relationships between species of *Geonemertes* and *Pantinonemertes*. Character state changes are indicated by transverse bars. Characters represented by the numbers are listed in Table 1. Redrawn, with permission, from Sundberg (1989a: Fig. 1).

Table 1. Identification of characters indicated by numbers in Figure 1.

Character number	Character
1	Habitat
2	Length at maturity
4	Cephalic glands, presence or absence of granular proteinaceous components
6	Cephalic glands, other ventral glands present or absent
7	Cerebral organs open from lateral longitudinal or ventral transverse furrow
9	Cerebral organs small or elaborated
10	Eyes four or more than four
11	Accessory lateral nerve present or absent
12	Body wall musculature weak or strong
13	Parenchyma sparse or extensive
14	Proboscis small or large
15	Proboscis used or not used for locomotion
16	Oesophagus long or short
18	Blood vessels thick-walled with 'valves' and extra-vascular pouches, thin-walled with 'valves' and extra-vascular pouches, or thin-walled with 'valves' but no pouches
19	Cephalic blood system a capillary network or simple vascular loop
21	Excretory system with or without specialisation of terminal region
22	Excretory system with no, low or high degree of multiplication of ducts and pores
23	Flame cells simple mononucleate without support bars, binucleate with transverse bars, or binucleate with transverse and longitudinal bars
25	Flame cells scattered or aggregated around cephalic loop
26	Sexes separate or hermaphroditic
27	Reproduction ovoviviparous or otherwise

be essentially similar: the two types of flame cells could well have evolved from a common ancestral form.

The degree of multiplication of the excretory system (**22**) can be related to habitat as indicated above. On the scheme proposed, reversal must be invoked in a species supposedly secondarily marine. This character is not helpful.

Lengthening of the oesophagus (**16**) can have two separate causes: precerebral length increasing, as in *Pantinonemertes,* or great enlargement of the stomach which can then only fit behind the brain, as in *Geonemertes*. This character does not unite the genera.

What unites P. enalios *and* P. californiensis*?*

26 Separate sexes replace hermaphrodity.

Comment: This is a fragile solitary character for uniting these two species as separate from the rest. Either condition may obtain in closely related terrestrial species (e.g., three of the four Australian land species have separate sexes and one is hermaphrodite). Here the change to separate sexes has to be recorded as a convergence on the line leading to *P. winsori* and *P. mooreae*.

What unites the remaining species?

1 Habitat becomes more terrestrial.
13 Parenchyma increases.
18 Blood vessels become thin-walled.

Comment: Terrestrial habitat (**1**) has to be reversed on one of the lines. Blood vessels are thin-walled (**18**) in the outgroup genus *Prosorhochmus* but thick-walled in the highly modified blood system of *Prosadenoporus* (Moore & Gibson, 1988a). It is thus by no means certain that 'blood vessels thin-walled' represents a derived character state.

What unites P. winsori *and* P. mooreae *with* Geonemertes *species?*

9 Elaboration of the cerebral organs.

Comment: Since the cerebral organs are elaborated in completely different ways (Moore & Gibson, 1981; Gibson, 1982), this union entirely fails.

What unites P. winsori *with* P. mooreae*?*

1 Habitat becomes less terrestrial.
26 Separate sexes replace hermaphrodity.
18 Blood vessel walls thicken.

Comment: Habitat (**1**) is a reversal, separate

sexes (**26**) a convergence, and blood vessels (**18**) are cited as reversal (but see above).

In summary, 32 character states are recorded. Four (starred in Fig. 1) are autapomorphies included for completeness. Excluding the characters concerning oesophageal length and cerebral organ elaboration, *26* character states are used. Eight of these involve convergence and four are reversals (four more are therefore involved in reversal) although this is the most parsimonious scheme which can be found, *i.e.*, with reversal and convergence minimised. Seven more of these character states concern the union of the two species of *Geonemertes,* which undoubtedly are very alike and very distinct from *Pantinonemertes.*

It is of course feasible to sort out these seven species in this way, but we are not convinced that it more accurately represents the relationships between these taxa. If the most parsimonious cladogram presents such a result, can a cladistic analysis possibly be on the right track?

Group II includes two freshwater genera, *Campbellonemertes* and *Potamonemertes.* Here cladistic classification (Sundberg, 1989b) illustrates even more strongly the particular difficulty posed by convergence. Sundberg's (1989b: Fig. 3) cladogram is shown in Fig. 2. Moore & Gibson's (1988b: Fig. 2) postulation of Group II relationships is given for comparison (Fig. 3) (Moore & Gibson's figure was incomplete in that it intentionally omitted the two non-Australasian genera *Leptonemertes* and *Katechonemertes*). The placing of the two freshwater genera constitutes the main difference between these figures.

For cladistic analysis it is difficult to find an outgroup for Group II. Use of *Oerstedia* species followed a tentative suggestion by Moore and Gibson but the suitability of this taxon is now very dubious. The three resulting character states at the base of the cladogram are not strong:

- **8** Ventrolateral cerebral organ openings become ventral.
- **16** Weak body wall musculature becomes strong.
- **22** Ciliated oesophagus becomes unciliated.

Each of these demands reversal at some later stage (reversal of **8** for the freshwater genera, where the cerebral organ openings are fully lateral, was omitted in error).

The first division separates *Plectonemertes sinensis,* the only fully marine species, from the rest. *Plectonemertes* is a monotypic genus with a number of specialised characters. The terrestrial and freshwater species are united by:

- **9** Cephalic glands discharge through single pore separate from the frontal organ.
- **28** Blood capillary network present in foregut region only.
- **29** 'Valves' become numerous in blood vessels (the term 'valve' has been retained, although electron microscopy studies by Jespersen & Lützen [1988b] have shown it to be quite inappropriate).

Comment: This combination of character states completely fails. **9** is a highly specialised condition which turns out to be an autapomorphy for *Potamonemertes percivali* (see below) and requires reversal for all other species. **28** is also used erroneously since it is an autapomorphy for *Plectonemertes sinensis* and does not apply to any other genus. **29** requires reversal for *Potamonemertes percivali.*

At the next division, *Potamonemertes* and *Campbellonemertes* are combined, and distinguished from the terrestrial species, on the following character states:

- **1** Increase in length.
- **3** Cephalic glands shorten.
- **11** Eyes are lost.
- **26** Blood vessels become lacunar.
- **27** Blood vessels lose cross connections.
- **37** Hermaphrodity.

Comment: These six character states do not establish a close relationship between the two freshwater genera. Increase in length (**1**) is correct for *Campbellonemertes johnsi* (100 mm) but not for *Potamonemertes percivali* (20–30 mm, like many terrestrial species). Shortening of the cephalic glands (**3**) is correct for *P. percivali* but not for *C. johnsi,* which has a well-developed extensive cephalic gland arrangement like all the terrestrial species. Loss of eyes (**11**) (except in most species of the tetrastemmid freshwater genus *Prostoma),* hermaphrodity (**37**) and absence of cross connec-

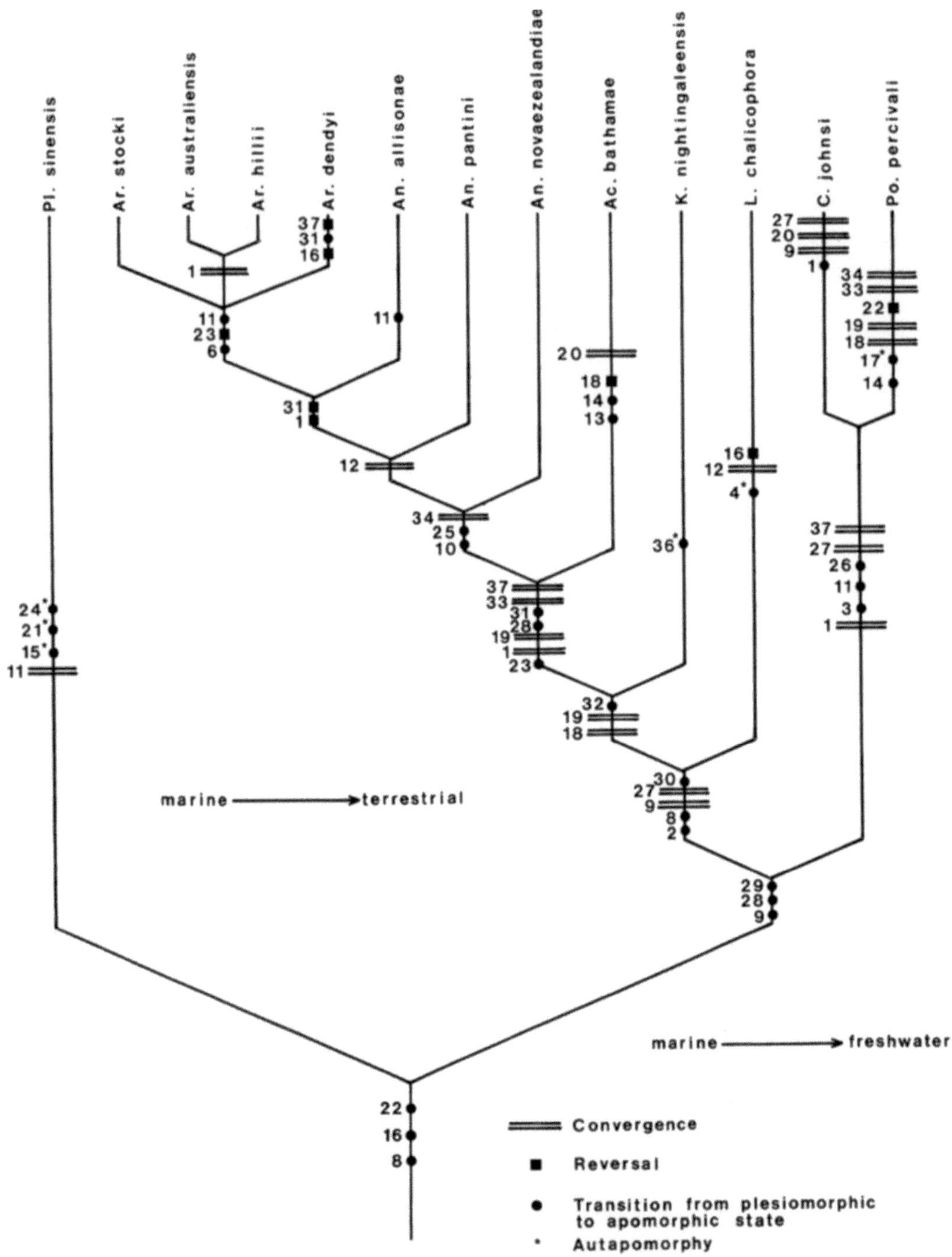

Fig. 2. Sundberg's cladogram showing the estimate of phylogeny within the Group II taxa; genera are *Acteonemertes* (Ac.), *Antiponemertes* (An.), *Argonemertes* (Ar.), *Campbellonemertes* (C.), *Katechonemertes* (K.), *Leptonemertes* (L.), *Plectonemertes* (Pl.) and *Potamonemertes* (Po.). Characters represented by the numbers are listed in Table 2. Redrawn, with permission, from Sundberg (1989b: Fig. 3).

Table 2. Identification of characters indicated by numbers in Figure 2.

Character number	Character
1	Length at maturity
2	Cephalic furrows lateral, horizontal and longitudinal or ventral and transverse
3	Cephalic glands entirely precerebral or extending behind brain
4	Cephalic glands either mainly basophilic or acidophilic
6	Cephalic glands with or without additional dorsal components
8	Cerebral organs opening ventrally, laterally or ventrolaterally
9	Cephalic glands discharging only via improvised pores, by single median pore separate from frontal organ, or through frontal organ supplemented by improvised pores
10	Cerebral organs small and simple, large with branching canal, or unique with two glandular regions
11	Eyes absent, less than four, four or more than four
12	Accessory lateral nerve present or absent
13	Accessory lateral nerve with or without contribution from dorsal cerebral ganglia
14	Ventral cerebral commissure with or without posterior dorsal projections
15	Peripheral nervous system normal or unusually well developed
16	Body wall musculature strong or weak
17	Parenchyma sparse or extensive
18	Proboscis small or large
19	Proboscis used or not used in locomotion
20	Stylet bulb region of proboscis with or without duct containing acidophilic glands
21	Oesophagus long or short
22	Oesophagus ciliated or unciliated
23	Stomach always in normal position or sometimes pushed forwards
24	Intestinal caecum with lateral pouches only or with lateral and anterior diverticula
25	Intestinal caecum with long anterior diverticula or with diverticula of variable length
26	Blood vessels distinct or lacunar
27	Blood vessels with or without transverse connectives in intestinal region
28	Post-cerebral blood system without capillary development, with capillary network in foregut region only or with capillaries present throughout body
29	'Valves' in blood system rare or numerous

Table 2. (Continued)

Character number	Character
30	Vascular plugs located on anterior projections of mid-dorsal vessel or on posterior of cephalic vessels
31	Cephalic blood system a simple vascular loop or capillary network
32	Excretory system of limited extent or very extensive
33	Nephridiopores numerous or exceptionally multiplied
34	Terminal region of excretory tubules not modified or thick-walled and specialised
36	Flame cells scattered, aggregated around cephalic loop or aggregated around lateral blood vessels
37	Sexes separate or hermaphroditic

tions between blood vessels (**27**) are conditions common to all known freshwater nemerteans, in cluding the four heteronemertean genera, and blood vessels often become lacunar (**26**) in addition to losing their cross connections.

Since Sundberg's paper, a new species of *Potamonemertes*, *P. gibsoni*, has been discovered in Tasmania (Hickman & Moore, 1990). This species is undoubtedly in the same genus but in general is less specialised than *P. percivali*. There are extensive cephalic glands opening through the frontal organ, and the following character states can now be seen as autapomorphies for *P. percivali*: **14**, the posterior knobs on the ventral cerebral commissure, **22**, the ciliated oesophagus, and **29**, the rarity of blood vessel 'valves' as well as **9**, the single pore separate from the frontal organ, used here at the stem of the cladogram. In all these respects *P. gibsoni*, by contrast, resembles the terrestrial species. Apart from these characters, the distinguishing features of *C. johnsi* and *P. percivali* consist of six character states showing convergence (the seventh, **27**, loss of blood vessel cross connectives, is repeated here in error. Increase in parenchyma (**17**) applies most strongly to other species and should be included as convergence). *Campbellonemertes johnsi* has other unique features, such as the structure of the cerebral organs (the unique condition referred to

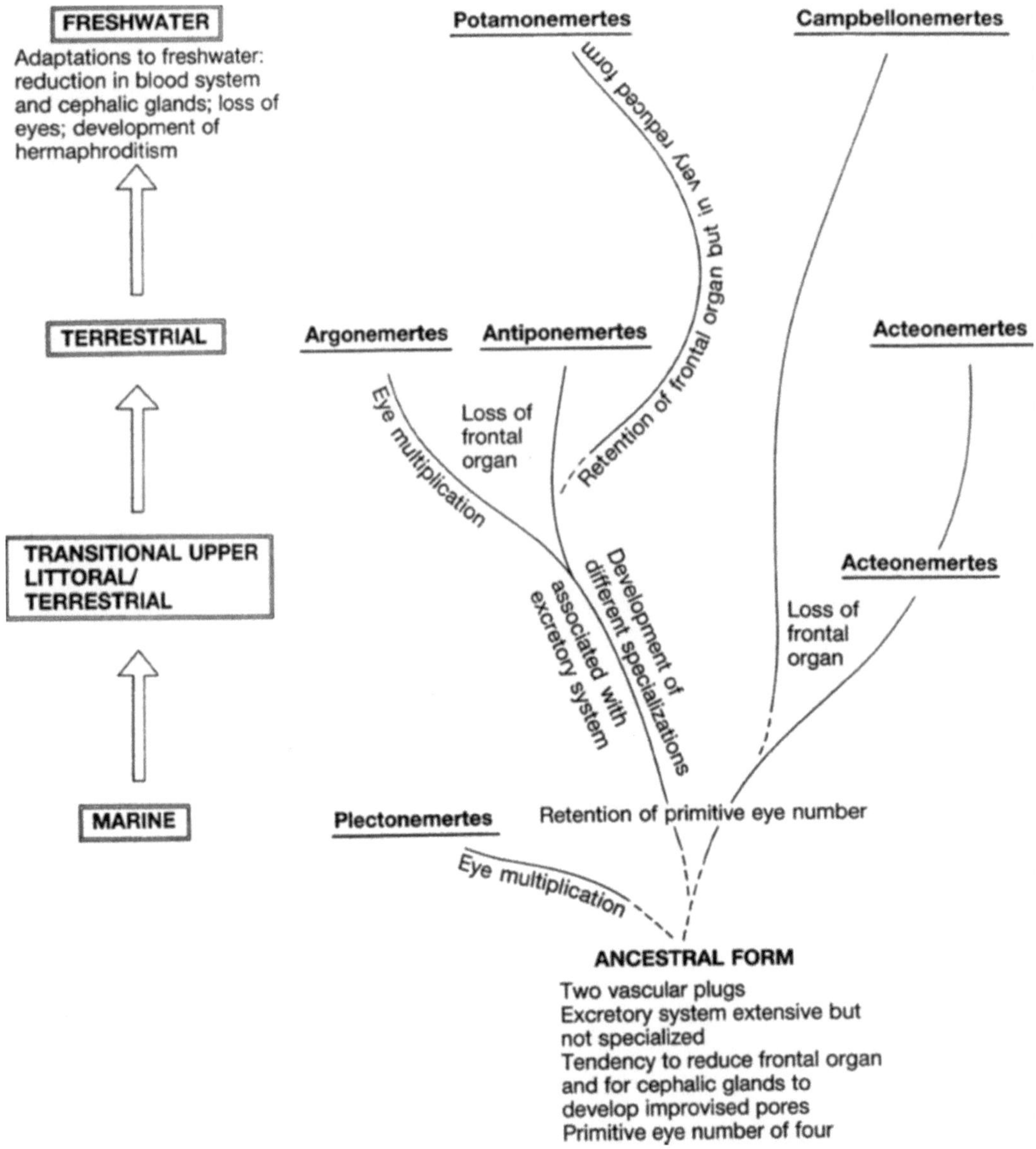

Fig. 3. The suggested evolutionary relationships between terrestrial, freshwater and marine Group II genera placed in the family Plectonemertidae; from Moore & Gibson (1988b: Fig. 2).

in character state **(10)**, the structure of the excretory ducts, the circular musculature within the connective tissue of the proboscis, and the ventral position of the testes. Whether the proboscis is used for locomotion cannot be determined (the species was described from three preserved specimens only and there is no record of behaviour in life).

The remainder of Sundberg's cladogram for Group II presents other difficulties, for example: character **8** is another aspect of **2**, character **30** similarly is a consequence of **27**, characters **14** and **36** both combine two separable characters, eye multiplication **(11)** should be recorded as convergent in *Argonemertes* species and *Plectonemertes*, **17**, extensive parenchyma, is characteristic of many of these species, especially *Acteonemertes bathamae*, for **19** there is no evidence about proboscis use in *Katechonemertes nightingaleensis*, **25**, the anterior diverticula, are variable in length only in *Antiponemertes* species and are completely missing from *A. allisonae*, **31**, *Argonemertes australiensis* and *A. hillii* both have the blood capillary network in the cephalic region, like *A. dendyi*. *Argonemertes stocki* combines such a network with a cephalic vascular loop. **32** *Leptonemertes chalicophora* has an extensive excretory system with many pores (and probably many flame cells too, but these are extremely hard to find in this species). **37** The change to hermaphrodity applies only to *Argonemertes dendyi*. For *L. chalicophora* no verdict on sexuality can be given, since all 66 specimens examined have been female.

Major problems, in addition to the problem of convergence, concern uncertainty about the primitive condition, for many characters, and possible linkage between some character states and small size: for example, the two small species of *Antiponemertes* and *Argonemertes* bring these two genera together in the diagram – the genera are in fact distinguished by more characters than are recorded here and the difference is strongly apparent when they are studied in life (Moore, 1989).

The classification of nemerteans in general

Terrestrial and freshwater nemerteans have received relatively thorough examination (Moore & Gibson, 1985). The very studies which established the prevalence of convergence have also provided the raw material for Sundberg's cladistic analyses. Do terrestrial and freshwater nemerteans constitute a special case, or is convergent evolution common among marine nemerteans also?

Papers presented at the two previous nemertean symposia (in Philadelphia, 1983, and Tjärnö, Sweden, in 1986), as well as others, provide many clues that convergence is widespread amongst marine nemerteans. Studies of the commensal egg predators of crustaceans demonstrate unequivocally that this way of life has evolved several times in these invertebrates. For long only a single genus, *Carcinonemertes*, was known, with all its species found as ectohabitants on crabs (Wickham & Kuris, 1985, 1988), but (as in the original terrestrial genus *Geonemertes*) this way of life is a product of convergence. Representatives of other genera have now been found living on egg masses of crustaceans, including *Alaxinus* on Alaskan king crabs (Gibson *et al.*, 1990), species of *Nemertopsis* from within barnacles in Hong Kong (Gibson, 1990b) and *Pseudocarcinonemertes* from lobsters off eastern Canada (Fleming & Gibson, 1981); without doubt specialised egg predation linked with commensalism has evolved separately in different enoplan families and provides a classic example of convergence. Norenburg (1988) studied another special habitat, the marine interstitial, where nemerteans from four of the five distinguished orders show convergent adaptations. A remarkable convergence here, both within and outside the phylum, is the pseudosegmentation of the specialised *Annulonemertes* (Berg, 1985) which is found posteriorly in *Arenonemertes* also. Riser (1988) described a new hoplonemertean from the supra-littoral zone in New Zealand, *Notogaeanemertes folzae*. This has features found both in *Prosorhochmus* and in *Acteonemertes* from the same habitat, but does not fit into either group of terrestrial and semi-terrestrial nemerteans.

Convergence is not confined to colonisation of particular habitats, as was shown at these symposia by Norenburg's (1985) study of the nemertean integument and also by Kem's (1988) studies of the distribution of pyridine alkaloids: the same toxin may occur in species which are not at all closely related, for example anabasine in various nemerteans and the unique toxin shared by *Pantinonemertes californiensis* and *Nipponnemertes pulcher*. There is a large number of similar examples: on any classificatory scheme, the taxonomist finds character states which contradict each other, for example those concerning the presence or absence of the accessory lateral nerves, hermaphrodity or separation of the sexes, extension of the cephalic glands, eye multiplication and many more. A particularly clear example of this contradiction arises from Gibson's (1988) proposed reclassification of the Hoplonemertea. This was intentionally controversial and has been further pursued in two papers published in sequence

(Sundberg, 1990; Gibson, 1990a) but one point is abundantly clear: *either* the fundamentally different condition of the rhynchocoel wall *or* the possession of a single stylet must have evolved at least twice.

Convergence between nemerteans and other phyla may be at the root of two especially controversial topics, the relationship of nemerteans to coelomate worms on the one hand and to chordates on the other hand. The nemertean blood system is formed by splitting of the mesoderm in very much the same way as occurs in annelids (Turbeville & Ruppert, 1985; Turbeville, 1986). Turbeville deduced that nemerteans were derived from coelomate worms. From studies of the excretory system, Bartolomaeus (1988) disputed this hypothesis. P. Willmer (1990) surveyed the origin of the coelom more widely and concluded that postulation of convergence was inescapable. Similarly, Jensen (1988) gave an account of resemblances between hoplonemerteans and early chordates. There is a convergence of four authors, including Jensen and E. N. Willmer (1970, 1974) who regard these resemblances as proof of direct ancestry of chordates from nemerteans: a more widely held view is instead to find some examples which represent convergence.

In conclusion, it would be dangerously misleading to assume that convergent evolution was a rare phenomenon in any nemertean group. We must therefore choose a method of classification which does not implicitly assume that resemblance is more likely to be due to common ancestry than to convergence. We believe that the traditional method, emphasising the assessment of all available characters, is the one which has the best chance of revealing true evolutionary relationships.

Acknowledgements

We should like to thank Dr Per Sundberg by paraphrasing his 1989b acknowledgement to ourselves – our study has gained considerably through many discussions with him, which is not to imply that he shares our views and conclusions. We are also grateful to the Zoological Society of London for permission to reproduce Fig. 2 of our 1988b paper.

References

Ashlock, P., 1974. The uses of cladistics. Annu. Rev. Ecol. Syst. 5: 81–90.

Bartolomaeus, T., 1988. No direct contact between the excretory system and the circulatory system in Prostomatella arenicola Friedrich (Hoplonemertini). Hydrobiologia 156: 175–181.

Berg, G., 1985. Annulonemertes gen. nov., a new segmented hoplonemertean. In S. Conway Morris, J. D. George, R. Gibson & H. M. Platt (eds), The origins and relationships of lower invertebrates. Clarendon Press, Oxford: 200–209.

Charig, A. J., 1982. Systematics in biology: a fundamental comparison of some major schools of thought. In K. A. Joysey & A. E. Friday (eds), Problems of phylogenetic reconstruction. Academic Press, London: 363–440.

Darwin, C., 1899. The origin of species by means of natural selection, or the preservation of favoured races in the struggle for life. John Murray, London, 432 pp.

Disney, R. H. L., 1983. A synopsis of the taxonomist's tasks, with particular attention to phylogenetic cladism. Fld Stud. 5: 841–865.

Eldredge, N. & J. Cracraft, 1980. Phylogenetic patterns and the evolutionary process. Columbia University Press, New York, 349 pp.

Fleming, L. C. & R. Gibson, 1981. A new genus and species of monostiliferous hoplonemerteans, ectohabitant on lobsters. J. exp. mar. Biol. Ecol. 52: 79–93.

Friday, A. E., 1987. Models of evolutionary change and the estimation of evolutionary trees. Oxf. Surv. Evol. Biol. 4: 61–88.

Gibson, R., 1982. Nemerteans of the Great Barrier Reef. 5. Enopla Hoplonemertea (Monostilifera). Zool. J. linn. Soc. 75: 269–296.

Gibson, R., 1988. Evolutionary relationships between mono- and polystiliferous hoplonemerteans: Nipponnemertes (Cratenemertidae), a 'missing link' genus? Hydrobiologia 156: 61–74.

Gibson, R., 1990a. Classification of the Nemertea Enopla: response to Sundberg's critique and cladistic analysis. Zool. Scr. 19: 141–142.

Gibson, R., 1990b. The macrobenthic nemertean fauna of Hong Kong. In B. Morton (ed.), Proceedings of the second international marine biological workshop: the marine flora and fauna of Hong Kong and southern China. University Press, Hong Kong 1: 33–212.

Gibson, R., D. E. Wickham & A. M. Kuris, 1990. A new genus and species of monostiliferoidean nemertean (Nemertea: Enopla) found on an egg mass of the anomuran decapod Paralithodes camtschatica. Zool. J. linn. Soc. 98: 185–198.

Halstead, L. B., 1978. The cladistic revolution – can it make the grade? Nature, Lond. 276: 759–760.

Hennig, W., 1966. Phylogenetic systematics. University of Illinois Press, Urbana, 263 pp.

Hickman, J. & J. Moore, 1990. A new species from Tasmania of the freshwater hoplonemertean, Potamonemertes. Pap. Proc. r. Soc. Tasm. 124: 51–59.

Hull, D. L., 1979. The limits of cladism. Syst. Zool. 28: 416–440.

Jensen, D. D., 1988. Hubrecht, Macfarlane, Jensen and Willmer: on the nature and testability of four versions of the nemertean theory of vertebrate origins. Hydrobiologia 156: 99–104.

Jespersen, Å. & J. Lützen, 1988a. The fine structure of the protonephridial system in the land nemertean Pantinonemertes californiensis (Rhynchocoela, Enopla, Hoplonemertini). Zoomorphology 108: 69–75.

Jespersen, Å. & J. Lützen, 1988b. Ultrastructure and morphological interpretation of the circulatory system of nemerteans (phylum Rhynchocoela). Vidensk. Meddr dansk naturh. Foren. 147: 47–66.

Joysey, K. A., 1988. The use of amino acid sequences in phylogenetic analysis. In T. W. Broadhead (ed.), Molecular evolution and the fossil record. Paleontological Society of America, Knoxville, Tennessee 1: 34–48.

Kem, W. R., 1988. Pyridine alkaloid distribution in the hoplonemertines. Hydrobiologia 156: 145–151.

Mayr, E., 1969. Principles of systematic zoology. McGraw-Hill, N.Y., 428 pp.

Mayr, E., 1982. The growth of biological thought. Belknap Press of Harvard University, Cambridge (Mass.), 974 pp.

Moore, J., 1985. The distribution and evolution of terrestrial nemertines. Am. Zool. 25: 15–21.

Moore, J., 1989. A record of the behaviour and distribution of New Zealand land nemertines. N.Z. nat. Sci. 16: 79–86.

Moore, J. & R. Gibson, 1981. The Geonemertes problem (Nemertea). J. Zool., Lond. 194: 175–201.

Moore, J. & R. Gibson, 1985. The evolution and comparative physiology of terrestrial and freshwater nemerteans. Biol. Rev. 60: 257–312.

Moore, J. & R. Gibson, 1988a. Marine relatives of terrestrial nemerteans: the genus Prosadenoporus Bürger, 1890 (Hoplonemertea). Hydrobiologia 156: 75–86.

Moore, J. & R. Gibson, 1988b. Further studies on the evolution of land and freshwater nemerteans: generic relationships among the paramonostiliferous taxa. J. Zool., Lond. 216: 1–20.

Norenburg, J. L., 1985. Structure of the nemertine integument with consideration of its ecological and phylogenetic significance. Am. Zool. 25: 37–51.

Norenburg, J. L., 1988. Remarks on marine interstitial nemertines and key to the species. Hydrobiologia 156: 87–92.

Panchen, A. L., 1982. The use of parsimony in testing phylogenetic hypotheses. Zool. J. linn. Soc. 74: 305–328.

Pantin, C. F. A., 1969. The genus Geonemertes. Bull. Br. Mus. nat. Hist. (Zool.) 18: 261–310.

Patterson, C., 1980. Cladistics. Biologist 27: 234–240.

Platnick, N. I., 1980. Philosophy and the transformation of cladistics. Syst. Zool. 28: 537–546.

Riser, N. W., 1988. Notogaeanemertes folzae gen. n., sp. n., an additional ecologically restricted hoplonemertean from New Zealand. Hydrobiologia 156: 125–133.

Simpson, G. G., 1961. Principles of animal taxonomy. Columbia University Press, New York, 247 pp.

Sneath, P. H. A. & R. R. Sokal, 1973. Numerical taxonomy. W. H. Freeman, San Francisco, 573 pp.

Sundberg, P., 1987. Classification and nemertean taxonomy. Ph.D. Dissertation, Göteborgs Universitet.

Sundberg, P., 1989a. Phylogeny and cladistic classification of terrestrial nemerteans: the genera Pantinonemertes Moore & Gibson and Geonemertes Semper. Zool. J. linn. Soc. 95: 363–372.

Sundberg, P., 1989b. Phylogeny and cladistic classification of the paramonostiliferous family Plectonemertidae (phylum Nemertea). Cladistics 5: 87–100.

Sundberg, P., 1989c. Classification of the family Plectonemertidae (Nemertea): a phenetic comparison. Zool. J. linn. Soc. 97: 57–68.

Sundberg, P., 1990. Gibson's reclassification of the enoplan nemerteans (Enopla, Nemertea): a critique and cladistic analysis. Zool. Scr. 19: 133–140.

Turbeville, J. M., 1986. An ultrastructural analysis of coelomogenesis in the hoplonemertine Prosorhochmus americanus and the polychaete Magelona sp. J. Morph. 187: 51–60.

Turbeville, J. M. & E. E. Ruppert, 1985. Comparative ultrastructure and the evolution of nemertines. Am. Zool. 25: 53–71.

Wickham, D. E. & A. M. Kuris, 1985. The comparative ecology of nemertean egg predators. Am. Zool. 25: 127–134.

Wickham, D. E. & A. M. Kuris, 1988. Diversity among nemertean egg predators of decapod crustaceans. Hydrobiologia 156: 23–30.

Wiley, E. O., 1981. Phylogenetics: the theory and practice of phylogenetic systematics. Wiley-Interscience, J. Wiley & Sons, New York, 439 pp.

Willmer, E. N., 1970. Cytology and evolution. Academic Press, N.Y., 2nd edition, 649 pp.

Willmer, E. N., 1974. Nemertines as possible ancestors of the vertebrates. Biol. Rev. 49: 321–363.

Willmer, P., 1990. Invertebrate relationships: patterns in animal evolution. University Press, Cambridge, 400 pp.

Hydrobiologia **266**: 103–113, 1993.
R. Gibson, J. Moore & P. Sundberg (eds), Advances in Nemertean Biology

Phylogeny, natural groups and nemertean classification

Per Sundberg
University of Gothenburg, Department of Zoology, Medicinaregatan 18, S-413 90 Göteborg, Sweden

Key words: Nemertea, cladistics, systematics

Abstract

Contemporary practice in the classification of nemerteans (phylum Nemertea) is critically discussed. It is argued that basing higher taxa on the existence of a unique combination of characters in a species (or genus) is unlikely to lead to monophyletic taxa, and that this approach should be abandoned in favour of a classification based on explicit hypotheses of phylogeny. These hypotheses should be based on all available characters and characters should not be excluded before the analysis. The classification should be based on a reconstruction of the phylogeny and reflect this phylogeny in an unambiquous way.

Introduction

Systematists have long strived to classify animals and plants into 'natural' groups. The meaning of 'natural' has, however, changed through time and reflects shifting views on genesis and nature (see e.g. Hull, 1989). Aristotle, following Plato, classified organisms into groups that shared the same 'essence'. Since essence was considered real and not a human invention, such classes were consequently said to be natural. A genus was defined by its essential character, the one that uniquely defined the group. Linnaeus also declared that his genera were 'natural' and that his classification represented a 'natural system'. By contrast with Aristotle, he considered sets of characters rather than single characters and the importance of a character was determined by its covariation with other characters in the complex. To Linnaeus, no single character was of dominant importance and a character might be missing from a species within a genus without the species being excluded. In this particular aspect, Linnaeus' classification can be called natural even in a modern sense (Cain, 1958).

A priori principles for forming natural classes were common among early taxonomists and played an important part in classification before Darwin's Origin of Species. One influential taxonomist, Cuvier, based his taxonomic theory on *a priori* weighting of characters, on the basis of their importance for the organism. He advocated a principle of 'subordination of characters' wherein certain physiological systems were of such major importance that they should govern the classification. By the 'admirable fecundity of the principle of the subordination of characters, and the beautiful laws to which it leads us' Cuvier classified invertebrates into six 'natural' divisions; 'natural' since it made it possible to make positive general statements that would be true for all members of the group. Groups were thus defined by the existence, and not the arrangement, of physiological systems (Winsor, 1976) and the classification was therefore artificial in a modern sense. The situation changed, however, and taxa were more and more established on empirical assessments of the totality of characters instead of a limited number of characters. The term 'natural' lost the Linnean meaning and came to signify unbiased *a priori* considerations of all available characters.

Even before the origin of evolutionary theory, zoologists in the nineteenth century came to believe that a hierarchy of groups, divided into subgroups, best represented existing relationships of 'natural affinity'. Darwin argued that the arrangement of living organisms in such hierarchies was consistent with his theory of descent and he begins his chapter which deals with classification (Darwin, 1859) with the statement:

> 'From the first dawn of life, all organic beings are found to resemble each other in descending degrees, so that they can be classed in groups under groups. This classification is evidently not arbitrary like the grouping of the stars in constellations'.

Darwin also emphasized that a natural grouping depicts the phylogeny of the group:

> 'the arrangement of the groups within each class, in due subordination and relation to the other groups, must be strictly genealogical in order to be natural; but that the amount of difference in the several branches or groups, though allied in the same degree in blood to their common progenitor, may differ greatly, being due to the different degrees of modification which they have undergone; and this is expressed by the forms being ranked under different genera, sections or orders'.

Biological classification before Darwin was ahistoric in the sense that it was not based on evolutionary history, as now recognized. Darwin changed this view and most taxonomists now agree that the hierarchical structure of homologous similarities found in nature reflects phylogeny. In theory, they also agree that the classification should mirror this phylogeny, but this conviction is not always reflected in the actual results. For many biologists taxonomy is something else than envisioned by Darwin. Meglitsch (1972), for example, wrote in a textbook on invertebrate zoology:

> 'all of the categories [genus etc.] are fairly artificial, being merely a convenient system of pigeonholes. However, they are not entirely arbitrary, for it is convenient to use this hierarchy of groups to reflect our ideas about relationships of animals. From this point of view, a genus should contain a group of species so similar that they appear to be closely related. Assigning a species to a genus is really the first step in organizing our information. It is useful in cataloguing, and also in systematizing'.

Thus, classification is mainly a matter of organizing our knowledge, although 'convenient' if it reflects ideas about evolutionary relationships. Abbott *et al.*, (1985), in a book on taxonomic analysis in biology stated that:

> 'We believe that the connection between taxonomic and evolutionary groups is by no means automatic and that it is important to understand that evolutionary interpretations are not the primary purpose of much work in taxonomy. Some would agree with us that the various aspects of [phylogenetic] reconstruction ... are secondary studies that utilize the classification as a starting point. There are others, with whom we disagree, who believe that a phylogenetic reconstruction, no matter how speculative, should be used to shape the final classification and information system.'

The question whether phylogeny can be reconstructed has been a source of major controversy among systematists since the mid 1960's. Many has claimed that since convergent and parallel selection is so extensive, all attempts to estimate the phylogeny are doomed from the outset. Recent advances in logic (especially Hennig's (1966) conclusion that only synapomorphies provide evidence of monophyly), combined with better methods for analyzing character data (e.g. Farris, 1970, 1983; see Felsenstein, 1983 for a review), have revived interest in phylogenetic reconstruction and classification (Wiley, 1981; Eldredge & Cracraft, 1980; Nelson & Platnick, 1981). Sneath & Sokal (1973) describe earlier approaches to phylogenetic reconstruction, and compare them to phenetic strategies for classification.

Here I discuss how supraspecific taxa, especially genera, are formed in the phylum Nemertea (nemerteans) and argue that most nemertean taxonomy has been mainly a matter of making classes, of organizing knowledge rather than aiming to discern natural groups. Hitherto, most classifications have been based on arbitrary recognition of 'important' characters, or character combinations of characters, chosen to be constant within and between groups. I will argue that this procedure is not guaranteed to lead to natural groups in the Darwinian sense, except in the trivial case of monotypic taxa. Finally, I discuss recent attempts to classify nemerteans according to cladistic principles, and I compare these results with other classifications.

Nemertean classification: the past

Nemerteans are acoelomate worms often (e.g. Hyman, 1951; Gibson, 1972) considered to be descendend from, or closely related to, the turbellarian flatworms; a view based largely on the acoelomate condition. This, and other oft-mentioned similarities are probably plesiomorphies (Norenburg, 1985) and Turbeville & Ruppert (1985) suggested a coelomate, protostomous ancestry, which would exclude the flatworms as ancestors. The presence of a circulatory system and a complete digestive tube adds support to this hypothesis.

Most nemerteans are marine, predominately benthic but some are pelagic; a few species are found in freshwater and terrestrial habitats. There are now about 900 described species arranged into 38 families and 177 genera (Gibson, 1985); a nemertean was mentioned as early as 1555 (Cedhagen & Sundberg, 1986), but early zoologists were seemingly unfamiliar with the group and the first comprehensive treatment of nemertean anatomy was published by Schultze (1851).

Between 1890 and 1965, Bürger, Coe, Stiasny-Wijnhoff, and Friedrich were the most influential and prolific authors on nemertean taxonomy. There are no indications that they aimed to produce natural groups (in a Darwinian sense), even though it is evident that they were aware of phylogenetic and evolutionary reasoning. For example Stiasny-Wijnhoff's (1930) conclusion that: 'Die Gattung *Oerstedia* ist existenberechtigt' [the genus *Oerstedia* is valid] is stated without any further discussion of why she thinks so. A common query in taxonomic papers from this period was whether a character is taxonomically 'reliable', meaning whether it was constant within a proposed taxon. There was no attempt to find characters identifying evolutionary units; the search for constant characters was similar to the pre-Darwinian search for 'essential characters'. Early nemertean taxonomy was basically typological and non-evolutionary and the question now asked is whether this has changed?

Nemertean taxonomy: the current situation

The most obvious change in modern nemertean taxonomic literature is the publication of detailed species descriptions. Recent revisions of dubious genera and families have been made possible because of the increased number of anatomical characters used. In consequence, diagnoses of the supraspecific taxa now include more morphological and ecological information. Has this, however, led to classifications better reflecting the phylogeny? Gibson & Moore (e.g. Gibson & Moore, 1976; Moore & Gibson, 1985) have had the largest influence on the structure of modern nemertean classification and an examination of their methods is necessary to understand the current systematics. I here argue that their approach is flawed and that they (and others) follow procedures which reflect tradition rather than contemporary taxonomic theory.

I concentrate below on how characters are chosen, and how they are used to erect and diagnose supraspecific taxa. I argue that major problems exist at both stages if the intention is to find natural groups, and that the resulting taxonomy is artificial.

The characters

Moore & Gibson (1973: 152) write that (see also e.g. Wilfert & Gibson (1974) and Gibson & Moore (1976)):

> 'The grouping of species into genera and of genera into families is particularly difficult in nemerteans ... Of fundamental importance, structures cannot easily be assessed for taxonomic value when their function is so little understood. Criteria that must be applied include such questions as: (a) Is the feature related to a particular environment or is there any other possible reason for convergence? (b) Does the feature occur in a number of apparently unrelated species? (c) How much variation does the structure show in other species possessing it?'

Criteria (a) and (b) aim crucially to exclude adaptive characters to ensure that groups are not defined by convergent similarity. The function is considered crucial and Gibson (1988: 68), for example, concluded: 'An assessment of these features indicates ... that we know insufficient about their function ...; the systematic value of such characters is accordingly limited'.

I agree that we must distinguish homologous from convergent similarity, but disagree that we have to know the function of an organ to reach conclusions about homology. Function may give clues to adaptation but this does not have to be known for a character to be used in taxonomic analysis. Studies of character evolution and function often rely upon correlations between the character and some environmental variable(s). This ahistorical approach may, however, give misleading results (e.g. Clutton-Brock & Harvey, 1984; Felsenstein, 1985) and before such correlations are taken as evidence for convergent selection, historical explanations have to be rejected (see e.g. Coddington, 1988; Donoghue, 1989). I can see no justification for dismissing characters whatever their function. Phylogenetic analysis will give clues to those characters leading to non-homologous similarity; to exclude characters before the analysis introduces unnecessary subjectivity.

The third criterion (c) above does not include function, but involves finding what taxonomists call 'reliable' characters, *i.e.* characters that are invariant within a group, with an intent to form groups where all the members possess this character. It is, however, important to distinguish between two kinds of variation with separate systematic implications. Intraspecific variation hampers all taxonomic analysis above species level. This problem is correctly identified by, for example, Gibson & Moore (1976: 187) where it is concluded that the position of the cerebral organ pore, relative to the eyes, varied within any single species and that 'such variability does not permit this feature to be used as a taxonomic aid'. Characters, however, also vary between taxa and this variation can be used to reconstruct phylogeny. Moore & Gibson (e.g. Moore & Gibson, 1988: 5), however, also dismiss characters displaying this kind of variation. They concluded for example that the 'degree of intrageneric variation [in kind of pre-cerebral septum] ... indicates that the nature of the pre-cerebral septum cannot be used in an assessment of their intergeneric relationships'. Similarly, Moore & Gibson (1988: 10) conclude when assessing reproductive strategies as taxonomic characters: 'More generally, the degree of variation between the reproductive strategies of the different species make it very unlikely that any features of their reproductive system are significant in higher classification of nemerteans'.

In these cases, there is confusion between characters that are important in phylogenetic reconstruction and classification and characters useful in keys. The important attributes of a taxonomic key are clarity and convenience and characters should therefore not vary within taxa. However, classification must precede the key and it is illogical to exclude variable characters before the classification has been carried out. There are no 'bad' or 'good' phylogenetic characters, all are significant in the phylogenetic inference, even when homoplastic. Ambiguous, heterochronic, or characters susceptible to artifacts (see Sundberg, 1979), are not 'bad'. It is rather a matter of how these characters are handled; some are less easy to observe and more prone to artifacts, but future work may eliminate these problems.

Supraspecific taxa

A common procedure in nemertean systematics is to use a 'unique combination of characters' to erect supraspecific taxa. In this section I show by citation example and then argue that the flaws in this procedure should lead to its abandonment in favour of an approach that aims to find monophyletic groups. Moore & Gibson (1972: 187) write that specimens of a freshwater nemertean (*Campbellonemertes johnsi)*: 'possess a combination of characters which excludes them from any other currently recognized genus'. Gibson (1984: 87) concludes that a particular set of characters is: 'sufficient justification for the establishment of the new genus *Craticulineus*; no other lineid heteronemertean is known at present with this combination of characters'. Or in Gibson *et al.* (1990):

> 'Table 1 shows that this combination of characters enables the present form to be excluded from all the known taxa in which the body wall longitudinal musculature is divided anteriorly, and it is accordingly placed in the new genus, *Alaxinus*'

The same reasoning is applied at the family level by Moore & Gibson (1988) in their discussion of marine relatives of terrestrial nemerteans, concluding that the two genera *Prosorhochmus* and *Prosadenoporus* have a: 'combination of features ... not known for any other monostiliferous taxon and fully justifies the two genera being included in the same family [Prosorhochmidae]'

Genera and families are thus formed on character combinations. There are two fallacies in this procedure. First, inclusion of any other character may lead to a combination that includes the species in another, or a new, genus. The classification will be unstable, and at its extreme producing monotypic genera equal in number to the described species. Including intraspecific variation could lead to that individual populations would be placed in separate genera. Second, this procedure does produce natural groupings except in the trivial cases of monotypic taxa. If we accept that the same character can appear in independent taxa (parallelisms/convergences) and that character reversals occur, it is trivial to find cases where the 'unique-combination' approach does not lead to natural groups. Take for example the phylogeny for eight hypothetical species with 11 characters (Fig. 1) (Table 1). A classification

Table 1. Eight hypothetical species described by 11 characters. The true phylogeny for these species is supposed to be known (Fig. 1). One of the two most parsimonious Wagner tree on the basis of these characters will be the same as the true tree, if we assume that 0 is the plesiomorphic state for all characters. Species B and C are placed together in the second tree, viewing character 7 as a synapomorphy for these two species, instead of a homoplasi as in the true tree. Only the relevant autapomorphies are included, it is assumed that the terminal taxa are monophyletic.

Character	1	2	3	4	5	6	7	8	9	10	11
Species A	1	0	0	0	1	0	0	1	1	0	0
B	1	0	0	1	0	0	1	1	0	0	0
C	1	0	0	1	0	0	1	1	0	0	1
D	1	0	0	1	0	0	0	1	0	0	1
E	0	1	1	0	0	0	1	1	0	0	0
F	0	1	1	0	0	1	0	0	0	1	0
G	0	1	1	0	1	1	0	0	0	1	1
H	0	1	1	0	1	0	0	0	0	1	1

based on 'unique combinations' produces a classification where all species form separate monotypic genera:

Genus I
  species A
Genus 11
  species B
Genus III
  species C
Genus IV
  species D

etc.

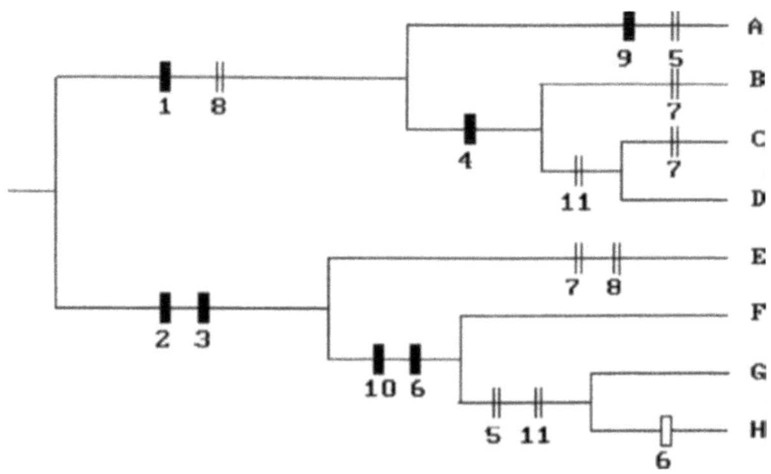

Fig. 1. A phylogenetic tree for eight hypothetical species, A-H. The character changes for 11 characters (Table 1) are assigned to the branches. Solid bars indicate a change from the plesiomorphic to the apomorphic state, parallel lines convergences, and open bars character reversals. Eight genera will be formed by applying the 'unique-combination-of-character' concept (see text), whereas the phylogeny would suggest maybe two (A–D, and E–H) monophyletic and genera.

This classification does not portray phylogenelic relationships among the species. Nor is it evolutionary in the Mayr-Simpson sense with both anagenetic and cladogenetic information retained in the classification.

A phylogenetic classification of these species can be constructed which describes the topology of the tree and the internested monophyletic groups. One possible classification would be to form two genera: one consisting of species A, B, C, and D (recognized by the synapomorphic state '1' in character 1); and the other of E, F, G, and H (recognized by the synapomorphies in characters 2 and 3). Using phyletic sequencing (Nelson, 1972), a classification which retains the information from the tree is:

Genus I
  Species A
    Species B
      Species C
      Species D

Genus II
  Species E
    Species F
      Species G
      Species H

Discussion

'The Darwinian Revolution' (e.g. Ruse, 1979) had an immense impact on biology in general. Darwin was the first to demonstrate that the most useful classifications are those where 'natural' means phylogenetic. He has, however, have had little influence on nemertean classification even though natural selection and evolution are accepted concepts among nemertean taxonomists; taxonomists like Coe and Wijnhoff were well aware of evolutionary thinking. Wijnhoff (1912: 22), for example, criticized Bürger for not producing a phylogenetic classification: 'Bürger's Systematik ist nur eine Entwicklung des Centralnervensystems, nicht eine Phylogenie der Nemertinen' [Bürger's systematics is just about the evolution of the central nervous system, and not a phylogeny of the nemerteans]. And she continued by declaring how a phylogenetic classification is achieved: 'Diese soll Rucksicht nehmen auf sämtliche Bauverhältnisse in den Nemertinenordnungen und in Anschluss daran die verschiedenen Entwicklungsrichtungen feststellen' [It should consider all the relationships between characters among the nemertean taxa and in connection to this also establish the evolution of the characters]. She discussed characters she thought more important than those used by Bürger, but her classification was no more phylogenetic than Bürger's. There are no taxonomic study on nemerteans from the first half of this century in which taxonomic conclusions are based on reconstruction of phylogeny. It is rather the reverse and classifications have been established on morphological similarity and phylogenetic interpretation has been subsidiary, a situation basically unchanged in contemporary nemertean systematics.

The systematic procedure outlined above is deficient because: (1) characters are excluded on questionable *a priori* grounds, (2) there is a conceptual mingling of diagnostic characters and those that identify monophyletic groups, and (3) the formation of supraspecific taxa is based on unique combinations of characters. Use of functional criteria for selecting characters in taxonomic studies has a long history in taxonomy, both to identify important characters for use in classification, and to avoid non-homologous similarity. The rationale behind this, and its validity, has been amply discussed (e.g. Cracraft, 1981; Gosliner & Ghiselin, 1984), and I will just add that the function of organs in nemerteans is so little known that it is difficult to justify their exclusion, whatever the merits of such criteria. Moore and Gibson (e.g. Moore & Gibson, 1985) also conclude that little is known about the function of various organs, but use the lack of knowledge as an argument to exclude characters (e.g. Gibson, 1988: 68) rather than to keep them. To use constancy of characters within proposed taxa as a ground for keeping or excluding characters must also be dismissed; phylogeny estimates should be based on all information available.

The 'unique-combination-of-characters' approach will lead to non-monophyletic groups. This kind of artificial classification is not confined to nemertean systematics but is a rather common practice among invertebrate taxonomists. For example, Martin *et al.* (1990) erected a new polychaete genus because 'the suite of characters of *Miscellania dentata* although variously shared with those of several other genera, justify the erection of a new genus'. Zenkevitch (1934) defined the genus *Fabriciella* on similar criteria and Ghiselin (1985) reviewing Parker's (1982) synopsis and classification of living organisms noted that many of the included groups were merely listed in alphabetical order, without indication of relationships. Many biologists and historians of science opine that biological taxonomy has changed little from pre- to post-Darwinian times (de Queiroz, 1988), and e.g. Dobzhansky (1937) wrote that:

> 'The fact is that the classification of organisms that existed before the advent of evolutionary theories has undergone surprisingly little change in the time following it. ... The phylogenetic interpretation has been simply superimposed on the existing classification'.

de Queiroz (1988) wrote that resistance to apply evolutionary theory is caused by failure to recognize the important conceptual distinction between classification and systematization (*sensu* Griffiths, 1974). Classification is ordering entities into classes, groups defined by the characters of the members, while systematization involves ordering species into systems whose existence depends on some natural process through which their parts are related. Most nemertean taxonomists treat taxa as if they are classes rather than systems, with higher taxa defined by a list of characters. The taxa are based on phenotypic similarity which is one-dimensional in that it lacks an evolutionary component.

Why then should we not accept artificial taxa, or classes *sensu* Ghiselin (e.g. Ghiselin, 1987)? If systematic biology is merely epistemological description and cataloguing of nature, and we consider that sufficient, the answer is that we should. If we wish to describe some underlying process, or describe a fact of nature that exists independent of the observer, the answer is no. Several authors (e.g. Griffiths, 1974; Cracraft, 1983) argue that only monophyletic taxa have explicit ontological status and that we need to identify such groups if our classification is to be natural. Furthermore, the information most commonly sought from a classification is phylogenetic relationship and only a classification based on the estimated phylogeny contains this information. It will therefore have a much wider application. Evolutionary taxonomists claim that other information, such as degree of morphological divergence, should be included in the classification. The only information, however, that can be drawn from a Linnean classification is the branching sequence. To group organisms according to degree of divergence introduces ambiguity and obscures phylogeny.

Only homologies indicate phylogenetic relationships and only shared homologies (synapomorphies) provided evidence of monophyly (Hennig, 1966). The identification of synapomorphies is thus paramount; unique derived characters (autapomorphies) are useful in diagnoses and keys, but cannot be used in classification. Homoplastic similarity, caused by either convergent or parallel evolution, leads to classifications which are inconsistent with true phylogeny. Cladistics recognizes that the problems of finding synapomorphies and identifying homoplastic similarity, is minimized by including many characters and not by empirical observations of similarity.

The classification of nemerteans depends upon histological studies of internal morphology and any taxonomic approach depends ultimately on accurate and comprehensive description of individual species. However, the handling of the animals prior to the fixation, and the fixation itself, unavoidably introduces artifacts (see Sundberg, 1979, 1984). In addition, any taxonomic study at a higher taxonomic level depends in part upon published descriptions, which in many cases are inadequate by modern standards. These problems have been considered an obstacle to a phylogenetic approach to nemertean taxonomy (Moore, *in litt.*; Gibson, *in litt.*; Gibson, 1990), in favour of the approach described above. I hold, however, that cladistic analysis is feasible in nemerteans and that it leads to better hypotheses about phylogeny. Furthermore, a cladistic analysis highlights problems in characters and their definitions, leading to a better understanding of the morphology, and pointing to directions in which new evidence should be sought, *i.e.* the reciprocal illumination of Hennig (1966). Two recent cladistic analyses (Sundberg, 1989a; 1990) demonstrate these points.

The finding of several features in common between Antarctic and Subantarctic *Nipponnemertes* species and reptantic Polystilifera led Gibson (1988) to reconsider parts of the traditional classification of the class Enopla (Table 2) (names of higher categories follow Sundberg (1991)). The separation of the subclass Hoplonemertea in two superorders: the Monostilifera with one stylet on the stylet basis and two separate muscle layers in the rhynchocoel, and the Polystilifera with several stylets and interwoven muscles in the rhynchocoel, follows earlier suggestions by Brinkmann (1917) and Stiasny-Wijnhoff (1923). Gibson (1988, and references therein), however, noted that there are a number of hoplonemertean species that combine a monostiliferous type of proboscis armature with a wickerwork arrangement of the rhynchocoel musculature. Thus, a classification based on these characters implies convergent evolution of either the proboscis armature, or the wickerwork rhynchocoel wall musculature.

Gibson (1988) emphasized rhynchocoel musculature and suggested that the Enopla should be divided into two major taxa, Distromatorhynchocoela (with a two-layered rhynchocoel) and Urichorhynchocoela (with a wickerwork musculature in the rhynchocoel). He considered the monostiliferous arrangement of the proboscis armature to have arisen at least twice. This hypo-

Table 2. The traditional classification of the class Enopla Nemertea (from Sundberg, 1991), and the reclassification in Gibson (1988).

Traditional	Gibson's classification
Subclass Hoplonemertea	Order Distromatorhynchocoela
Superorder Monostilifera	Suborder Monostiliferoidea
Superorder Polystilifera	Suborder Bdellonemertoidea
Order Reptantia	Order Urichorhynchocoela
Suborder Eureptania	Suborder Pelagonemertoidea
Suborder Archireptantia	Tribe Archipelagica
Order Pelagica	Tribe Eupelagica
Suborder Archipelagica	Suborder Herponemertoidea
Suborder Eupelagica	Tribe Paramonostilifera
Subclass Bdellonemertea	Tribe Polystilifera
	Subtribe Archireptantia
	Subtribe Eureptantia

thesis, however, can only be tested by including further characters in the analysis. A cladistic analysis (Sundberg, 1990) based on a larger number of characters could not corroborate Gibson's hypothesis.

Gibson (1990) criticized this analysis and doubted 'whether a cladistic analysis can be meaningfully applied to nemertean taxonomy given the present state of information' (Gibson, 1990: 141). Lack of knowledge hampers *any* taxonomic analysis, but there is no reason why the information used by the traditional taxonomists cannot be used for cladistic analysis.

Analysis of the two genera *Pantinonemertes* and *Geonemertes* (Sundberg, 1989a)* exemplifies how our understanding of character evolution can be increased by a cladistic analysis, and how it can point to directions for further investigation. Most nemerteans have excretory systems with flame cells of a basal type with a mononucleate cell with a single ciliary 'flame'. The flame cells, however, are specialized in *Geonemertes* and *Pantinonemertes,* with binucleate cells reinforced with cuticular bars. In *Geonemertes,* this reinforcement consists of both parallel transverse and longitudinal bars, while in *Pantinonemertes* only irregularly arranged transverse bars are present. Moore & Gibson (1981), and Moore (pers. comm.) considered these differences evidence for two monophyletic genera, thereby implying a transformation series where an unreinforced mononucleate flame cells evolved along two independent lines into the two type of reinforced cells. The cladistic analysis (Sundberg, 1989a) indicated that *Pantinonemertes* was paraphyletic, and that the more complicated reinforcement in *Geonemertes* evolved from the plesiomorphic state of only transverse bars as in *Pantinonemertes*. Later, Jespersen & Lützen (1989) concluded from ultrastructural evidence that the: 'fundamental cellular architecture of [the flame cells in the] land nemertean *Pantinonemertes californiensis* ... is the same as that of *G. pelaensis*. The similarities are so great that they must indicate a true and very close relationship'. Jespersen & Lützen (*in litt.*) considered flame cell organization in *Pantinonemertes* to be the plesiomorphic state, and the kind found in *Geonemertes* the apomorphic. They regarded the dissimilarities as modifications of an essentially identical structure, thus supporting the hypothesis proposed by the cladistic analysis.

This study also exemplifies the importance of clearly defining characters used in taxonomic description. Moore & Gibson (1981) described the cerebral organs as 'elaborated' in two *Geonemertes* (*sensu* Moore & Gibson) species, and *Pantinonemertes winsori* and *P. mooreae*. This state was also used in cladistic analysis, and found to be a synapomorphy for these four species. 'Elaborate', however, does not mean the same for these species (Moore pers. comm; Moore & Gibson, 1993) and does not indicate a relationship.

These examples demonstrate that cladistic analysis is viable also in the case of nemerteans, despite our lack of knowledge. Hypotheses can only be substantiated through complete set of characters, and by objective analyses carried out as unbiased as possible. Nemertean systematics will only become evolutionary if taxa are based on phylogeny and cladistics is the only formal approach to taxonomy which aim to produce genealogical classifications.

Different approaches to classification can only be assessed in the light of their aim. Cladistic analysis aims to reconstruct the phylogeny and to classify the taxa in accordance with the estimated

---

* Unfortunately, a substantial number of coding errors was detected after the publication of this study (Moore in. litt.). The purpose of the study in Sundberg (1989a) was twofold: to classify the included species in monophyletic groups, and to examine whether the evolution of the terrestrial habitat had been via ancestral species living in habitats intermediate between the terrestrial and the marine. A reanalysis of the species based on corrected characters (from Moore) did, however, not change any of the major conclusions made in the paper. But, the coding errors make it impossible to use the published cladogram for tracing character evolution for many of the features.

Similar problems were encountered in the study of Sundberg (1989b), but a subsequent reanalysis did not change the main conclusions made in the study. That is, that *Argonemertes* is paraphyletic, and that the two freshwater genera *Potamonemertes* and *Campbellonemertes* has evolved from a marine ancestor. Care should be taken, however, when tracing character transformations along the cladogram published in Sundberg (1989b).

phylogeny. Moore & Gibson (1993) concluded that 'the traditional method, emphasising the assessment of all available characters, is the one which has the best chance of revealing true evolutionary relationships'. Their main criticism of cladistic analysis is that it fails 'through [its] fundamental assumption that convergent evolution is a rare occurrence'. However, this is a criticism of the parsimony criterion and not of cladistic analysis. Parsimony fails in revealing the true phylogeny if convergent similarity is more common than homologuous similarity but the crucial point is that we do not know why organisms are similar and a first approximation is that it is due to a common ancestor. (If Moore & Gibson are right in their claim than convergent similarity is more common than genealogical, the traditional nemertean classification must be abolished as nonevolutionary since it is in all its major parts based on some groups being morphologically more similar than others.) Homology can only be corroborated by including other characters and the analysis should therefore be based on all available data, without *a priori* exclusions of characters. It should aim to classify taxa based on phylogeny and a procedure that fails in this aim should be abandoned.

Acknowledgements

The following persons, who obviously represent different points of view, have been kind to read and comment upon the manuscript: Christer Erséus, Mats Envall, Ray Gibson, Bill Kem, Jon Norenburg, Janet Moore, Pam Roe, and Mikael Svensson. I am very grateful to their comments. This research is supported by the Swedish Natural Science Research Council.

References

Abbott, L. A., F. A. Bisby & D. J. Rogers, 1985. Taxonomic analysis in biology. Columbia University Press, New York, 336 pp.

Brinkmann, A., 1917. Pelagic nemerteans from the 'Michael Sars' North Atlantic deep-sea expedition 1910. Reports of the Scientific Results Michael Sars North Atlantic deep Sea Expedition, 3: 1–18.

Cain, A. J., 1958. Logic and memory in Linnaeus's system of taxonomy. Proc. linn. Soc., London 169: 144–163.

Cedhagen, C. & P. Sundberg, 1986. A previously unrecognized report of a nemertean in the literature. Arch. Nat. Hist. 13: 7–8.

Clutton-Brock, T. H. & P. H. Harvey, 1984. Comparative approaches to investigating adaptation. In J. R. Krebs & N-B. Davies (eds), Behavioural ecology: an evolutionary approach. Blackwell, Oxford: 7–29.

Coddington, J., 1988. Cladistic tests of adaptational hypotheses. Cladistics 4: 3–22.

Cracraft, J., 1981. The use of functional and adaptive criteria in phylogenetic systematics. Am. Zool. 21: 21–36.

Cracraft, J., 1983. The significance of phylogenetic classifications for systematic and evolutionary biology. In J. Felsenstein (ed.), Numerical taxonomy. Proceedings of the NATO Advanced Study Institute, Ser. G (Ecological Sciences), No 1. Springer-Verlag, New York: 1–17.

Darwin, C., 1859. On the origin of species by means of natural selection, or the preservation of favored races in the struggle for life. John Murray, London, 502 pp.

De Querioz, K., 1988. Systematics and the Darwinian revolution. Phil. Sci. 55: 238–259.

Dobzhansky, T., 1937. Genetics and the origin of species. Columbia University Press, New York, 364 pp.

Donoghue, M. J., 1989. Phylogenies and the analysis of evolutionary sequences with examples from seed plants. Evolution 43: 1137–1156.

Eldredge, N. & J. Cracraft, 1980. Phylogenetic patterns and the evolutionary process. Columbia University Press, New York, 349 pp.

Farris, J. S., 1970. Methods for computing Wagner trees. Syst. Zool. 19: 83–92.

Farris, J. S., 1983. The logical basis of phylogenetic analysis. In N. Platnick & V. Funk (eds), Advances in cladistics, II. Columbia University Press, New York: 7–36.

Felsenstein, J., 1983. Parsimony in systematics: biological and statistical issues. Annu. Rev. Ecol. Syst. 57: 379–404.

Felsenstein, J., 1985. Phylogenies and the comparative method. Am. Nat. 125: 1–15.

Ghiselin, M. T., 1985. Narrow approaches to phylogeny: A review of nine books of cladism. Oxf. Surv. Evol. Biol. 1: 209–222.

Ghiselin, M. T., 1987. Species concepts, individuality, and objectivity. Biol. Phil. 2: 127–143.

Gibson, R., 1972. Nemerteans. Hutchinson, London, 224 pp.

Gibson, R., 1984. *Craticulineus novaezealandiae* gen. et sp. nov., a new lineid heteronemertean from the South Pacific (Nemertini). Zool. Anz., Jena 213: 77–88.

Gibson, R., 1985. The need for a standard approach to taxonomic descriptions of nemerteans. Am. Zool. 25: 5–14.

Gibson, R., 1988. Evolutionary relationships between mono-

and polystiliferous hoplonemerteans: *Nipponnemertes* (Cratenemertidae), a 'missing link' genus? Hydrobiologia 156: 61–74.

Gibson, R., 1990. Classification of the Nemertea Enopla: response to Sundberg's critique and cladistic analysis. Zool. Scr. 19: 141–142.

Gibson, R. & J. Moore, 1976. Freshwater nemerteans. Zool. J. linn. Soc. 58: 177–218.

Gibson, R., D. E. Wickham & A. M. Kuris, 1990. A new genus and species of monostiliferoidean nemertean (Nemertea: Enopla) found on an egg mass of the anomuran decapod *Paralithodes camtschatica*. Zool. J. linn. Soc. 98: 185–198.

Gosliner, T. M. & M. Ghiselin, 1984. Parallel evolution in opistobranch gastropods and its implications for phylogenetic methodology. Syst. Zool. 33: 255–274.

Griffiths, G. C. D., 1974. On the foundations of biological systematics. Acta Biotheor. 13: 85–131.

Hennig, W. 1966. Phylogenetic systematics. University of Illinois Press, Urbana, 263 pp.

Hull, D. L., 1989. Metaphysics of evolution. SUNY Press, New York, 331 pp.

Hyman, L. H., 1951. The invertebrates. Vol. II: Plathyleminthes and rhynchocoela. McGraw-Hill, New York, 550 pp.

Jespersen, Å. & J. Lützen, 1989. Ultrastructure of the protonephridial terminal organ in the terrestrial nemertean *Geonemertes pelaensis* (Rhynchocoela, Enopla, Hoplonemertini). Acta Zool. 70: 157–162.

Martin, D., C. Alos & R. Sarda, 1990. *Miscellania dentata* gen. et sp.n. (Polychaeta: Syllidae) from the Spanish Mediterranean coast. Zool. Scr. 19: 169–172.

Meglitch, P. A., 1972. Invertebrate Zoology, 2nd edition. Oxford University Press, New York, 834 pp.

Moore, J. & R. Gibson, 1972. On a new genus of freshwater hoplonemertean from Campbell Island. Freshwat. Biol. 2: 187–202.

Moore, J. & R. Gibson, 1973. A new genus of freshwater hoplonemertean from New Zealand. Zool. J. linn. Soc. 52: 141–157.

Moore, J. & R. Gibson, 1981. The *Geonemertes* problem (Nemertea). J. Zool., Lond. 194: 175–201.

Moore, J. & R. Gibson, 1985. The evolution and comparative physiology of terrestrial and freshwater nemerteans. Biol. Rev. 60: 257–312.

Moore, J. & R. Gibson, 1988. Further studies of the evolution of land and freshwater nemerteans: generic relationships among the paramonostiliferous taxa. J. Zool., Lond. 216: 1–20.

Moore, J. & R Gibson, 1993. Methods of classifying nemerteans: an assessment. Hydrobiologia 266: 89–101.

Nelson, G. J., 1972. Phylogenetic relationship and classification. Syst. Zool. 21: 227–230.

Nelson, G. & N. Platnick, 1981. Systematics and biogeography: cladistics and vicariance. Columbia University Press, New York, 567 pp.

Norenburg, J., 1985. Structure of the nemertine integument with consideration of its ecological and phylogenetic significance. Am. Zool. 25: 37–51.

Parker, S. P. (ed), 1982. Synopsis and classification of living organisms. McGraw-Hill, New York, 2,398 pp.

Ruse, M., 1979. The Darwinian revolution. University of Chicago Press, Chicago, 320 pp.

Schultze, M. S., 1851. Beiträge zur Naturgeschichte der Turbellarien. C. Koch, Greifswald, 78 pp.

Sneath, P. H. A. & R. R. Sokal, 1973. Numerical taxonomy. W. H. Freeman, San Francisco, 573 pp.

Stiasny-Wijnhoff, G., 1923. On Brinkmann's system of the Nemertea Enopla and *Siboganemertes weberi*, n. g. n. sp. Quart. J. Micros. Sci. 67: 627–669.

Stiasny-Wijnhoff, G., 1930. Die Gattung *Oerstedia*. Zoölogischer Mededelungen, Leiden, 13: 226–240.

Sundberg, P., 1984. Multivariate analysis of polymorphism in the hoplonemertean *Oerstedia dorsalis* (Abildgaard, 1826). J. exp. mar. Biol. Ecol. 78: 1–22.

Sundberg, P. 1979. Statistical analysis of variation in characters in *Tetrastemma laminariae* (Nemertini), with a redescription of the species. J. Zool., Lond. 189: 39–56.

Sundberg, P., 1989a. Phylogeny and cladistic classification of terrestrial nemerteans: the genera *pantinonemertes* Moore & Gibson and *Geonemertes* Semper. Zool. J. linn. Soc. 95: 363–372.

Sundberg, P., 1989b. Phylogeny and cladistic classification of the paramonostiliferous family Plectonemertidae (phylum Nemertea). Cladistics 5: 87–100.

Sundberg, P., 1990. Gibson's reclassification of the enoplan nemerteans (Enopla, Nemertea): a critique and cladistic analysis. Zool. Scr. 19: 133–140.

Sundberg, P., 1991. A proposal for renaming the higher taxonomic categories in the phylum Nemertea. J. nat. Hist., Lond. 25: 45–48.

Turbeville J. M. & E. E. Ruppert, 1985. Comparative ultrastructure and evolution of nemertines. Am. Zool. 25: 53–71.

Wijnhoff, G., 1912. Die Systematik der Nemertinen. Zool. Anz. 40: 337–341.

Wiley, E. O., 1981. Phylogenetics: the theory and practice of phylogenetic systematics. John Wiley & Sons, New York, 439 pp.

Wilfert, M. & R. Gibson, 1974. A new genus of hermaphroditic freshwater heteronemertean (Nemertinea). Z. Morph. Tiere 79: 87–112.

Winsor, M. P., 1976. Starfish, jellyfish, and the order of life. Yale University Press, New Haven, 228 pp.

Zenkevitsch, L. A., 1934. Uber das Vorkommen der Brackwasser-polychaete *Manavunkia* (*M. polaris* n.sp.) an der Murmansküste. Zool. Anz. 109: 195–203.

Hydrobiologia **266**: 115–140, 1993.
R. Gibson, J. Moore & P. Sundberg (eds), Advances in Nemertean Biology

Major characters and enoplan systematics

Frank B. Crandall
Turkey Run Research Institute, 900 Turkey Run Road, McLean, Virginia 22101–1700, USA

Key words: Nemertea, Monoxiphidia, Mictosemia, classification, taxonomic characters

Abstract

The history of enoplan systematics is briefly reviewed. Various taxonomic characters significant at the generic level and above are discussed, with comments on their application to traditional classifications and to recently propounded theories and new approaches to enoplan classification schemes. Several new characters are introduced and their potential significance examined. Finally, a revised classification scheme is presented which leaves open some of the phylogenetic questions about which there is presently much debate but, at the same time, provides secure higher taxa to contain the many forms and groups currently undergoing revision.

Introduction

Most characters used in the earliest descriptions of nemerteans were features of external morphology, many of which are now recognized as having negligible or minor importance in differentiating groups at the generic level and above. With the advances in microtomy and histological technique in the late 1800's came greater understanding of the internal morphology of the phylum. Over the past century, and particularly the last fifty years, numerous internal features have come to be recognized as characters of major significance in developing a phylogenetic understanding of the group and a more secure basis for classification of species into genera and higher taxa. It is with those characters significant at the generic and higher levels that this discussion is primarily concerned. However, a number of the features given in older descriptions, now regarded as minor characters, are still valuable in distinguishing between species within genera. Recently studied details of a few of these characters will also be discussed.

This study grew out of earlier work on the genus *Amphiporus* (Gibson & Crandall, 1989, 1991) which relegated more than 70 forms (approximately 10 percent of the Enopla) to *species inquirendae* status. In working out proper placements for many of these species, particularly the pending transfer of some of them to the Cratenemertidae, it became apparent that several underlying problems of enoplan systematics required re-examination. The need for further discussion has been heightened by an alternative theory of enoplan systematics (Gibson, 1988) which has stimulated critical re-evaluation of a scheme which had been in place for some 70 years, during which time the number of recognized enoplan species approximately doubled and many new genera and higher taxa were erected within the existing framework.

Although this paper presents a revised approach to the classification of the Enopla, its underlying purpose is to focus attention upon a number of difficult problems and issues which must be addressed by any classification scheme if it is to lead to a better phylogenetic understanding of this group and become a useful tool in taxonomic practice. It presents a case for recognition of intermediate taxonomic groups which are internally consistent but leaves open the question of how these groups may interrelate at higher levels.

Materials and methods

Material consisted of slide series in the collections of the U.S. National Museum of Natural History, the British Museum (Natural History), the Cambridge University Zoology Museum and Pantin collections, and of alcoholic specimens from the USNM collections, Bouin-Hollande fixed specimens from the collection of J. L. Norenburg, and Bouin and formalin fixed material from my personal collection.

Whole specimens and fragments were embedded in 56 °C m.p. paraffin, sectioned at 6 μm, and processed with an unpublished technique developed to provide adequate staining and differentiation in very old and often indifferently fixed museum material. The procedure consists of two stages of pre-mordanting followed by staining with an adaptation of the Mallory, Gomori, and Gurr-MacConaill trichrome methods. In some cases specimens or fragments were cleared in benzyl benzoate (refractive index 1.5681) after dehydration to permit study of gross internal morphology prior to embedding in paraffin.

Some figures are identified by a three-digit slide series number rather than by name or USNM number. This number refers to a particular specimen or fragment and is used for one or more of the following reasons: (1) a new USNM number will be assigned to the section series, (2) it is not USNM material, (3) the specimen is as yet unnamed, or (4) the generic placement under which it was originally listed is now dubious.

Brief historical review of enoplan classification

Schultze (1851, 1853) is generally regarded as the first to have a good understanding of nemertean morphology. He treated the Nemertinea as an order of the Turbellaria, dividing them into the suborders Anopla and Enopla, though Johnston (1827-1828), who thought it a feature of the gut, was the first to mention the existence of the armature apparatus in one group and its absence in the other. McIntosh (1873-1874) likewise divided the order Nemertinea into suborders Anopla and Enopla and thence into families. Hubrecht (1879) divided the ordo Nemertini into subordo Paleo-, Schizo-, and Hoplonemertini and from there into families. None of these authors discussed the position of *Malacobdella*. Bürger (1891, 1895, 1897, 1907, 1904) divided the Klasse Nemertini, primarily on nervous system-body wall relationships, into the orders Proto-, Meso-, Hetero-, and Metanemertini (which included *Malacobdella*). Carus (1885) regarded the family Malacobdellidae as a fourth suborder of Hubrecht's scheme, but without giving it a name. Verrill (1892) named Bdellomorpha as a separate suborder. Coe (1905) discussed the prior schemes in some detail, rejecting portions of Bürger's classification as containing unwarranted groupings and separations (*i.e.,* for not being phylogenetic) and drawing on the older schemes of Schultze, Hubrecht, and Verrill, he regarded the Nemertea as a class of the Platyhelminthes, recognizing four orders: Paleo-, Hetero-, Hoplo-, and Bdellonemertea. Wijnhoff (1912) objected to some features of Bürger's classification as being unphylogenetic and divided the Nemertinen into Anopla, containing the orders Paleo- and Heteronemertini, and Enopla containing the Hoplo- and Bdellonemertini.

After taxonomists turned from the scheme of Bürger toward that evolved by Verrill, Coe, and Wijnhoff, nemertean systematics remained stable except for further division of the Hoplonemertea as a large number of new genera and species were added to this group in the early years of this century. Brinkmann (1917) recognized the Anopla and Enopla as subclasses and divided the Order Hoplonemertea into suborders Monostilifera and Polystilifera, and the latter into Tribus Reptantia and Tribus Pelagica. Stiasny-Wijnhoff (1923) again criticized Bürger's approach as not being phylogenetic and followed Brinkmann's scheme while further subdividing the Reptantia into Subtribus Archireptantia and Eureptantia, then later

(1936) subdividing the latter into Cladus Aequifurcata and Inaequifurcata. Unaccountably, neither Brinkmann nor Stiasny-Wijnhoff referred to the work of Coe or Verrill. The scheme resulting from the studies of these authors is shown in Fig. 1.

More recently, Iwata (1960, 1972, 1985), based on embryological studies showing great similarity in the early stages of development of *Malacobdella* and the Hoplonemertea, proposed division of the Enopla into three suborders: Monostiliferoidea, Polystiliferoidea, and Bdellonemertiodea. However, because of the great differences in the adult morphology, most workers preferred to retain Bdellonemertea as a separate order on a par with Hoplonemertea. Gibson (1988) proposed a very different theory for classification of the Enopla based on whether the rhynchocoel wall musculature is bi-layered or interwoven as the primary character and postulated convergent evolution of monostiliferous armature in the bilayered and interwoven groups, as shown in Fig. 2.

Discussion of some taxonomic characters

The following sections discuss characters that have figured prominently in recent debates on enoplan systematics and present new information on characters explored in recent investigations.

Proboscis structure and armature

The enoplan proboscis is characteristically divided into three distinct regions: anterior chamber, bulb, and posterior chamber. Even *Malacobdella* exhibits this basic structure, whether one regards its lack of armature as primitive or as secondary loss. The only features which have been sufficiently recorded to serve as widely applicable taxonomic characters are number of proboscis nerves and details of armature.

The number of longitudinal nerves in the anterior proboscis wall has been widely used as a taxonomic character and was earlier thought to be constant, or nearly so, for a given species.

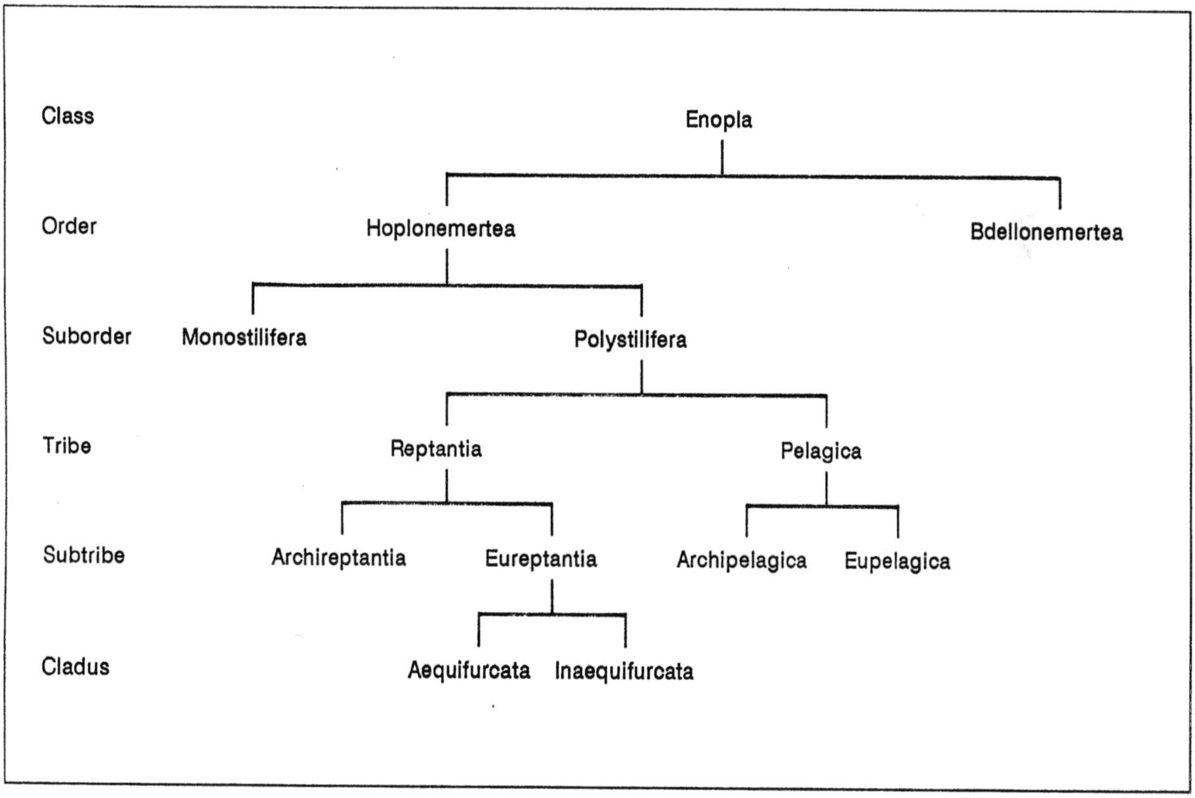

Fig. 1. Classification scheme evolved from the work of Verrill, Coe, Brinkmann and Stiasny-Wijnhoff.

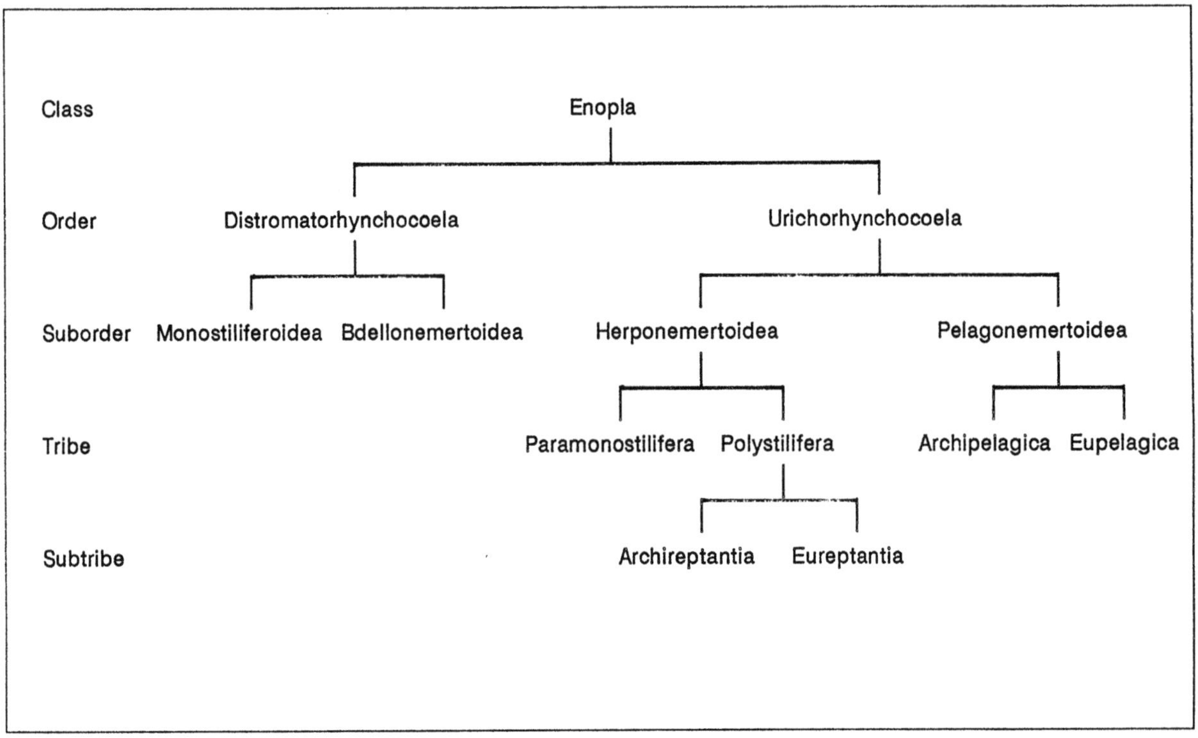

Fig. 2. Classification scheme proposed by Gibson (1988).

However, examination of a large number of specimens frequently reveals considerable intraspecific variation (Berg, 1972, 1985; Gibson *et al.*, 1982; Norenberg, 1986), but with frequency distributions (usually strongly leptokurtotic and often skewed) which indicate the modal number as the underlying pattern. Norenburg (1986) has noted that despite intrageneric variability different genera also tend to have characteristic modal values for nerve number. The number of nerves may not be constant at all levels of the proboscis in a single individual (Norenburg, 1986). The condition occurs often enough that species descriptions, especially those based upon single specimens, should report examination for this feature.

Most authors in this century have regarded the polystiliferous condition as primitive and armature in the Pelagica as more primitive than in the Reptantia, a view supported by the gradient of developmental complexity within the Polystilifera. For example, *Balaenanemertes* and *Nectonemertes* in the Archipelagica have proboscides barely differentiated into distinct regions, with the bulb demarcated only by a thin, incomplete septum of muscle fibers and the armature consisting of a lump-like basis on one side of the lumen. Still within the Archipelagica, *Pelagonemertes* shows differentiation of the bulb region somewhat more reminiscent of monostiliferous forms and in one of its species a basis comprised of the characteristic tongue-shaped portion joined to an ovoid posterior portion resembling a monostiliferan basis. As one moves through the Eupelagica to the Reptantia there is a general progression in the level of differentiation and organization, with an increasing tendency toward a radial symmetry in the placement of the armature components within the proboscis structure.

The number of accessory stylet pouches has been much used over the years, at least in the Monostilifera, as a character at the specific level and, in combination with other characters, can have implications at the generic level as well. In the Polystilifera there tend to be a rather large number of accessory pouches, each containing only one or two tacklike stylets, arranged in non-coplanar fashion as a radial belt around the wall of the anterior bulb region. The preponderance of monostiliferous forms have two accessory pouches, directly opposite one another at the periphery of the mid-bulb region. Where several pouches are present they are not equally spaced from one another around the bulb, rather two will be found opposite one another with the others tending to cluster more or less near to these. In all cases, they are arranged in essentially a single plane normal to the axis of the proboscis. This coplanar arrangement holds even for the few monostiliferous species with more than six pouches, for example, *Amphiporus formidabilis* with 8–12 and the *Amphiporus cruciatus- spinosus-spinosissimus-multihastatus* complex with 3–30. In such cases the number of accessory pouches is always significant specifically and sometimes generically. The number of species with more than two pouches decreases as the number of pouches increases, and forms with but a single pouch are conspicuously absent, suggesting that two pouches, biradially arranged, is plesiomorphic for monostiliferous species. The majority of monostiliferous forms have 2–4 stylets per pouch, with the number tending to decrease as the number of pouches increases. The Cratenemertidae typically have two pouches, usually with 3 – 8 + stylets each and mostly tend toward the high end of this range but otherwise do not show notable differences from other monostiliferous forms in proboscis structure.

Stricker (1985) provided persuasive arguments, supported by ultrastructural studies, for a general theory of stylet and pouch development from much simpler antecedents and endorsed the prevailing view that monostiliferous armature is the more highly evolved and was derived from the polystiliferous type by reduction. The morphologic gradient in the polystiliferous groups supports the concept of progression of armature from a patch of tissue, representing a crude basis studded with several small simple stylets, on the side wall of an almost open lumen, to a radially symmetrical, non-coplanar structure with highly differentiated tissues and a nearly solid bulb region. However, the morphology of monostiliferous armature, with its bilateral (or biradial) coplanar architecture, predominance of two pouches, and absence of forms with a single pouch, forces us to examine possible alternatives to this reduction hypothesis, namely that there was a much earlier divergence where, in one group, Stricker's primitive antecedents underwent reduction to bilaterality before their evolution into stylet pouches or that there was an altogether separate development of armature from a primitive bilateral origin (*i.e.*, from different precursors). Either possibility suggests that occurrence of more than two pouches in monostiliferous forms represents secondary proliferation. Unfortunately, comparable ultrastructural data for polystiliferans and cratenemertids is lacking, thus a number of important questions remain unresolved.

Rhynchocoel wall structure

Enoplan rhynchocoel wall musculature (with limited exceptions) is either bilayered (Fig. 3) or a single layer of interwoven circular and longitudinal fibers (Fig. 4). The interwoven type has been known since Hubrecht (1879) for *Drepanophorus* and subsequently for a few species of *Amphiporus*. Noted particularly by Brinkmann (1914-15, 1917), Wijnhoff (1914), Stiasny-Wijnhoff (1923, 1926, 1936), Coe (1905, 1926) and Korotkevich (1955, 1964), its occurrence throughout the monostiliferous families Cratenemertidae and Plectonemertidae, the Reptantia, and in some 40 percent of the Pelagica, about one-third of all enoplan species, mark it as a very important major character. It has been regarded as significant at the familial or higher level by Friedrich (1968), Berg (1972, 1985), Gibson (1988, 1989), Moore & Gibson (1988a, b), Gibson & Crandall (1989), and Crandall (1993).

Pouches and diverticular appendages to the rhynchocoel are found throughout the Reptantia and in a number of monostiliferous species. In the Reptantia they occur along most of the body length, are metameric lateral thin-walled diverticula which extend, in most cases, for considerable

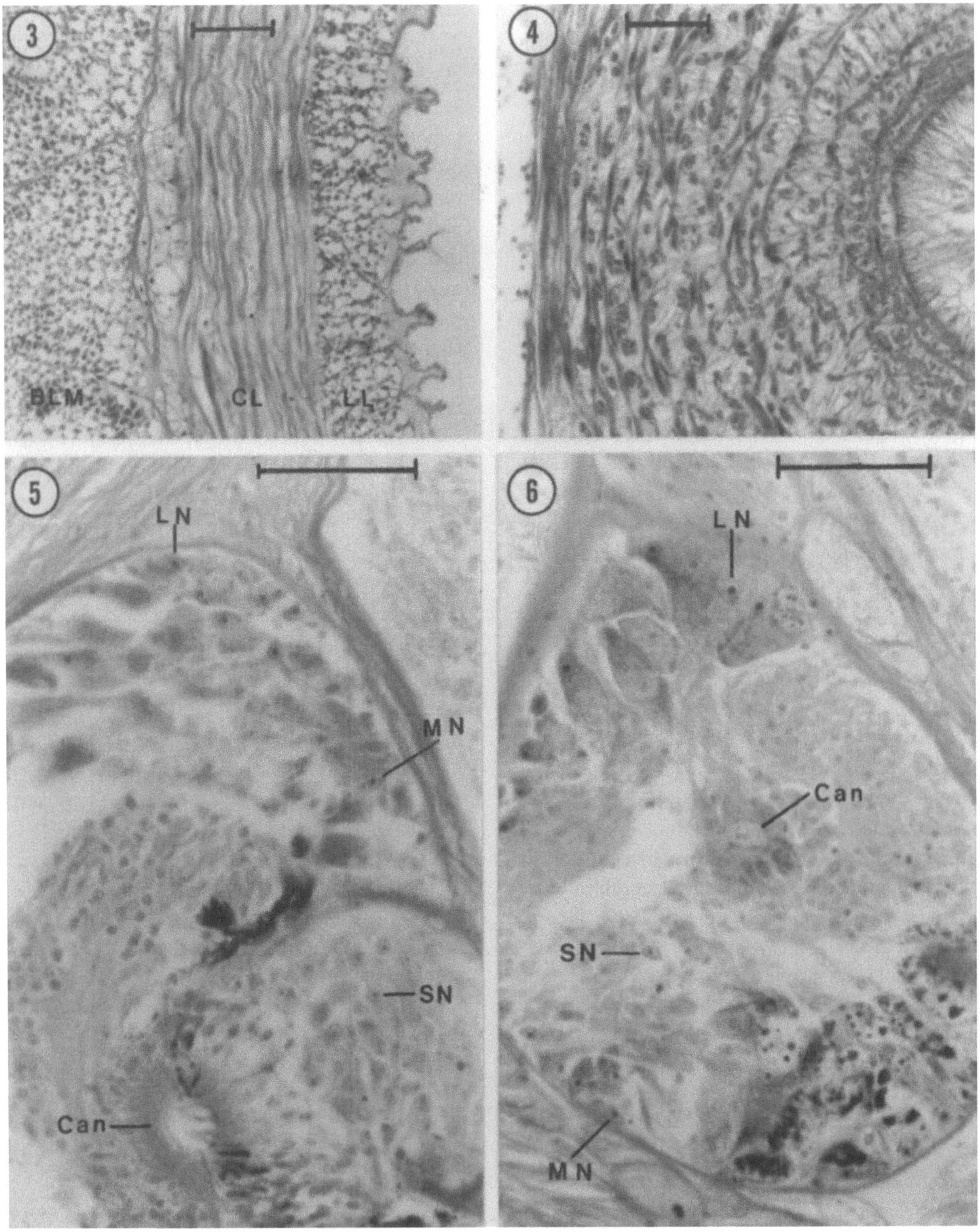

Figs 3–6. All sections transverse. Scale bars = 0.1 mm. 3. Bilayered rhynchocoel musculature, amphiporid (007). 4. Interwoven rhynchocoel musculature, *Nipponnemertes punctatulus* (010). 5. Cerebral organ gland cells, cratenemertid (027). 6. Cerebral organ gland cells, amphiporid (017). BLM = bodywall longitudinal muscle; Can = canal; CL = circular layer; LL = longitudinal layer; LN = large nucleolus; MN = medium nucleolus; SN = small nucleolus.

distances toward the lateral body margins and have, usually, a sphincter-like junction with the rhynchocoel proper. By contrast, diverticula or pouches in non- reptantic forms are usually limited to the foregut region with walls essentially the same as the rhynchocoel itself, *i.e.*, containing muscle fibers in addition to the membraneous lining layer. Such pouches occur as unpaired dorsal or ventral appendages in several monostiliferous forms and as a few paired lateral structures in *Valdivianemertes.* In all cases cratenemertid pouches have an open nonsphincter-like connection to the rhynchocoel lumen. In all but one or two cases such pouches have been mentioned as a generic-level character. Two specimens belonging to different cratenemertid species exhibited a pouchlike portion of the posterior rhynchocoel; however, closer examination and muscle fiber tracing indicates these are probable contraction artifacts. Coe (1943) reached a similar conclusion with regard to the ventral rhynchocoelic pouches described for *Proneurotes multioculatus* by Montgomery (1897), but Friedrich (1940) found them to be valid structures in *P. baltica.*

Whether Gibson (1988) was justified in using interwoven musculature as the pre-eminent character for primary division of the Hoplonemertea has been the subject of heated debate. I am of the opinion that it should be 'one' of the characters used for this purpose but not the only one. Moreover, I conclude that Gibson's proposal would have been on safer ground had some critical investigations been performed to test his concept. For example, when one looks at the rhynchocoel musculature one is actually looking at only half of a biodynamic system; the other half is the innervation of the muscle fibers. The interweaving of the muscle fibers has profound consequences for the way in which innervation can be accomplished, *i.e.*, routing of nerve fibers, neuromuscular junctions, etc. If these features occur in only one pattern, it would not definitely prove a single evolutionary path, but would suggest that this was the case. On the other hand, if two (or more) distinct patterns were found, it would provide justification for believing that there were two (or more) evolutionary pathways by which this condition had been achieved. There are, no doubt, other tests that could be used to garner evidence applicable to this problem.

A further difficulty with Gibson's classification arises over inclusion of the Pelagica as a main branch of the Urichorhynchocoela, since only about 40 percent of the pelagic species have interwoven musculature. Among archipelagic species 84 percent are bilayered and the remaining 16 percent trilayered with approximately 11 percent of the total showing some interpenetration of fibers between layers. *Obnemertes* has a few fibers interweaving between layers, *Mesarmaueria* has three layers, and *Proarmaueria* three layers with a few interwoven fibers. Some 30 percent of the Eupelagica have bilayered musculature; although, nearly half of these show some small degree of interweaving near the brain. Three of the six families have genera which are bilayered: *Dinonemertes, Planonemertes, Plionemertes*, and *Tononemertes* are entirely bilayered, while *Chuniella* is bilayered with a few fibers interweaving.

An additional complication in attempting to formulate an unambiguous scheme for understanding the interwoven rhynchocoel musculature is the fact that the prevalent condition is roughly even distribution of circular and longitudinal fibers throughout the thickness of the layer. But, some species have a predominance of longitudinal fibers toward the inside of the layer while in others it is toward the outside. Unfortunately, recorded data for this feature is too fragmentary to permit any generalizations at present.

There is no clear substantiating evidence either for or against Gibson's hypothesis. Perhaps the likelihood of parallel or convergent evolution of interwoven muscle is greater than that of evolving monostilifery twice, indicating that Gibson's classification should be held in abeyance pending further investigation.

Cerebral organs

Size, structure, and relative position of cerebral organs have been recognized as significant at the generic level or above (e.g., Stiasny-Wijnhoff,

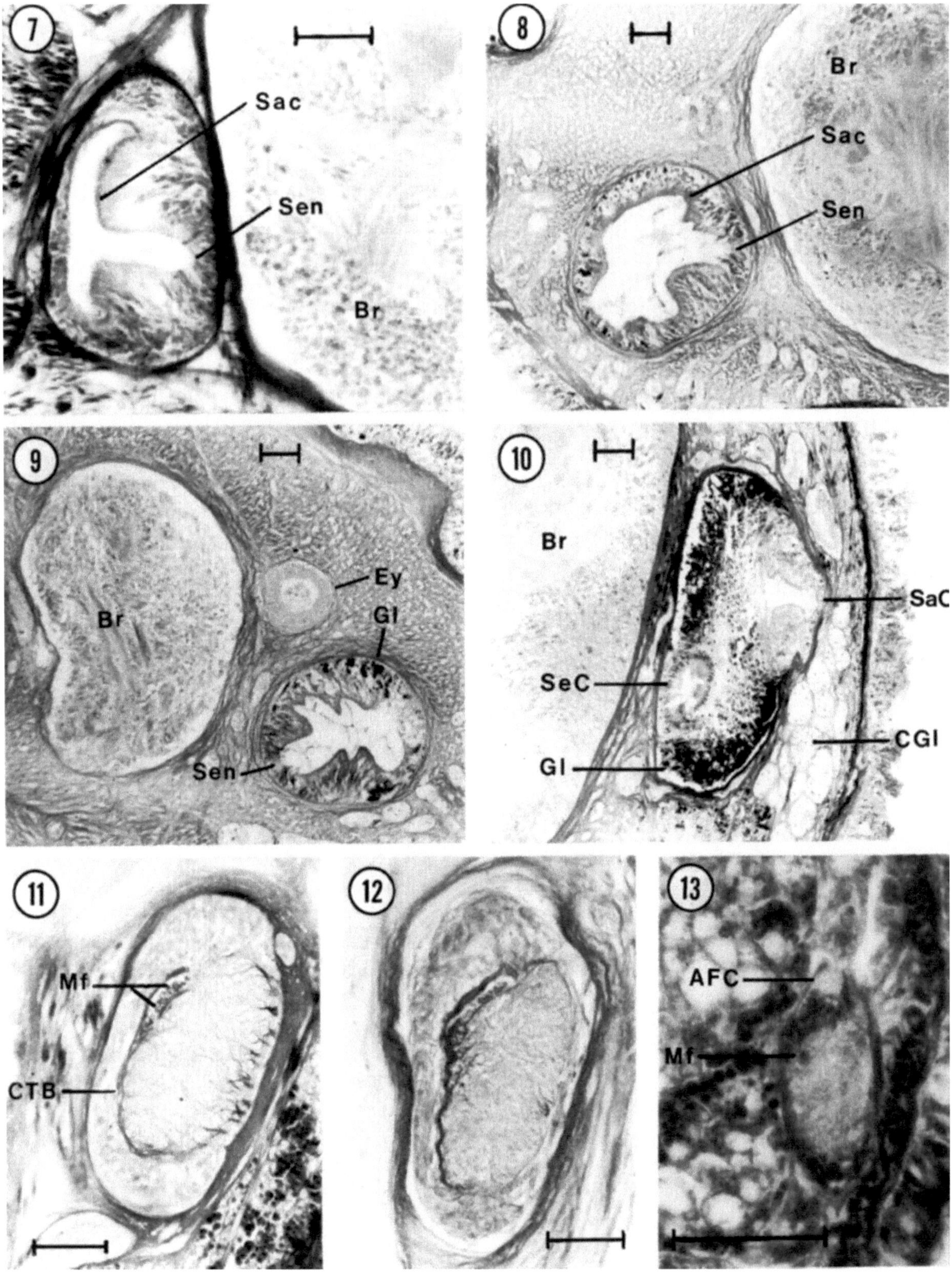
7
Sac
Sen
Br
8
Br
Sac
Sen
9
Ey
Gl
Br
Sen
10
Br
SaC
SeC
CGl
Gl
11
Mf
CTB
12
13
AFC
Mf

1923, 1926, 1936; Friedrich, 1955, 1968; Berg, 1972, 1985). However, relative size and position can appear to vary as a consequence of distortion of the cephalic region during fixation (Sundberg, 1979; Norenburg, 1986). Moreover, statements as to size are often subjective, varying considerably with different authors and even with the same author depending upon the presumed norm for the genus to which a species is assigned. Statements regarding position can also be misleading, as in the case of the older German literature where the phrase 'weit vor dem Gehirn' often occurs but examination of the animal shows the cerebral organ just in front of the brain, although distinctly separated from it.

The detailed histology and cytology of the cerebral organs is a valuable taxonomic tool which has been underutilized. Most of the detailed examinations of the fine structure of the cerebral organs have been done by physiologists studying their possible neuro-endocrine functions and few by systematists looking for useful taxonomic characters. My recent studies of the Cratenemertidae reveal, with ordinary histological techniques, what appear to be no less than four, and possibly as many as seven, different kinds of gland cells in a single cerebral organ and certainly two (or more probably three or four) kinds of canal epithelial lining cells. These observed differences in cell types are based not only on staining results but upon less mutable features such as nuclear shape, size and position within the cell; particle size and density of granular material within certain types of gland cells; pronounced differences in size and position of nucleoli in some types of gland cells; and length, diameter, and density of cilia in epithelial cells (Fig. 5). One monostiliferous specimen, a putative amphiporid, has been found to possess what appear to be most of the gland cell types found in the cratenemertids; however, the gross morphology of the organ is less complex and there is only a simple canal with an entirely sensory epithelium (Fig. 6). The cerebral organs in the Cratenemertidae and Reptantia are far more complex in structure with much greater cellular differentiation than is found, with a few rare exceptions, in monostiliferous forms and thus, in my view, represent a more highly evolved condition than that present in all the rest of the Enopla.

Branching of the cerebral organ canal was used by Stiasny-Wijnhoff as a diagnostic character for the Eureptantia and later for its division into the Aequifurcata, in which both branches are lined with sensory epithelium, and the Inaequifurcata, where one branch is lined with sensory epithelium and the other is lined with an epithelium which is histologically much different and usually enlarges to end in a blind sac. Gibson (1988) stated he had observed such a branching of canals and the presence of sacs in a number of Antarctic members of *Nipponnemertes* and used this as one of the features for arguing a close relationship between the Cratenemertidae and the Reptantia. Stiasny-Wijnhoff (1936) in discussing such sacs noted several species originally placed in *Amphiporus* which subsequently have been transferred to *Nipponnemertes* or have recently been found to belong among the Cratenemertidae and pointed out that in these species there was not a true sac but only a widening of an unbranched canal. She described this condition as forming a 'T-shaped slot' canal cross section. A typical example of this configuration is shown in Fig. 7.

I have examined some of the same Antarctic material examined by Gibson, additional unde-

Figs 7–13. All sections transverse. Generic names in quotes are no longer valid, but new generic names for these species have not yet been published. Scale bars = 0.1 mm. 7. T-slot cerebral organ canal, *Nipponnemertes sp.* (USNM 98358). 8. Complex cerebral organ canal, cratenemertid (027). 9. Complex cerebral organ canal, opposite side of cratenemertid (027). 10. Branched cerebral organ canal, *'Amphiporus' bimaculatus* (006). 11. Lateral nerve cord with myofibrillae, *'Amphiporus' rubellus* (029). 12. Lateral nerve cord with myofibrillae, *'Amphiporus' bimaculatus* (006). 13. Lateral nerve cord with single large myofibril, *Argonemertes dendyi* (001). AFC = accessory fiber core; Br = brain; CGl = cephalic gland; CTB = connective tissue band; Ey = eye; Gl = gland cells; GlC = gland cells; Mf = myofibrillae; SaC = sack type canal; Sac = sack type epithelium; SeC = sensory type canal; Sen = sensory type epithelium.

scribed Antarctic cratenemertid specimens from the Smithsonian Oceanographic Sorting Center, and a number of cratenemertids from the Pacific and Atlantic coasts of North America. Only a few of these species possessed what could truly be called a blind sac like that found in the Inaequifurcata, but nearly all have a much expanded region of the canal just after it enters the cerebral organ. Sometimes this enlargement is of complex configuration and occupies more than half of the cross-sectional area of the organ (Fig. 8, 9). *Amphiporus bimaculatus* Coe (1901, 1904, 1905), now known to be a definite cratenemertid of undetermined genus, possesses a canal that divides, just after entering the organ, into two branches, *i.e.*, a sensory branch having a uniform elliptical cross section throughout its length and a shorter branch of similar size but complex shape, lined with characteristic 'sac-like' epithelium (Fig. 10).

A wide range of conditions prevails in the Plectonemertidae. *Argonemertes dendyi* has a 'T-slot' configuration, *A. australiensis* has a short branch of elliptical cross section lined with sac-type epithelium, while *A. hillii* has a branch of elliptical cross section, lined with sactype epithelium, extending to the rear of the organ. *Antiponemertes novaezealandiae* has separate sensory and sac-type branches of comparable cross section. The plane of the sections in the specimens examined made it impossible to determine accurately the configuration of the sac-type epithelium in *Antiponemertes pantini* and *Acteonemertes bathamae,* only that it was present.

Whether or not Gibson's (1988) contention regarding actual bifurcation of canals or presence of blind sacs is strictly applicable in all cases, the important point is that the Cratenemertidae and Plectonemertidae possess both a sensory and a sack-type epithelium which are histologically very different as well as canal goemetry much more elaborate than other monostiliferous forms, a condition they share with the Inaequifurcata. By contrast, the rest of the monostiliferous forms and the Aequifurcata possess only a sensory type of epithelium, and while the canal may be bifurcated in the latter, the configuration is otherwise relatively simple. At the same time, there appear to be significant differences between the Cratenemertidae and Plectonemertidae in the number and complexity of cell types forming the glandular components of the cerebral organs. In both respects the Plectonemertidae more closely resemble the other monostiliferous forms, while the Cratenemertidae more closely resemble the Inaequifurcata.

On the basis of structural histology and details of cytology, there is available a rich array of potential cerebral organ characters. Therefore, I am inclined to disagree with the contention of Moore & Gibson (1988b) that in the family Plectonemertidae, 'the clear correlation between habitat and cerebral organ elaboration ... indicates that their form cannot be used in assessing higher taxonomic affinities.' This statement may be true in terms of gross size, shape, and position and perhaps even in respect of proportions of different cell types; but it is unlikely to be true of the cytological features of the cells represented.

I have equal disagreement with Sundberg's (1990) statement, '... but their homologous states must be corroborated by other characters'. This would seem to imply that if we find a strong enough mathematical correlation between two groups for a given set of characters it necessarily demonstrates homology for a particular character. However, such an approach is highly dependent on selection of the set of characters used in the analysis, and this factor alone can materially skew the result. Moreover, if it fails to consider the natural history of the characters in the set, how and why they might have gotten to their present states, it may yield an implausible result. Thus, I feel it more prudent to look to the organ itself, and all of its parts, for evidence of homologies.

Myofibrillae

Gibson (1988) noted the occurrance of myofibrillae in the lateral nerve cords of several genera of the Reptantia, several genera of the Monostilifera, all of the terrestrial genera and in several Antarctic members of *Nipponnemertes* and made the further comment that their occurrance in the rest of *Nipponnemertes* was not recorded in the literature. In examining a number of species of this genus and other cratenemertids from a variety of locations, I have found myofibrillae in every specimen and in some the myofibrillae and associated connective tissue form a feature of extraordinary prominence.

There appear to be definite differences in the arrangement of the myofibrillar structure in the different groups. The cratenemertids have a band of dense connective tissue lying along the entire mesial margin of the fiber core of the nerve cord and the myofibrils are found within a vacuolar space in the dorsal one-fourth to one-half of this band (Fig. 11, 12; see also Fig. 30, 31). In at least some of the Reptantia the myofibrillae occur in an oval configuration in the dorsal portion of the lateral nerve cord. I am less certain of their presence in some of the terrestrial specimens I have recently examined, but this may have been due to the staining of many of these specimens. However, my impression is that the connective tissue band and numerous myofibrils, so characteristic of the cratenemertidae, are replaced by only one or, at most, a few unensheathed myofibrils. Figure 13 shows a single large, bare myofibril in the dorsomesial portion of the main fiber core of *Argonemertes dendyi*.

In the cratenemertids the myofibrillae and connective tissue band arise immediately behind the ventral brain commissure and pass into the ventral ganglion just as it is beginning to form the lateral nerve cord (Fig. 14). The connective tissue band assumes its characteristic position on the mesial surface of the fiber core before the lateral nerve cord becomes distinct from the brain.

I am inclined to think the position and detailed structure of the myofibrillae may constitute a character consistent at the generic level and, perhaps in some cases, at the family level, though their functional significance remains obscure.

Body wall structure

Arrangement of anterior longitudinal body wall musculature as a single layer or divided into an inner and outer layer separated by gland cells or parenchyma has been recognized as a generic level character by Friedrich (1955), Kirsteuer (1967, 1973, 1974), Moore & Gibson (1988a), and Norenburg (1986).

The pre-cerebral septum, whether absent, closed, split, or dissolved, has been discussed by Friedrich (1935, 1955, 1970) and Kirsteuer (1974) as a generic character. Moore & Gibson (1988b) found it to be of variable intergeneric applicability, and Norenburg (1986) has shown that proboscis insertions formed only from the inner longitudinal layer exhibit significant intraspecific variability. I have found in a genus of very closely related forms with undivided anterior longitudinal musculature one species where the entire longitudinal layer contributes muscle fibers to the insertion and a sibling species where only the inner 40 percent of the longitudinal layer contributes fibers to the insertion. Certainly the divided anterior longitudinal musculature seems well established as a character at the generic level and, in some cases, perhaps is applicable at the family level. I think the conclusion must be that the details of the septum, while very useful at the specific level, should be used with caution at higher levels.

Body wall diagonal muscle layer

Diagonal muscles, arranged as fibers more or less at right angles to each other, lying between, but obliquely to, the circular and longitudinal layers of the body wall, are not unique to the Enopla, being known also from paleo- and heteronemerteans such as *Carinoma* and *Cerebratulus*. Early accounts of diagonal muscle included Bürger (1895), who noted its presence in a number of enoplan species and discussed it in some detail, and Coe (1905), who noted that in *Amphiporus punctatulus* it was fairly easy to see because it was separated from the circular layer by a thin sheet of connective tissue. More recently the presence or absence of a diagonal layer and, if present, its thickness, whether it forms a complete layer, and its extent posteriorly has been noted at the generic

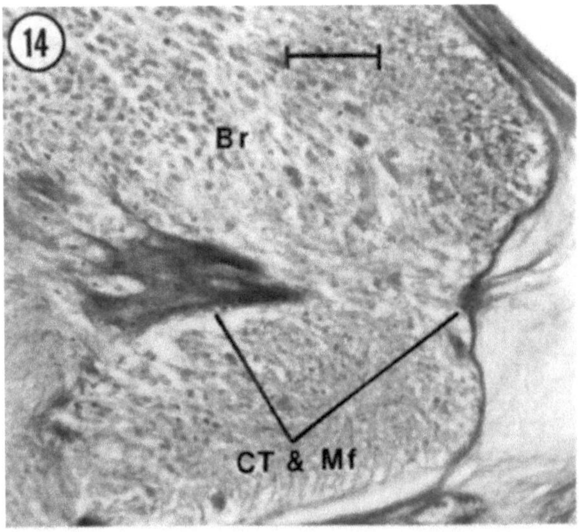
14
Br
CT & Mf

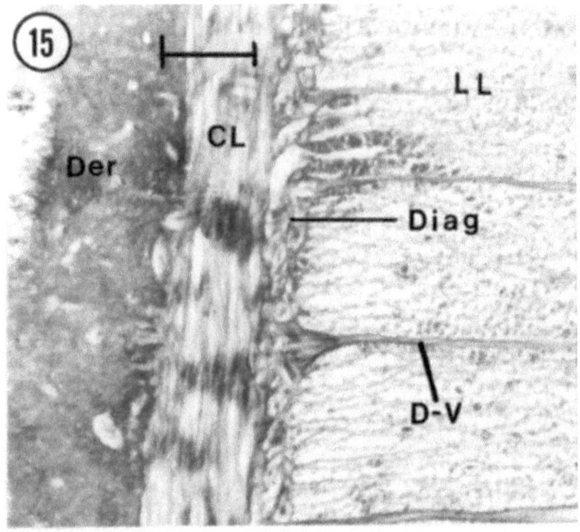
15
LL
CL
Der
Diag
D-V

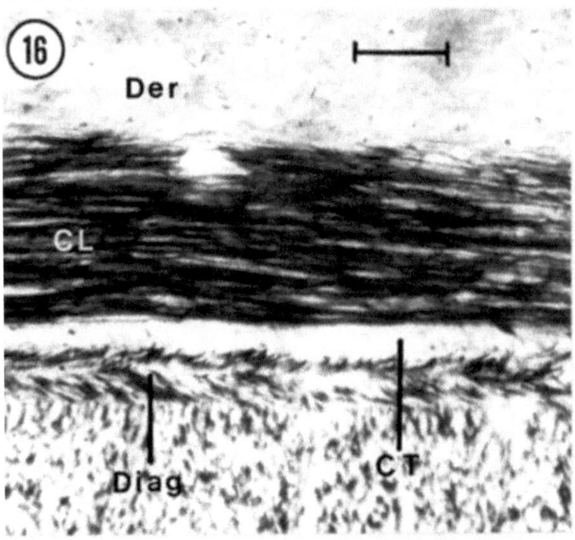
16
Der
CL
Diag
CT

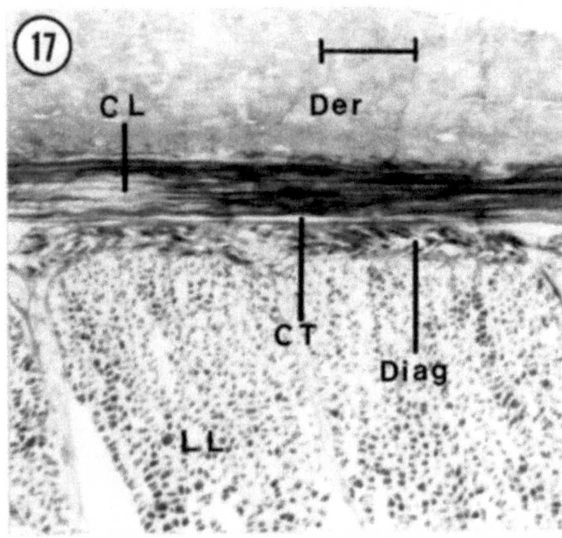
17
CL
Der
CT
Diag
LL

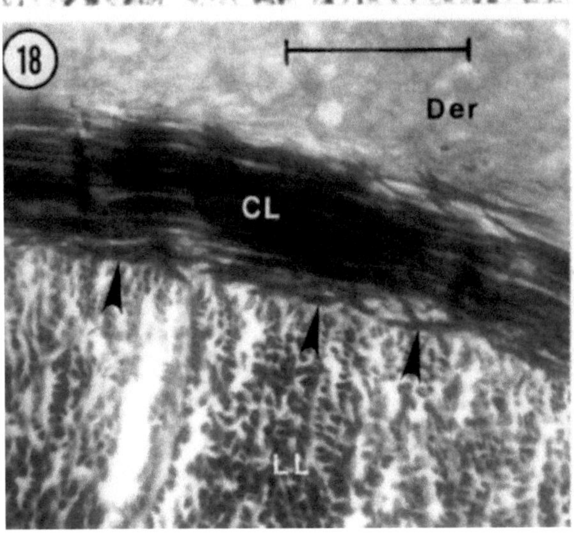
18
Der
CL
LL

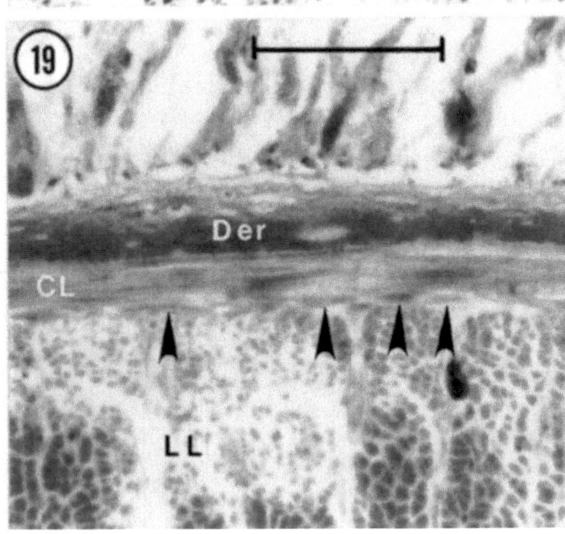
19
Der
CL
LL

level by Friedrich (1955), Kirsteuer (1967, 1974), and Norenburg (1986) who also notes that it may be very difficult to see.

Recent studies of a variety of enoplan material show that diagonal muscle is of wider and more frequent occurrence than perhaps previously thought and, more important from a taxonomic point of view, that there are two kinds of diagonal muscle layers. The first kind usually is formed of units consisting of 2–4 muscle fibers enclosed in a tube of fascia-like connective tissue. A series of these units lie side by side, at about a 45° angle to, and just inside the circular layer. A second layer of such units lies just inside the first and at approximately right angles to it but the fascia-like covering is distinct from that enclosing the muscle bundles of the longitudinal layer. Because there is usually more than one muscle fiber per diagonal unit, because the layer is usually several fibers thick, and because of the amount of strongly-staining connective tissue, this type of diagonal muscle is, at worst, not difficult to detect and, at best, extremely obvious (Fig. 15, 16, 17). The thickness of the connective tissue elements differs between species but there is always identifiable connective tissue between the circular and diagonal layers. Hereafter, I shall refer to this type as fasciated diagonal muscle.

The second type of diagonal muscle consists of single muscle fibers of about the same diameter as those of the circular layer, without a fascial covering, lying from five to twenty fiber diameters apart, directly against the circular layer. A second layer of these widely-spaced fibers lies at roughly right angles to the first, and both lie obliquely to the circular and longitudinal layers. Diagonal muscle of this type is very difficult to observe in normal transverse or sagittal sections, except under optimal conditions (Fig. 18, 19) and is best verified in tangential sections (Fig. 20, 21). I have designated it non-fasciated diagonal muscle.

In the groups studied thus far, the state of the diagonal muscle character (*i.e.*, absent, non-fasciated, fasciated) seems to hold constant within genera, except in those groups which are already felt, on other grounds, to be heterogeneous. For example, some species presently in *Nipponnemertes* (e.g., *N. punctatulus* and *N. magnus*), or soon to be transferred to the Cratenemertidae, have prominent fasciated diagonal muscle, while other species currently assigned to this genus have been found to possess the non-fasciated type. This may well explain the differing accounts of this character by Gering (1912, 1913), Berg (1972, 1985), and others when examining different species. Both types have been found in Cratenemertid and Amphiporid species and the plectonemertid *Acteonemertes bathamae* has an extremely well-developed non-fasciated diagonal layer but, unfortunately, not enough material has been examined to generalize more broadly at this time.

Body wall inner circular muscle layer

Although of common occurrence in the Paleonemertea (*Tubulanus, Carinoma. Carinomella)* and in some Heteronemertea (*Zygeupolia, Micrura*), circular muscle inside the longitudinal layer of the body wall occurs in very few enoplan species. This very rarity in the Enopla led Gibson (1990b) to refer to this character in naming his new genus *Ischyronemertes* containing three species which are closely united by this and numerous other shared character states. Iwata (1954) described as *Amphiporus insolitus* a form with a thin layer of inner circular muscle in the anterior portion of the body but, in light of new findings of the last two decades, the generic placement of this taxon should now be re-assessed.

Wijnhoff (1914) and Stiasny-Wijnhoff (1936)

Figs 14–19. All sections transverse. Generic names in quotes are no longer valid, but new names have not yet been published. Scale bars = 0.1 mm. 14. Myofibrillae and connective tissue band entering brain, *'Amphiporus' bimaculatus* (006). 15. Fasciated diagonal muscle layer, *Nipponnemertes punctatulus* (007). 16. Fasciated diagonal muscle layer, *Nipponnemertes magnus* (CUZM). 17. Fasciated diagonal muscle layer, amphiporid (019). 18. Non-fasciated diagonal muscle fibers, *'AmphiPorus' spinosus* (USNM 98904). 19. Non-fasciated diagonal muscle fibers, amphiporid (018). Br = brain; CL = circular layer; CT = connective tissue; Der = dermis; Diag = diagonal muscle; D-V = dorso-ventral muscle; LL = longitudinal layer; Mf = myofibrillae.

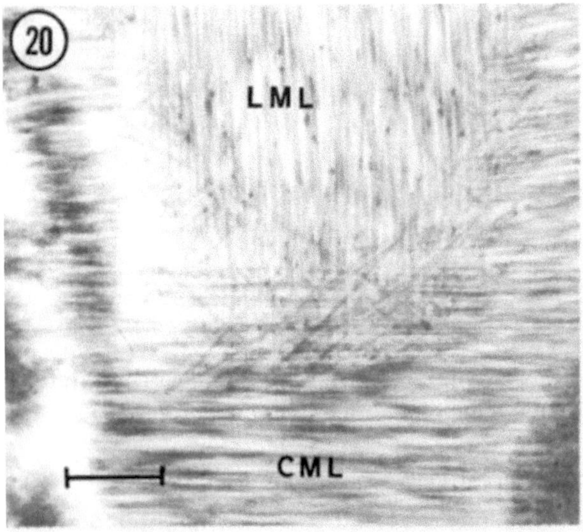
20
LML
CML

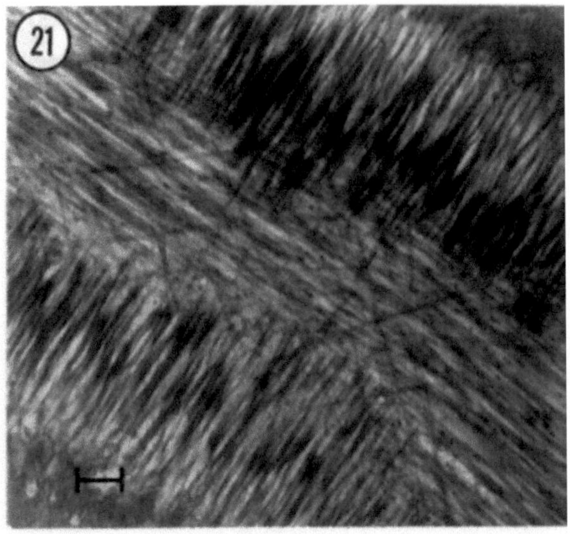
21

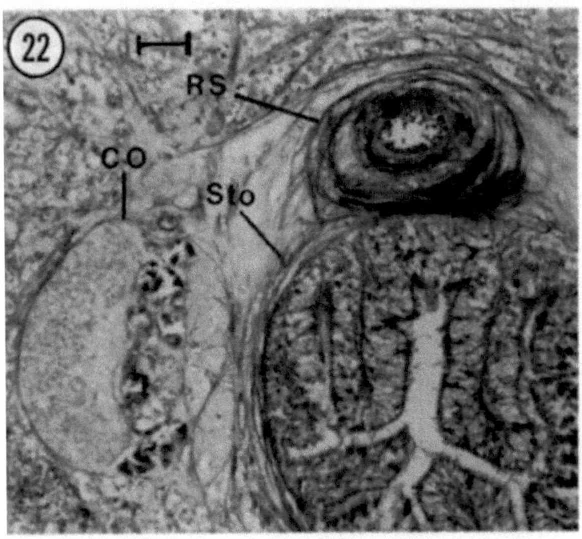
22
RS
CO
Sto

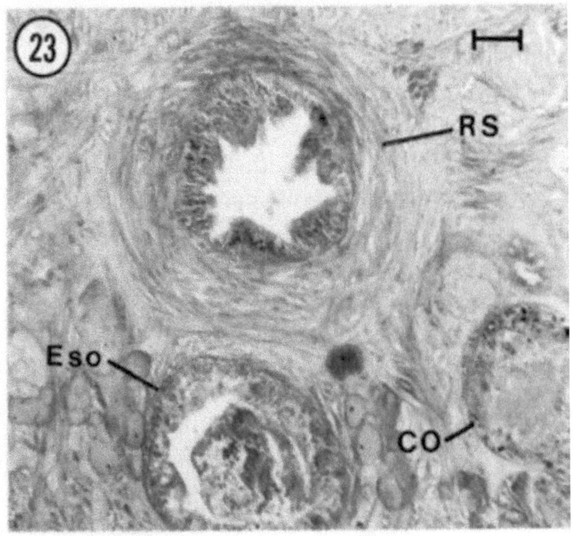
23
RS
Eso
CO

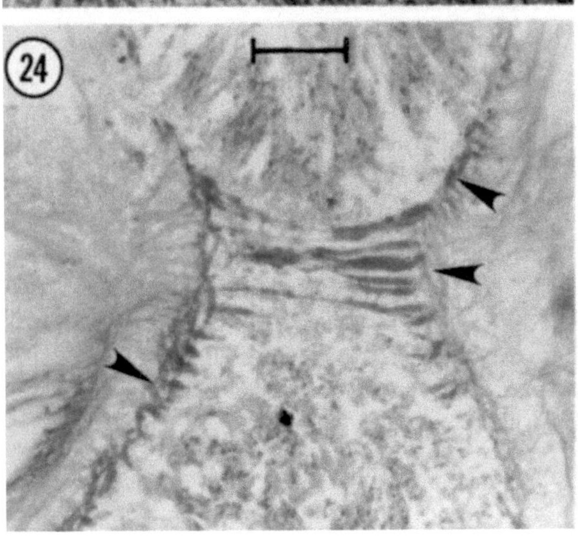
24

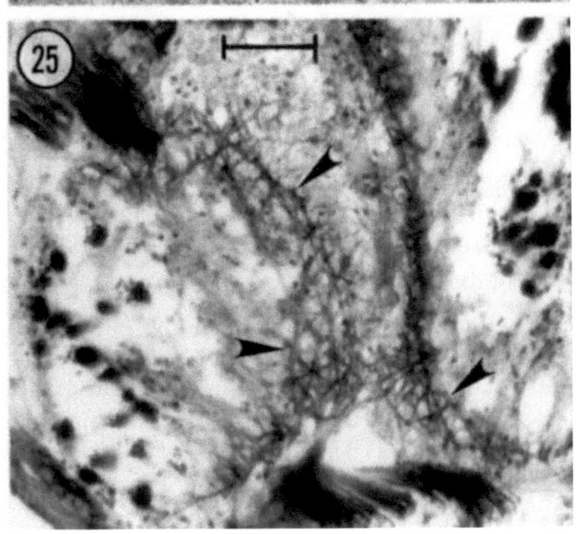
25

advance arguments for regarding the circular muscle fibers surrounding the gut and rhynchocoel in some of the Reptantia as an inner circular layer, and Gibson (1982) lists among the Reptantia two monospecific families and a third with four species as having an inner circular layer. However, the interpretation that these muscles are homologous with the inner circular layer of other higher taxa is difficult to accept since Stiasny-Wijnhoff showed them essentially as dorso-ventral muscle bands straddling the gut rather than associated with the body wall. Moreover, the lateral nerves, blood vessels, nephridia, gonads, and rhynchocoelic diverticula lie between this layer and the regular body wall musculature, except in *Drepanophoresta* where the circular layer lies outside the lateral nerve cords.

In *Ischyronemertes*, on the other hand, this layer exists dorsally and ventrally as a single layer just inside the longitudinal layer but laterally is divided into an inner band alongside the gut and an outer band which lies outside all organs except the nephridia. Clearly, this character appears to have utility at the generic, and possibly higher, levels, but more intensive investigation is required properly to assess its potential.

Esophageal and rhynchodaeal sphincters

Some enoplans have a sphincter-like investment of circular muscle fibers around the posterior esophagus just ahead of the brain. Coe (1905) mentions such a feature also in groups as diverse as *Micrura, Zygeupolia*, and *Carinoma*. However, it is inconsistently reported and may represent no more than a specific level character in the Hoplonemertea.

The presence or absence of sphincters of circular muscle fibers surrounding the rhynchodaem is often useful in distinguishing between closely related species. For example, a well- developed sphincter surrounds the rhynchodaeum just anterior to the brain in *Amphiporus spinosissimus* (Fig. 22) but is largely absent in the very closely related *A. spinosus*. These sphincters are sometimes massively developed (Fig. 23). Since this character has been only infrequently recorded in the literature, its possible usefulness beyond the specific level is difficult to estimate.

Basic gut plan

The simple straight tube is putatively the primitive condition in bilateria with separate mouth and anus as has been demonstrated in many embryological studies. Numerous times in the evolution of various bilaterian groups the intestine has elaborated its morphology in response to the need for greater absorptive surface area. Two basic avenues have been followed: (1) adding diverticula to a straight tube, or (2) lengthening the tube with respect to the body wall thereby forcing it to become sinuous, although some higher groups, the mammals for example, have combined the lengthened tube and diverticula, but in this case the diverticula are in the form of villi projecting into the lumen.

The straight, diverticulated intestine is characteristic for nemerteans. Departure from this plan in *Malacobdella*, with its sinuous non-diverticulated intestine, has customarily been regarded as due to lifestyle adaptation. Perhaps one reason for assuming that the gut of *Malacobdella* represents secondary specialization is that there has not been a good 'outgroup' to use in comparing it with the hoplonemerteans. Riser's (1988, 1989) discovery of *Arhynchonemertes* with a largely enoplan bauplan and its several presumably primitive

Figs 20–25. All sections transverse unless otherwise noted. Generic names in quotes are no longer valid, but new generic names for these species have not yet been published. Scale bars = 0.1 mm. 20. Tangential section, bodywall showing non-fasciated diagonal muscle fibers, amphiporid (017). 21. Tangential section, bodywall showing non-fasciated diagonal muscle fibers, amphiporid (017). 22. Rhynchocoel sphincter, *'Amphiporus' spinosissimus* (024-025). 23. Rhynchodaeal sphincter, amphiporid (018). 24. Nearly tangential view of muscle in intestinal wall, *Nipponnemertes punctatulus* (007). 25. Nearly tangential view of muscle in intestinal wall, *Nipponnemertes magnus* (CUZM). CML = circular muscle layer; CO = cerebral organ; Eso = esophagus; LML = longitudinal muscle layer; RS = rhynchodaeal sphincter; Sto = stomach.

characters, including a straight tubular intestine, provides us with such an 'outgroup' and forces us to confront a more profound possibility, *i.e.*, that this feature of *Malacobdella* is not secondary specialization at all but, in fact, a very ancient and fundamental divergence from the rest of the Enopla. Pursuing this reasoning farther, we see that the Archipelagica, a group widely presumed to be rather primitive, generally have relatively few intestinal diverticula, lending support to the theory that the straight non-diverticulated intestine is the plesiomorphic condition for the ancestral Enolpa. At the same time, the condition in the archipelagic family Buergerellidae, where the intestinal diverticula are very few but very extensively branched, further attests to the underlying concept of gut elaboration being due to the need for increased absorptive surface area.

Mouth-proboscis relationships

Relationships between mouth and proboscis openings and the corresponding association between esophagus and rhynchodaeum have been long used at generic and higher levels. In most of the monostiliferous forms the esophagus opens into the rhynchodaeum internally whereas *Malacobdella* is sharply distinguished by the rhynchodaeum opening into an esophagus of unique histological structure. These great differences both of gross morphology and of histology once again raise questions as to the relationship of *Malacobdella* to the rest of the Enopla.

The hoplonemerteans have a range of conditions from the esophagus opening into the rhynchodaeum shortly in front of the brain to completely separate openings on the body surface. Most monostiliferous forms have the esophagus opening into the rhynchodaeum a short distance behind the proboscis pore and some distance anterior to the brain. Some of the Cratenemertidae and many of the Reptantia exhibit an intermediate condition in which both esophagus and rhynchodaeum open into a common atrium represented by an indentation in the body wall having a somewhat differentiated epithelium. Stiasny-Wijnhoff (1936) noted that even in Reptantia with separate openings on the body surface the strip of epithelium between the two pores is histologically differentiated from the surrounding integument. An atrium has sometimes been used as a definitive generic character, for example by Friedrich (1955) in the case of *Communoporus*, and by various authors more loosely as a familial character. However, in discussing the varied reporting of this character in *Valdivianemertes*, Crandall (1993) notes the somewhat subjective nature of this character and Norenburg (pers. comm.) points out that a strongly contracted cephalic region or some eversion of an atrium can give the appearance of separate openings. Therefore it seems clear that detailed histological examination must be employed in assessing the status and significance of this character, especially in one-of-a-kind specimens.

Foregut arrangement and appendages

Caecael or diverticular appendages to the esophagus and the stomach have been mentioned by many workers, notably Friedrich (1955, 1957, 1968, 1970), Kirsteuer (1967, 1974), and Berg (1972, 1985) who have discussed their use as generic characters and emphasized the importance of distinguishing between natural morphological structures and preservation artifacts. Kirsteuer (1963) observed that careful tracing of muscle fibers would usually make possible discrimination between normal and artifactual pouches. Another effective method is overlaying scaled drawings of different specimens where an anterior and posterior landmark (*e.g.*, proboscis pore and pyloric junction) are drawn the same distance apart, providing an opportunity to assess the not only the genuineness of unusual morphology but also its range of variation. However, the numerous variations on the 'standard' foregut plan and intraspecific variability in preserved specimens warrant great caution in using these features as firm taxonomic characters. Moreover, there should always be sound histological support for any description of gross morphology.

Caecum

An intestinal caecum anterior to the pyloric junction is characteristic for most enoplans, and various details of its morphology, such as girth and forward extent of the caecum proper; number, position, and elaboration of lateral diverticula; and particularly the forward extent and elaboration of anterior pouches, have been invoked as taxonomic characters at the generic level (Friedrich, 1955, 1968, 1970; Kirsteuer, 1967, 1974; Gibson, 1972, 1982; Berg, 1972, 1985; Norenburg, 1986). In some cases a particular configuration of these features seems to hold, with great consistency, for all of the species of a genus, while in other genera, variations in configuration form a reliable criterion for distinguishing between species. Given this degree of variability between groups, caecal morphology must be used with considerable caution. Even reported absence of a caecum must be evaluated carefully, since the terminology, especially in the earlier literature, is often unclear as to whether the reference is to the caecum as a whole or just its anterior pouches.

Intestinal musculature

In some cratenemertid species, the intestinal walls, or at least those of the intestinal diverticula, are invested with a loose meshwork of rather widely separated muscle fibers. In some species these fibers are as thick as those in the body wall layers, while in others they are quite fine and delicate. In species where gonadal muscle fiber meshworks are also present (see discussion of this character later in this article) their fibers tend to be of comparable size. Relatively thick fibers have been found in *Nipponnemertes punctatulus* and *N. magnus* and in *Amphiporus rubellus* and *A. bimaculatus* (both now known to be cratenemertids) (Fig. 24, 25). Fine, delicate fibers have been found in two Antarctic cratenemertids for which detailed descriptions are pending (Fig. 26). The functional significance of this character is not clear, but its presence seems to correlate with other relatively uncommon characters. Thus, it may be taxonomically useful, at least at the generic level.

Rectum and anus

Many species have a rectum occurring as a relatively narrow tube running from the gut to the anus and lined with an epithelium histologically differentiated from the gut (Fig. 26). Other species appear to lack this structure and the terminal part of the gut opens directly through the anus. Frequently, the entire rectum, if it is short, or at least the terminal portion, if it is long, is invested with circular muscle fibers forming a definite sphincter. These fibers are a separate structural element and not part of the body wall musculature.

The position of origin and opening of the rectum may vary but, whether dorsal, ventral or terminal, they are usually consistent with one another. The position of the rectum and whether or not it is invested with a sphincter of circular muscle fibers appears to be constant for those genera studied. Whether these characters are useful beyond this level remains to be determined.

The position of the anastomoses of the lateral vessels and of the lateral nerve cords with respect to the rectum has been regarded as important. For example, one of the key diagnostic characters given for *Proneurotes* Montgomery (1897) is anastomosis of the lateral nerve cords below the [ventral] rectum in contrast to the diagnosis for *Amphiporus* (as understood at that time) of a dorsal anastomosis.

Vascular system

The number of vascular plugs penetrating the rhynchocoel wall, has been recognized as a significant character by Pantin (1969) and Gibson (1972), more recently at the generic level by Moore & Gibson (1981), and at the familial level by Gibson (1988) and Moore & Gibson (1988b). Some of the Cratenemertids have been found to possess vascular plugs which extend for considerable dis-

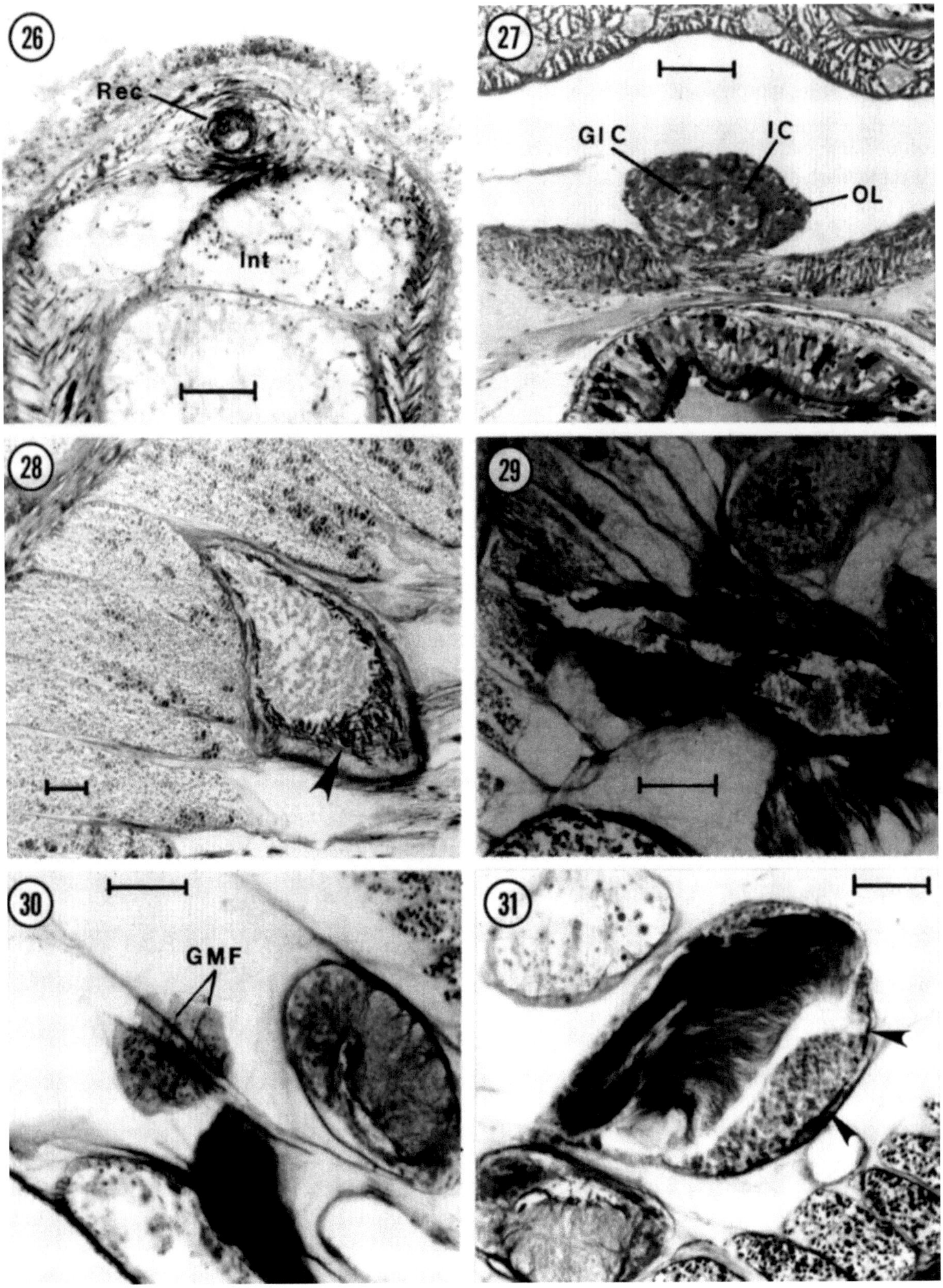
26
Rec
Int
27
Gl C
IC
OL
28
29
30
GMF
31

tances in the lumen of the rhynchocoel and are of great size and complexity, with a thick outer layer and a distinct inner core with what appear to be large gland or secretory cells (Fig. 27). A reasonable presumption is that the function of the vascular plug is to control the volume and/or chemical composition of the rhynchocoelic fluid. Why these species have elaborated the vascular plug to such an extent is unknown.

Extracerebral vessels, when present, have been recognized as a generic character by Friedrich (1955) and Kirsteuer (1974). Lacunae in the vascular system have been recognized by Moore & Gibson (1981, 1985), among others, but Norenburg (1986) cautions against over-reliance on the 'lacunar' nature of some vascular structures due to fixation-induced variability. Both of these characters are potentially useful but require much more thorough study in a wider array of species before their significance can be fully assessed.

Recent ultrastructural work by Jespersen & Lützen (1988) sheds new light on nemertean vascular structure generally and offers some insights into the possible significance of lacunar versus thick-walled vessels. Their study also indicates that the endothelial lining of the vessels in *Malacobdella* differs significantly from that in all of the hoplonemerteans they studied. This finding tends to further confirm the separation of the Bdellonemertea from other Enopla.

Excretory system

The structure and extent of the excretory system, has been noted by Bürger (1895) and Coe (1901, 1904, 1905) at both specific and generic levels, and more recently by Pantin (1969), Moore & Gibson (1981, 1985) at the generic level. The absence of an excretory system in the Pelagica has long been an important character state above the family level. The presence of nephridia in presumably primitive *Arhynchonemertes* and absence in the pelagic cratenemertid *Korotkevitschia* suggest that the pelagic habitat has resulted in secondary loss. The histology of the nephridia in the bdellonemerteans differs from that in the hoplonemerteans in ways that are more suggestive of ancient divergence than reduction. In recent years, the morphology of the nephridia in the terrestrial prosorhochmids and plectonemertids has become a well-studied and consistently reported feature and, therefore, an extremely useful taxonomic character (or set of characters) for these groups. Similarly consistent study of other groups should yield equally useful information.

Ocelli

The number, arrangement, size, and structure of ocelli have been widely used over time but accorded varying degrees of taxonomic significance by different workers. Gibson (1972, 1982, 1988) has pointed out their great variability within a number of genera and species; however, Norenburg (1986) has noted their usefulness in establishing relationships when used in conjunction with other characters.

Gibson (1988) noted the high degree of organization and complexity in the eye structure of the Reptantia and *Nipponnemertes* and used this as one of the defining features for his Herponemertiodea (*cf.* Fig 9). My own studies have found striking histological similarity of the eyes in these two groups, and among the cratenemertids there is great consistency of eye structure from species to species. (It should be noted that the Plectone-

Figs 26–31. All sections transverse unless otherwise noted. Generic names in quotes are no longer valid, but new names have not yet been published. Scale bars = 0.1 mm. 26. Rectum, *'Amphiporus' spinosissimus* (024-025). 27. Vascular plug, cratenemertid (027). 28. Nearly tangential view of gonadal wall muscle, discharged ovary of *Nipponnemertes punctatulus* (007). 29. Nearly tangential view of gonadal wall muscle, nearly discharged ovary of *'Amphiporus' rubellus* (002). 30. Nearly tangential view of gonadal wall muscle, testis of *Nipponnemertes sp.* (USNM 98358). 31. Transverse view of gonadal wall muscle, testis of *Nipponnemertes sp.* (USNM 98358). GlC = gland cells; GMF = gonadal muscle fibers; IC = inner core; Int = intestine; OL = outer layer; Rec = rectum.

mertidae do not exhibit comparable structural organization and complexity).

However, as Sundberg (1990) noted, eyes are not universally present in the Cratenemertidae. They are absent in *Valdivianemertes* (*cf.* Crandall, 1993), *Korotkevitschia*, and *Nipponnemertes scoresbyi.* In the Reptantia, eyes are missing in *Siboganemertes* and *Uniporus*. At the same time, I disagree with Sundberg's statement, 'One obvious problem with the character 'eye complexity' is that it cannot be unambiguously defined', since there must be some combination of descriptive elements which could define it rather precisely. His objection is valid only insofar as no such adequate definition, which will satisfy the needs of both cladists and traditional morphologists, has, thus far, been formalized.

It is my belief that eye structure and complexity is a valid character (or, more accurately, collection of characters) at the generic and family level and perhaps beyond. Moreover, presence or absence of eyes seems a valid character at the generic level and has been so used by a number of workers.

Gonadal musculature

In general, alternation of paired, thin-walled gonads and paired intestinal diverticula is the prevailing condition in the Enopla, with the gonads occupying the spaces lateral to the main intestinal canal between adjacent diverticula. Departures from this generality have usually been accorded some degree of taxonomic significance. For example, the far-anterior location of the testes in the Pelagica is a diagnostic feature. Some of the characteristics often noted include how far forward the first pair of gonads is located; how many separate sacs occur on each side; whether they open dorsally, ventrally, or, in an extreme case, into the gut.

There is a feature in some species which, as far as I am aware, has not previously been discussed. During examination of the original specimens from which Coe (1905) described *Amphiporus punctatulus* (since transferred to *Nipponnemertes* by Friedrich (1955)) and *Amphiporus rubellus* (a definite cratenemertid but of still unresolved generic status), it became apparent that the ripe gonads were invested with a meshwork of powerful muscle fibers (Fig. 28, 29). These fibers are about the same diameter to those of the body wall musculature and considerably larger than those lying in the parenchyma in the vicinity of the gut. The spacing and apparent diameter of these fibers obviously depends upon whether the gonad is distended with mature gametes or has been largely discharged.

Most of the species with large muscle fibers also have large arrays of dorso-ventral muscle in the intestinal region and frequently are dorsoventrally flattened in the posterior regions, *i.e.*, adapted for swimming. Coe (1905) stated that his specimens of *A. punctatulus* were taken in a surface net with electric light, (implying that they were swimming at the surface at night) and that they were sexually immature. This swimming behavior, as discussed by Gering (1913) and Berg (1972, 1985), is consistent with spawning and, indeed, every one of the original *A. punctatulus* specimens I have sectioned not only were sexually mature but had partially discharged gonads and, in the case of females, ova in the large ducts and on the body surface, indicating that the animals almost certainly were taken in the very act of spawning. I believe the probable purpose of the muscular meshwork investiture of the gonad in these forms is rapidly to discharge the gonads during spawning, *i.e.*, an ejaculatory mechanism. The advantage of this capability in forms which spawn while swimming is obvious, as such a mechanism greatly increases the concentration of gametes in a local volume of water thereby enhancing the probability of successful fertilization.

Gonadal muscle meshwork has also been found in *Amphiporus bimaculatus, Nipponnemertes magnus,* and at least two Antarctic cratenemertids of undetermined genus. In the latter specimens, however, the muscle fibers are rather fine and delicate, much like those of the muscle meshwork in the intestinal wall (Fig. 30, 31). This type of gonadal wall meshwork musculature may be common among the cratenemertids and has also been found in an amphiporid of indeterminate genus. My speculation is that this character may prove useful at the generic level and possibly even at a family level.

Discussion of enoplan classification problems

Gibson's (1988) suggested classification groups the Monostiliferoidea and Bdellonemertoidea together into the Distromatorhynchocoela as one main branch of the Enopla. His Urichorhynchocoela, as the other main branch, bifurcates into the Pelagonemertoidea and the Herponemertoidea, with the latter comprising the Paramonostilifera, for the families Cratenemertidae and Plectonemertidae, and the Polystilifera as a new name for the Reptantia (*cf.* Fig. 2). I have several major objections to this scheme.

First, Gibson attributes the wide array of differences between the Monostilifera and Bdellonemertea to secondary loss in the latter of numerous characters present in the former. Such features as basic gut plan, foregut arrangement, vascular differences, and absence of armature may represent quite ancient divergences and, if so, suggest that the 'missing' characters may never have evolved in the bdellonemertean line.

Second, inclusion of the Pelagonemertoidea among the Urichorhynchocoela is questionable. This retains the historic association of the Pelagica and Reptantia under the Coe-Brinkmann-Stiasny-Wijnhoff scheme based on commonality of proboscis armature. However, the Archipelagica have layered rhynchocoel musculature and nearly one-third of the Eupelagica are bilayered. The forms comprising the Pelagica have continued to be grouped together on the basis of a complex of shared characters including polystilifery, absence of cerebral organs, absence of nephridia, arrangement of gonads, and general body plan despite the fact that there might be legitimate speculation as to polyphyletic origin.

Third, introduction of the name Herponemertoidea simply replaces one problem with another, since both it and Reptantia mean crawling, creeping or slithering and are indicative of the primitive state of the Nemertea generally and not of any unique feature or features of the groups so designated. This is true whether the crawling condition is recognized under the rubric of patristic similarity (Cain & Harrison, 1960), symplesiomorphy (Hennig, 1966), or primitive similarity (Sneath & Sokal, 1973). Reptantia is legitimized as a descriptive taxonomic name only by its contrast to the Pelagica in its original usage to divide the Polystilifera into two groups.

Fourth, use of the name Polystilifera for what has been customarily designated the Reptantia could perhaps be justified on the grounds that removing Polystilifera as a higher group name frees it for reuse and that it contrasts conveniently with Paramonostilifera. However, this replacement is very undesirable because of the inversion of levels from past usage with the resulting confusion that would ensue.

Fifth, the designation Paramonostilifera makes a definite phylogenetic statement, which seems unwarranted, given that no evidence is adduced to substantiate the argument that monostiliferous proboscis armature has arisen more than once. But, neither has there been presented any solid evidence that rhynchocoel musculature differentiated into different states more than once.

All of this having been said, Gibson's scheme recognized the crucial point that the division of the Hoplonemertea into Monostilifera and Polystilifera as defined by Brinkmann and Stiasny-Wijnhoff is no longer satisfactory in view of the number of similarities between the Cratenemertidae and the Reptantia, contrasts between the Cratenemertidae and most monostiliferous forms in respect of these same characters, and numerous differences between Reptantia and Pelagica. The difference between the Brinkmann-Wijnhoff scheme and the Gibson scheme rests on whether stilifery or rhynchocoel musculature should be regarded as the pre-eminent character and the other as subordinate. The problem for many taxonomists in deciding whether to use one scheme or the other is that of an intuitive estimate of the probabilities attached to each, in the absence of substantive direct evidence.

Sundberg (1990) disagreed with Gibson's

scheme based on a 35-character cladistic analysis which he felt better supported retention of the old scheme. I have three objections to this analysis, as presented. First, a number of the characters selected seem rather arbitrary. Others of equal or greater importance could have been included and some of questionable utility could have been omitted. It must be recognized that in this type of analysis exclusion of a character automatically assigns it an *a priori* weight of zero and inclusion a weight of unity. Second, each family is assigned a single state for each character but, in a number of instances, two or even three character states are present within the family, and the coding scheme does not account for these differences. Also, there are several instances where I regard the character states attributed as erroneous. Third, the *a posteriori* character weighting performed by the computer program may be correct according to the tenets of its methodology but produces results at such variance with our total body of accumulated knowledge that it is difficult to give it credence. (Gibson (1990a) noted other specific objections.) In fairness, Sundberg was working with data which are fragmentary and lacking in needed precision. A future re-analysis, based on a revised set of data, may well prove very helpful, but the present analysis is too problematical to accept as making a case for retention of the old scheme.

Recently Sundberg (1991) proposed eliminating the tribus and cladus designations of Brinkmann and Stiasny-Wijnhoff as being inconsistent with accepted modern usage and proposed raising the established taxa one level, so that the Archi- and Eureptantia and the Archi- and Eupelagica are placed at the subordinal level. I agree that conformity with modern nomenclatural usage is long overdue. Moreover, his proposed elevation of the higher taxa by one level seems reasonable given the growing number of lower taxa and the need for dividing them into more coherent groups at intermediate levels. However, his dropping of the categories of Aequifurcata and Inaequifurcata within the Eureptantia seems unwarranted. His reason was, '...they would naturally have been at the level 'superfamily', which would have required that their names be changed

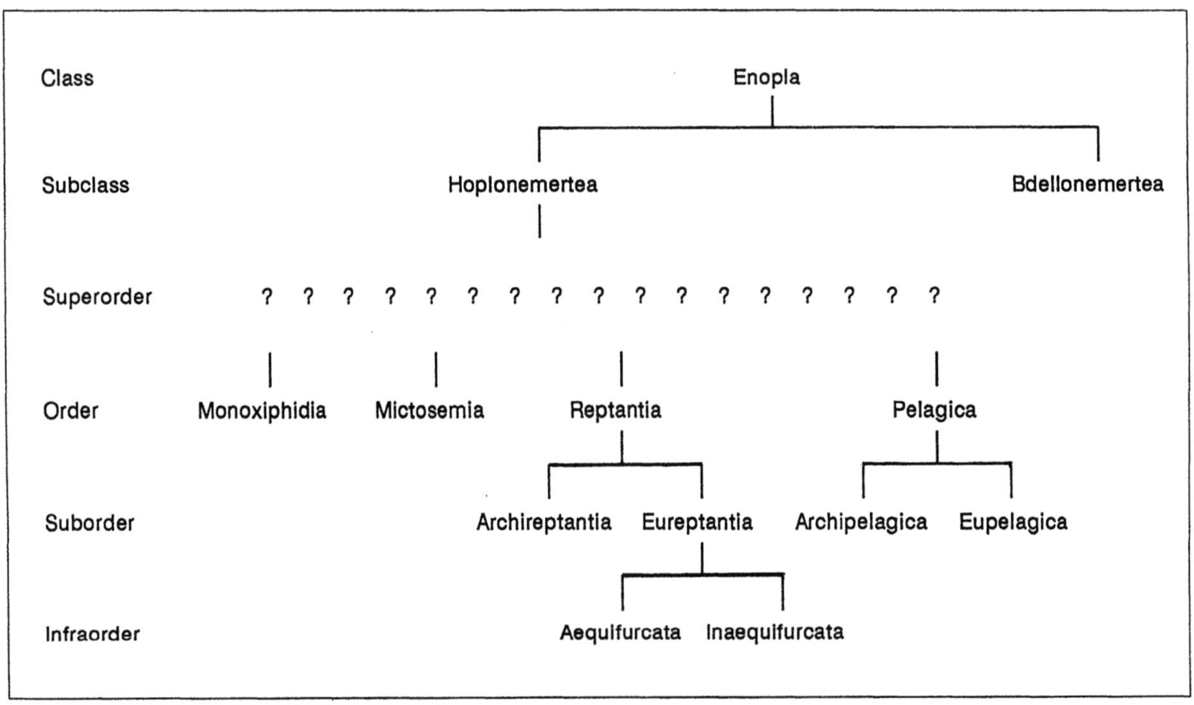

Fig. 32. Revised scheme for classification of the Enopla.

to something derived from the type genera', but the objection is readily overcome by recognizing them at the infraordinal level which seems just as natural as superfamily. For reasons discussed elsewhere, his superorders Monostilifera and Polystilifera are regarded as unwarranted.

How then are we to deal with enoplan classification so that we may proceed in an orderly way with taxonomic work? I believe this can be best accomplished by recognizing internally coherent groupings as secure taxonomic units and arranging them in a hierarchy insofar as we have reasonable grounds for so doing. This allows us to leave open such questions as whether proboscis armature or rhynchocoel musculature should be a pre-eminent classificatory character while still accommodating their use as major characters at lower levels.

First, we must deal with the relationship of the Hoplonemertea and Bdellonemertea. The embryological studies of Iwata and several features of adult morphology provide solid support for regarding the two groups as arising from a common ancestral stock, while the pronounced differences in features of the adult morphology discussed earlier support the argument for a quite ancient divergence of the two lines. On these grounds I would provisionally retain the Bdellonemertea as a subclass of the Enopla with the Hoplonemertea as the other subclass. How Riser's *Arhynchonemertes* relates to the rest of the Enopla is a question that must remain open for the time being. I am inclined to think it a relict form from an even earlier ancestral stock and, though it has certain features in common with the Enopla, it is problematical whether it should be included in that assemblage. However, it might well be used as an outgroup for cladistic analyses of the Enopla and is likely to prove a better candidate in that role than any member of the Anopla.

I see the Hoplonemertea as comprising four distinct and readily identifiable groups at the ordinal level. While I use type of proboscis arma-

Table 1. Some major character similiarities and differences in subgroups of the Enopla.

	Hoplonemertea							Bdellonemertea
	Monoxiphidia	Mictosemia	Reptantia			Pelagica		
			Archireptantia	Eureptantia		Archi pelagica	Eupelagica	
				Inaequifurcata	Aequifurcata			
Stilifery	mono	mono	poly	poly	poly	poly	poly	none
Rhynchocoel:								
Musculature	layered	woven	woven	woven	woven	layered	mixed	layered
Diverticula	no	no	yes	yes	yes	no	no	no
Cerebral organs	yes	yes	yes	yes	yes	no	no	no
Canals branch	no	mixed	mixed	yes	yes	NA	NA	NA
Epithelial types	1	2	mixed	2	1	NA	NA	NA
Gut:								
Straight or sinuous	straight	straight	straight	straight	straight	straight	straight	sinuous
Diverticulate or tubular	divertic.	divertic.	divertic.	divertic.	divertic.	divertic.	divertic.	tubular
Gonads metameric	yes	yes	yes	yes	yes	no	no	no
Nephridia	yes	yes	yes	yes	yes	no	no	different

ture and rhynchocoel musculature as major characters in arriving at these groupings, I shall not use the names Monostilifera, Polystilifera, Distromatorhynchocoela, or Urichorhynchocoela at any level, since they all carry connotations which could lead to much confusion and are better left available should concrete evidence someday be found for resurrecting any of them at the superordinal level. However, for now superorders seem unnecessary. The families Cratenemertidae and Plectonemertidae are separated from both the Reptantia and the old Monostilifera, with which groups they share both contrasts and similiarities, and united under the name Mictosemia (from *miktos* = mixed + *semion* = mark by which a thing is known, distinguishing feature) because of their mixed characters. Since it is necessary to select a new name for the remainder of the monostiliferous forms, they have been designated Monoxiphidia (from *mono* + *xiphidion* = dagger). Retention of the prefix mono- will help to minimize confusion but is not meant to imply that the stilifery character is regarded as more important than rhynchocoel musculature.

The Reptantia and Pelagica as taxonomic groupings remain unchanged. Therefore, these names are retained; although, for reasons noted earlier, Reptantia, as a name, is far from ideal. The subordinate levels under both of these groups are reasonably well understood and defined and are left intact. Sundberg's (1991) classification levels are used except for retention of the Aequifurcata and Inaequifurcata at the infraordinal level. The revised classification scheme I propose is shown in Fig. 32, and similiarities and differences in some, but by no means all, of the major characters are shown in Table 1.

Acknowledgements

I am indebted to: 1) the Division of Worms, U.S. National Museum of Natural History for the use of specimen collections and laboratory facilities, 2) the Smithsonian Oceanographic Sorting Center for access to recently collected specimens, 3) Dr Jon Norenburg for the use of specimens from his personal collection, 4) the British Museum (Natural History) for access to its collections and Dr Alex Muir for his kind hospitality, 5) the Cambridge University Zoology Museum and its Director Dr K. A. Joysey for access to its collections and for the loan of the original Punnet specimen of *Amphiporus magnus* and permission to restain it, 6) Dr Janet Moore of the Cambridge University Zoology Department for: providing access to the Pantin collections in her care, her gracious hospitality, her helpful discussions of the Plectonemertidae, and her careful reading of, and helpful comments on, the manuscript, and 7) the Registry of Tumors in Lower Animals, USNMNH, for the use of outstanding histology and photomicrography facilities.

Special appreciation and thanks are due Prof. Ray Gibson who, in our long collaboration, has been a constant source of enlightenment, stimulus, and encouragement. I am grateful to Drs Janet Moore, Jon Norenburg, Per Sundberg and Prof Nathan W. Riser for many discussions which have helped to focus my thinking, in a variety of ways, on issues addressed in this paper. However, the views expressed are peculiarly my own; I am sure these colleagues, and no doubt many others, will find reservations and disagreements with various conclusions reached. This is contribution number 91-04 of the Turkey Run Research Institute.

References

Berg, G., 1972. Studies on *Nipponnemertes* Friedrich, 1968 (Nemertini, Hoplonemertini). I. Redescription of *Nipponnemertes pulcher* (Johnston, 1837) with special reference to intraspecific variation of the taxonomical characters. Zool. Scr., 1: 211–225.

Berg, G., 1985. Studies on *Nipponnemertes* Friedrich (Nemertini, Hoplonemertini). II. Taxonomy of *Nipponnemertes pulcher* (Johnston) and some other species. Zool. Scr., 14: 239–246.

Brinkmann, A., 1914–15. *Uniporus*, ein neues Genus der Familie Drepanophoridae Verrill. Bergens Mus. Aarbok, No. 6: 1–29.

Brinkmann, A., 1917. Die pelagischen Nemertinen. Bergens Mus. Skr. ny Raek., 3, 1: 1–194.

Bürger, O., 1891. Nemertinen des Golfes von Neapel. Nachr. Königl. Gesellsch. Wiss. Gottingen. aus 1891: 286–301.

Burger, O., 1895. Die Nemertinen des Golfes von Neapel und der angrenzenden Meeres-Abschnitte. Fauna Flora Golf. Neapel, 22: 1–743.

Bürger, O., 1897–1907. Nemertini. In: Bronn's Klassen und Ordnungen des Tierreichs, Bd. 4, Supplement, Leipzig.

Bürger, O., 1904. Nemertini. Das Tierreich, 20: 1–151.

Cain, A. J. & G. A. Harrison, 1960. Phyletic weighting. Proc. zool. Soc. Lond. 135: 1–31.

Carus, J. V., 1885. Prodromus faunae Mediterraneae sive descripto annimalium maris Mediterranei incolarum quam comparata silva rerum quatenus innotuit adiectis locis et nominibus vulgaribus corumque auctoribus in commodum zoologorum. Vol. I. Coelenterata, Echinodermata, Vermes, Arthropoda. E. Schweizerbart'sche Verlagshandlung (E. Koch), Stuttgart.

Coe, W. R., 1901. Papers from the Harriman Alaska Expedition. XX. The Nemerteans. Proc. Wash. acad. Sci. 3: 1–110.

Coe, W. R., 1904. Nemerteans of the Pacific Coast of North America. Part II. Harriman Alaska ser. 11: 111–220.

Coe, W. R., 1905. Nemerteans of the west and northwest coasts of America. Bull. Mus. Comp. Zool. Harvard, 47: 1–318.

Coe W. R., 1926. The pelagic nemerteans. Mem. Mus. comp. Zool. Harv. 49: 1–244.

Coe, W. R., 1943. Biology of the nemerteans of the Atlantic coast of North America. Trans. Conn. Acad. Arts Sci. 35: 129–328.

Crandall, F. B., 1993. The genus *Valdivianemertes* Stiasny-Wijnhoff 1923 (Nemertea, Enopla, Hoplonemertea): nomenclatural status and proper systematic position. Hydrobiologia 266: 175–184.

Friedrich, H., 1935. Studien zur Morphologie, Systematik und Okologie der Nemertinen der Kieler Bucht. Arch. Naturgesch., 4: 293–375.

Friedrich, H., 1940. Einige neue Hoplonemertinen aus der Ostsee. Kieler Meeresforsch. 3: 233–251.

Friedrich, H., 1955. Beitrage zu einer Synopsis der Gattungen der Nemertini monostilifera nebst Bestimmungsschlussel. Z. wiss. Zool., 158: 133–192.

Friedrich, H., 1957. Beitrage zur kenntnis der Arktischen Hoplonemertinen. Vidensk. Meddr dansk naturh. Foren. 119: 129–154.

Friedrich, H., 1968. *Sagaminemertes*, eine bemerkenswerte neue Gattung der Hoplonemertinen und ihre systematische Stellung. Zool. Anz. 180: 33–36.

Friedrich, H., 1970. Nemertinen aus Chile. Sarsia 40: 1–80.

Gering, G., 1912. Neue Nemertinen der schwedischen Westkuste. Zool. Anz. 39: 520–523.

Gering, G., 1913. Neue Nemertinen der schwedischen Westkuste. Zool. Jb. Abt. Syst. 34: 187–202.

Gibson, R., 1972. Nemerteans. London, Hutchinson, 224 pp.

Gibson, R., 1982. Nemertea. In: Synopsis and Classification of Living Organisms, ed. S. P. Parker, Vol. 1, pp 823–846. New York, McGraw-Hill.

Gibson, R., 1988. Evolutionary relationships between mono- and polystiliferous hoplonemerteans: *Nipponnemertes* (Cratenemertidae), a 'missing link' genus? Hydrobiologia 156: 61–74.

Gibson, R., 1989. The macrobenthic nemertean fauna of Hong Kong. In: Morton, B. S. (ed.) Proceedings of the Second International Marine Biological Workshop: The Marine Flora and Fauna of Hong Kong and Southern China. Hong Kong Univ. Press, Hong Kong Vol. 1.

Gibson, R., 1990a. Classification of the Nemertea Enopla: response to Sundberg's critique and cladistic analysis. Zool. Scr. 19: 141–142.

Gibson, R., 1990b. The macrobenthic nemertean fauna of the Albany region, Western Australia. In: Proceedings of the Third International Marine Biological Workshop: The Marine Flora and Fauna of Albany, Western Australia, eds. F. E. Wells, D. I. Walker, H. Kirkman, and R. Lethbridge, Vol. I, pp 89–194. Western Australian Museum, Perth.

Gibson, R. & F. B. Crandall, 1989. The genus *Amphiporus* Ehrenberg (Nemertea, Enopla, Monostiliferoidea). Zool. Scr., 18: 453–470.

Gibson, R. & F. B. Crandall, 1991. *Amphiporus* Ehrenberg (Nemertea): Proposed designation of *Planaria lactiflorea* Johnston, 1828 as the type species. Bull. zool. Nomenclature, 48: 22–24.

Gibson, R., J. Moore & F. B. Crandall, 1982. A new semiterrestrial nemertean from California. J. Zool., Lond. 196: 463–474.

Hennig, W., 1966. Phylogenetic systematics. University of Illinois Press, Urbana, 263 pp.

Hubrecht, A. A. W., 1879. The genera of European nemerteans critically revised, with description of several new species. Notes Leyden Mus. 1: 193–232.

Iwata, F., 1954. Some nemerteans from the coasts of the Kii Peninsula. Publs Seto mar. biol. Lab. 4: 33–42.

Iwata, F., 1960. Studies on the comparative embryology of nemerteans with special reference to their interrelationships. Publs Akkeshi mar. biol. Stn No. 10: 1–51.

Iwata, F., 1972. Axial changes in the nemertean egg and embryo during development and its phylogenetic significance. J. Zool., Lond. 168: 521–526.

Iwata, F., 1985. Foregut formation of the nemerteans and its role in nemertean systematics. Am. Zool. 25: 23–36.

Jespersen, A. & J. Lützen, 1988. Ultrastructure and morphological interpretation of the circulatory system of nemerteans (Phylum Rhynchocoela). Vidensk. Meddr dansk naturh. Foren. 147: 47–66.

Johnston, G., 1827–28. Contributions to the British fauna. Zool. J. 3: 321–336, 486–491.

Kirsteuer, E., 1963. Beitrag zur Kenntnis der Systematik und Anatomie der adriatischen Nemertinen (Genera *Tetrastemma*, *Oerstedia*, *Oerstediella*). Zool. Jb. Abt. Anat. 80: 555–616.

Kirsteuer, E., 1967. New marine nemerteans from Nossi Be, Madagascar. Results of the Austrian Indo-Westpacific Expedition 1959/60 Part VIII. Zool. Anz. 178: 110–122.

Kirsteuer, E., 1973. A new polystiliferous hoplonemertean,

Curranemertes natans gen. et sp.n., from the Caribbean Sea (Nemertina, Polystilifera Reptantia). Zool. Scr., 2: 125–140.

Kirsteuer, E., 1974. Description of *Poseidonemertes caribensis sp.* n., and discussion of other taxa of Hoplonemertini Monostilifera with divided longitudinal musculature in the body wall. Zool. Scr. 3: 153–166.

Korotkevich, V. S., 1955. Pelagicheskie nemertinui dalnevostochnuikh morei SSSR. Izdav. Zool. Inst. Akad. Nauk S.S.S.R. 58: 1–131.

Korotkevich, V. S., 1964. Pelagic nemerteans of Antarctic and temperate waters of the southern hemisphere. Biol. Rep. Sov. Antartic Exped. 1955–58, 2: 134–172.

McIntosh, W. C., 1873–74. A Monograph of the British Annelids. Part I, Nemerteans. London, Ray Society. 1873–74.

Montgomery, T. H., 1897. Descriptions of new metanemerteans with notes on other species. Zool. Jahrb., Abt. Syst. 10: 1–14.

Moore, J. & R. Gibson, 1981. The *Geonemertes* problem (Nemertea). J. Zool., Lond. 194: 175–201.

Moore, J. & R. Gibson, 1985. The evolution and comparative physiology of terrestrial and freshwater nemerteans. Biol. Rev. 60: 257–312.

Moore, J. & R. Gibson, 1988a. Marine relatives of terrestrial nemerteans: the genus *Prosadenoporus* Bürger, 1890 (Hoplonemertea). Hydrobiologia 156: 75–86.

Moore, J. & R. Gibson, 1988b. Further studies on the evolution of land and freshwater nemerteans: generic relationships among the paramonostiliferous taxa. J. Zool., Lond. 216: 1–20.

Norenburg, J., 1986. Redescription of a brooding nemertine, *Cyanophthalma obscura* (Schultze) gen. et Comb. n., with observations on its biology and discussion of the species of Prostomatella and related taxa. Zool. Scr. 15: 275–293.

Pantin, C. F. A., 1969. The genus *Geonemertes*. Bull. Br. Mus. nat. Hist., Zool. 18: 263–310.

Riser, N. W., 1988. *Arhynchonemertes axi* gen.n., sp.n. (Nemertinea) – an insight into basic bilaterian organology. Fortschritte Zool. 36: 367–373.

Riser, N. W., 1989. Speciation and time-relationships of the nemertines to the acoelomate metazoan bilateria. Bull. mar. Sci. 45: 531–538.

Schultze, M., 1851. Beiträge zur Naturgeschichte der Turbellarien. Greifswald, C. A. Koch, 78 pp.

Schultze, M., 1853. Zoologische Skizzen. Z. wiss. Zool. 4: 178–195.

Sneath, P. H. A. & R. R. Sokal, 1973. Numerical taxonomy: the principles and practice of numerical classification. San Francisco, W. H. Freeman, 573 pp.

Stiasny-Wijnhoff, G., 1923. On Brinkmann's system of the Nemertea Enopla and *Siboganemertes weberi* n. gen. n. sp. Q. J. Microsc. Sci. 67: 627–669.

Stiasny-Wijnhoff, G., 1926. The Nemertea Polystilifera of Naples. Pubbl. Staz. zool. Napoli 7: 119–168.

Stiasny-Wijnhoff, G., 1936. Die Polystilifera der Siboga expedition. Monographie XXII aus: Uitkomsten op Zoologisch, Botanisch, Oceanographisch en Geologisch gebied verzameld in Nederlandsch Oost-Indie 1899–1900. Leiden, E. J. Brill. 1936.

Stricker, S. A., 1985. The stylet apparatus of monostiliferous hoplonemerteans. Am. Zool. 25: 87–97.

Sundberg, P., 1979. Statistical analysis of variation in characters in *Tetrastemma laminariae* (Nemertini) with a redescription of the species. J. Zool., Lond. 189: 39–56.

Sundberg, P., 1990. Gibson's reclassification of the enoplan nemerteans (Enopla, Nemertea): a critique and cladistic analysis. Zool. Scr. 19: 133–140.

Sundberg, P., 1991. A proposal for renaming the higher taxonomic categories in the phylum Nemertea. J. Nat. Hist. 25: 45–48.

Verrill, A. E., 1892. The marine nemerteans of New England and adjacent waters. Trans. Conn. acad. Arts Sci., 8: 382–456.

Wijnhoff, G., 1912. Die systematik der Nemertinen. Zool. Anz. 40: 337–341.

Wijnhoff, G., 1914. The proboscidian system in nemertines. Q. J. Microsc. Sci., 60, Part II: 273–312.

Hydrobiologia **266**: 141–157, 1993.
R. Gibson, J. Moore & P. Sundberg (eds), Advances in Nemertean Biology

Observations on the morphology of some North American nemertines with consequent taxonomic changes and a reassessment of the architectonics of the phylum

Nathan W. Riser
Marine Science Center, Northeastern University, Nahant, Massachusetts 01908, USA

Key words: Nemertea, North America, taxonomy, *Tenuilineus*, *Parvicirrus*, *Tarrhomyos*

Abstract

A majority of the nemertine species from the western North Atlantic were originally described from life in the nineteenth century. Many of these were established by A. E. Verrill who had 'an eye for species' no matter which phylum he was working with, and thus when living nemertines which he described are encountered, they can usually be recognized. The morphology of most of these species has never been reported; some may prove to be species described earlier from the eastern North Atlantic. Morphological observations on the nemertine species from the North East coast of the Pacific are inadequate and have prevented satisfactory comparison with species from Japan and eastern Russia. The morphology of some species collected in the Gulf of Maine and from Cape Cod, as well as a re-examination of some of the slides of type specimens of species from the west coast of North America, indicates that their generic placement must be re-examined. The architectonic plan of the heteronemertines postulated over 100 years ago can not be substantiated and is redescribed. Three new heteronemertean genera are described: *Tenuilineus* gen.n. *Parvicirrus* gen.n., *Tarrhomyos* gen.n.

Introduction

The present study began in 1950 in an effort to establish the morphology of New England nemertines so that preserved specimens could be identified. Satisfactory results were never obtained, but the morphology of the animals resolved a number of puzzling questions. Details of the anatomy of most nemertine species from the north east coast of the Pacific are inadequate and an examination of some of the slides of type specimens from the west coast of North America indicate that generic placement of some of these species must be reassessed. Specimens of *Lineus longissimus* (Gunnerus 1770) and *Micrura fasciolata* (Ehrenberg, 1831) from Spain were furnished by Drs N. Anadón and G. Vernet for sectioning and staining in order to determine generic parameters. Cultures of *Lineus lacteus* (Rathke, 1843), *L. pseudolacteus* Gontcharoff, 1951, *L. nigricans* (Bürger, 1892) and *L. sanguineus* (Rathke, 1799) from Dr J. Bierne and of *L. vegetus* (Coe, 1931) from Dr P. Roe were obtained for comparative studies, as was a single specimen of *L. bilineatus* (Renier, 1804) from the Isle of Man courtesy of A. Rogers.

The terms Lineidae and lineid are used in this text *sensu* Gibson (1982), *i.e.*, with the single autapomorphy of paired, lateral horizontal, cephalic furrows. Quotation marks around generic names indicate that the species does not belong in that genus and that generic placement is uncertain at the present time.

Observations

Amphiporus angulatus *(Müller, 1774)*
Gibson & Crandall (1989) considered the name of this species to be a *nomen nudum*, but accepted *Amphiporus angulatus sensu* Coe, 1901 as a *species inquirenda*. I agree with Verrill (1892) that the description of specimens from Greenland furnished to Müller by Fabricius is readily identifiable as this very common species in the Gulf of Maine and that the figure of the head, which Stimpson furnished to Girard for the description of *Omatoplea stimpsonii* Girard (1855), is an excellent depiction of the head of *A. angulatus* as seen with the naked eye or with the aid of a hand lense. The question of the authorship of the name, Müller or Fabricus, remains open. Well anaesthetized specimens of this species and of *Cyanophthalma cordiceps* (Jensen, 1878) are very flaccid and the relationship of the organs within them is as confusing as in contracted specimens. The description of *Amphiporus hastatus* McIntosh, 1873–74 by Oudemans (1885) could apply to either of these species. He referred to sections 1/40 mm thick and questioned many of his own observations. I would expect that re-examination of *A. hastatus* would show the internal anatomy to be similar to that of *A. angulatus* and *C. cordiceps*. The frontal glands of these two species are massively developed and occupy most of the space in the parenchyme lateral to the rhynchodaeum. The remaining space is packed with submuscular mucous cells (which stain less intensely than the frontal mucous glands) intermingled with longitudinal muscle bundles. The longitudinal muscles divide prior to the proboscideal insertion with the larger group of bundles continuing along the rhynchodaeum after giving off the proboscideal longitudinals. Figure 19 of Coe (1901) of the histology of the head is totally misleading. Submuscular mucous cells become more isolated behind the brain. They, like the frontal glands, are composite glands. Structures within this glandular mass are moved about during fixation (and in life) so that the cerebral organs may come to lie in front of, beside, or behind the brain. The anterior loop of the cephalic blood vessels is obvious if the apex of the body is not contracted straight back, but where the two vessels lie in the glandular tissue, they become contorted so that 1–5 sections of the vessel may be present on each side in a cross-section. In extreme situations, this gives the appearance of a plexus however, the appearance of the two sides meeting dorsally and/or ventrally has never been encountered, and Oudemans description of the blood vessels in the head of *A. hastatus* only postulates such a plexus since he states that he could not clearly see nor follow the system.

Amphiporus angulatus does not belong to the genus *Amphiporus*, and while it could be readily transferred to the genus *Cyanophthalma*, the potential value of certain morphological characters indicate that such a transfer should not be made until the morphology of *A. hastatus*, *A. nebulosus*, *A. gelatinosus* Coe, 1905, and other species of the 'hastatus' group with extensive development of frontal and submuscular glands is known.

Cyanophthalma cordiceps *(Jensen, 1878)*
L. Brunberg loaned material from Danish waters to compare with the *T. vittatum* material which I had collected in the Gulf of Maine. This allowed for the synonymizing of the two species by Norenburg (1986) and helped clarify some of the confusion in the descriptions by Friedrich (1933, 1935) and Brunberg (1964). The nephridial tissue is abundant in the parenchyme dorsally and ventrally to the nerve cords. The spread of the nephridial tissue in Friedrich, 1935, Abb. 4, indicates that his material was very strongly contracted. Efforts to duplicate Brunberg's figure (1964, Fig. 14) of the 'excretory system' were unsuccessful and optical sections of cleared whole mounts showed that the submuscular mucous glands produced comparable tracings. Friedrich (1933) described in detail the necks of the submuscular glands but did not recognize the glands as such and thought that they might be primordia

of gonads or have some relationship to the nephridial system. He referred to the glandular mass in the precerebral region as parenchyme and stated that if muscle fibers were present, they were not visible. In contracted specimens, nephric tissue may extend anterior to the brain. There are 16 nerves on the proboscideal plexus in which the fibers run circumferentially. Single cell bodies occur at intervals in the inner and outer halves of the nerves, the outer sending processes to the epithelium and the inner to the inner musculature of the proboscis.

Amphiporus ochraceus *(Verrill, 1873)*
Corrêa (1961) referred to the frontal glands ('cephalic glands' *sic*) as small. They appear as three composite units occupying about 1/3 of the area in mid-dorsal line above the proboscideal system. The nephric system is well developed and extends from a short distance behind the brain to the beginning of the intestine with one nephric duct passing to the body surface on either side near the middle of the anterior half of the system. Eleven proboscideal nerves reported by Corrêa (1961) in her description of Florida specimens, has been a constant number in the specimens which I have examined from Cape Cod.

Amphiporus frontalis *(Verrill, 1892)*
Three specimens were obtained from sand near the MLT line at Blacks Harbor, New Brunswick, Canada, 17 March, 1980. The eye pattern and color conformed to the description and Fig. 1, Pl. 34 of Verrill (1892). The largest specimen was slightly more than 8 cm long while gliding, but contracted to 5.5 cm after fixation. Sections showed it to be filled with tubules 6 μm in diameter that were initially assumed to be blood vessels since they had endothelial linings. Their meanderings filled all of the space between the longitudinal muscles and the internal organs. It was concluded that these tubules had been left behind by some parasite when the second specimen was sectioned and lacked these passages. The large specimen had degenerated ovaries, and the second specimen was immature. Neither series of sections permit a complete or satisfactory

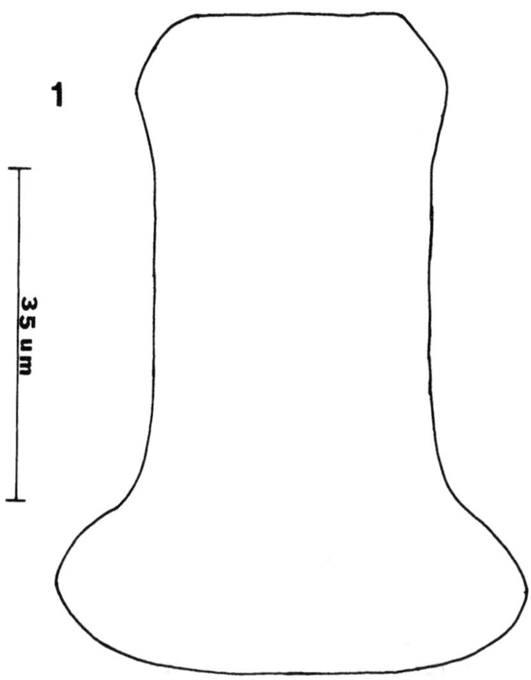

Fig. 1. Amphiporus frontalis. Basis of central stylet.

description. Stereocilia (bristles) at the apex of the head (Fig. 2) surround the frontal organ. The frontal glands are very small and extend a very short distance, back into head. The cerebral organs are large and contact the brain, reaching slightly behind in one specimen. The longitudinal muscles of the body wall divide in the brain region. Widely spaced fibers from the inner group form the proboscis insertion and a few continue forward along the ventral wall of the rhynchodaeum. The rhynchocoel extends almost to the posterior end of the body. The muscular wall of the proboscis is very thin and contains 11 proboscideal nerves. The basis (Fig. 1) of the central stylet has an unusual flair at its base where it is 5.5 μm wide and rapidly narrows for most of the length to 3 μm. The basis flairs again but not as markedly, about 4 μm wide at the top. The st/so ratio is 0.72. There are two accessory stylet pouches each with 3–5 accessory stylets. An intestinal diverticulum extends anteriorly almost to the brain on either side of the foregut. The nephridial system is restricted to a very short preintestinal area and is not extensively developed.

The eye pattern of the living animals and the

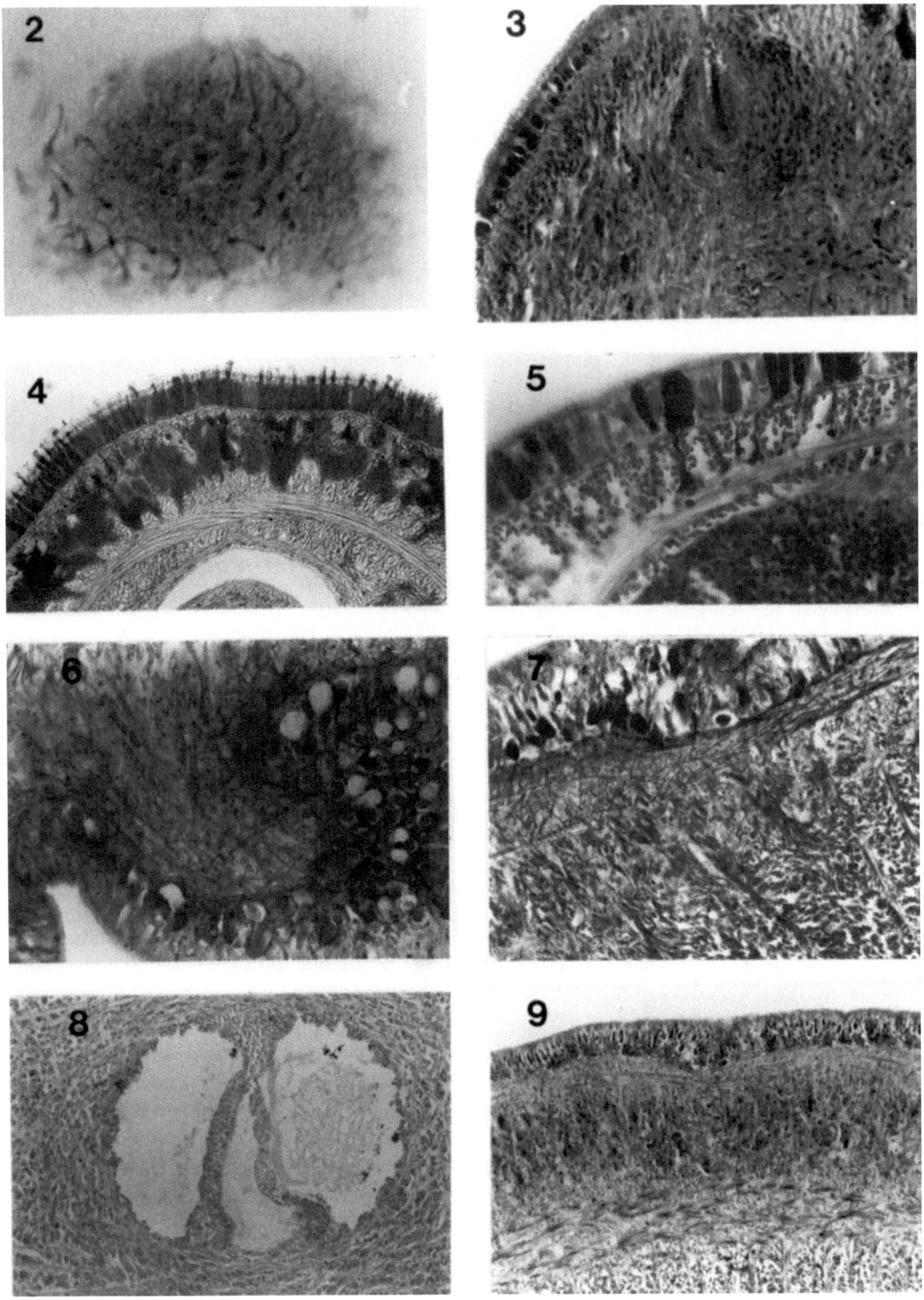
2
3
4
5
6
7
8
9

morphology of the stylet apparatus are the only features at the present time to distinguish the species.

Tenuilineus gen.n.
syn. *Lineus* Verrill 1892, *pro parte, Coe 1895, 1943, pro parte*

Lineidae *sensu* Gibson, 1982 with single dorsal/dorsolateral cephalic blood lacuna, subepidermal gland cells pass between longitudinal muscles; longitudinal muscles between rhynchocoel and foregut; subepithelial esophageal glands present, longitudinal muscles absent; transition to heterotype proboscis musculature, outer longitudinal muscles present for short region behind proboscis insertion; circular muscle cylinder around rhynchodaeal complex; pilidium larva in life-history; caudal cirrus absent.

Type species: *Tenuilineus bicolor* (Verrill, 1892) gen.n., comb.n. (USNM # 157042)

Etymology: L.; *tenuis*, thin; *linea*, line

Holotype: U.S.N.M. # 157042

Tenuilineus bicolor (Verrill, 1892) gen.n., comb.n.
syn. *Lineus bicolor* Verrill 1892:426; Coe 1895:499 1943:239, Riser 1991:435

Specimens were obtained from rubble dredged at 10 m depth in Buzzards Bay, near Woods Hole, Mass.

Apical sensory organs are absent. The subepidermal glands are not restricted by connective tissue and reach the internal circular muscle layer. Anteriorly the subepidermal glands crowd together so that almost all of the longitudinal muscle bundles are adjacent to the subepidermal circular muscles (Fig. 4). Further posteriorly, some longitudinal bundles abut the inner circular muscles and continuing posteriorly there is an irregular alternation with some individual muscle bundles extending between the two groups of circulars while other bundles are present in a subepidermal position. Ventrally and posteriorly, the subepidermal glands are few in number and the space between them is occupied by individual (*i.e.* single not fragmented) longitudinal bundles (Fig. 5). The nephric system extends on each side as a single tubule from the origin of the intestine to the region of the first gonads intimately bound to the intestine with each exiting at the posterior end of the system as a single efferent duct slightly dorsal to the lateral nerve cords. Coe (1943) briefly described the embryology and figured the pilidium larva, a life-history feature which along with the morphological characters herein noted, distinguish *Tenuilineus* from *Lineus*. Other species -- *Tenuilineus albocinctus* (Bergendal, 1903) n. com.; The description of *L. rubescens* by Coe (1905) and an examination of the type slides indicate that it too may belong to *Tenuilineus*. No significant morphological features distinguish *T. bicolor* from the descriptions of *T. albocinctus* by Brunberg (1964) and Cantell (1975).

Micrura affinis (Girard, 1855)
syn.: *Poseidon affinis* Girard, 1855:28.
Nemertes affinis Verrill, 1874a:39, 1874b:412.
Micrura affinis Verrill, 1879:186, 1892:428; Coe, 1895:492, 1943:248–249; Riser, 1990:604.

A large number of specimens has been obtained by dredging in Massachusetts Bay, and from beneath test panels and in the bryozoan turf on Ammen Rock, Cashes Ledge, at 31–33 m in the Gulf of Maine with the aid of divers. All living individuals were bright red with a cream colored ventral surface except males in July and August

Fig. 2. Amphiporus frontalis. Stereocilia around opening of frontal organ.
Fig. 3. Micrura affinis. Longitudinal section through one apical sensory organ.
Figs 4–5. Tenuilineus bicolor. 4. Cross section through anterior dorsal body wall. 5. Cross section through posterior body wall.
Fig. 6–7. Tarrhomyos luridus. 6. Oblique section through fold in body wall. 7. cross section through body wall.
Figs 8–9. Cerebratulus marginatus. 8. Cross section through rhynchodaeum. 9. Cross section through epidermis and adjacent tissue abutting the outer longitudinal muscles.

at which time, their dorsum is duller and the ventral surface dirty grey. Other color phases reported by Verrill (1892) and Coe (1943) were not encountered and may refer to specimens obtained south of Cape Cod. Specimens lacking gonads usually bear annuli around the body, but these do not differ in color from the rest of the body. The caudal cirrus is colorless, and adhesive as well as contractile. The tip of the head is similar to Fig. 5B of *M. fasciolata* Cantell (1975) except that the posterior intrusion of white is distinct and the two large eyes are at the tips of the two anterior red extensions. The smaller eyes are masked by the red pigment and require manipulation of the specimen in order to be seen. Comparison with a small specimen of *M. fasciolata* from Spain showed no morphological differences that could not be attributed to size or degree of contraction thus, the two species are only distinguishable on the basis of the color pattern of living specimens, a questionable character considering the variation reported for *M. fasciolata*. The apical sensory organs of *M. affinis* are very large (Fig. 3), almost as large as those on the sections of the type specimen of *Cerebratulus longiceps* Coe, 1905. Subepithelial esophageal muscles are absent and a weakly developed longitudinal muscle plate occurs beneath the rhynchocoel. The specimens from beneath the test panels at Ammens Rock were routinely associated with *Nipponemertes pulcher* (Johnston, 1837) and occasionally with *Amphiporus angulatus*.

Parvicirrus gen.n.
syn. *Lineus* Verrill 1879, 1892, *pro parte; Coe 1943, pro parte*

Lineidae sensu Gibson, 1982 with rhynchodaeum attached dorsally and ventrally producing paired rhynchodaeal blood lacunae; subepidermal glands penetrate between OLM bundles; proboscis with three muscle layers, ILM, CM, OLM; longitudinal muscle plate beneath rhynchocoel; subepithelial esophageal glands and longitudinal, caudal cirrus present; pilidium larva in life history.

Type species: *Parvicirrus dubius* (Verrill, 1879) gen.n., comb.n. (USNM # 157040)
Etymology: L.; *parvus*, small; *cirrus*, filament.
Holotype: U.S.N.M. # 157040

Parvicirrus dubius (Verrill, 1879) gen.n., comb.n.
syn. *Lineus dubius* Verrill 1879:186, 1892:426; Coe 1943:241, Riser 1987:21

I originally encountered this species under stones along the coast of southern Maine and have collected specimens from Cape Cod, Massachusetts to Black's Harbor, New Brunswick, Canada. It looked (and looks) like a small, soft, bleached variety of *Lineus viridis*. The white eyes, and presence of a caudal cirrus (not observed by Verrill) are distinctive. The subepidermal mucous glands do not penetrate between the bundles of the OLM, but the serous glands do but not to the ICM. The excretory system begins immediately behind the mouth and ends at the origin of the intestine with a single posterior efferent duct on either side. The outer longitudinal muscle layer of the proboscis forms a thin and frequently disrupted ring. The two muscle crosses are thin. Neurochord cells are present associated with the ventral ganglia. Specimens collected in July are ripe and eggs obtained by cutting females can be fertilized with diluted sperm from cut males. Late gastrulae break out of the membranes and swim by ciliary beat. By twenty-four hours, a small compact pilidium is attained.

One specimen was obtained by dredge at 16 m depth in Buzzards Bay, Massachusetts. All other specimens were collected intertidally beneath stones or in clean coarse sand. Except for those collected in moving intertidal sand dunes (Riser, 1987) there has been no regularity in encountering individuals at any given locale, and rarely have more than five been obtained at any given time.

Coe (pers. com.) never saw this species and based upon partial morphological details which I furnished to him, concluded that it belonged in the genus *Micrura*. It differs from *Micrura* in the penetration of the OLM by subepidermal glands, connective tissue enclosing the brain, presence of subepithelial longitudinal esophageal muscles, and the musculature of the proboscis.

Lineus socialis (Leidy, 1855)

Apical sensory organs are present. The rhynchodaeum is not attached dorsally, thus paired rhynchodaeal blood vessels are not formed. Longitudinal muscles are absent beneath the rhynchocoel. Longitudinal muscles are present beneath the epithelium of all surfaces of the esophagus and circular muscle bands are well developed especially ventrally. Subepidermal glands penetrate between the longitudinal muscles, frequently to the nerve plexus around the ICM. Esophageal glands are well developed and under certain conditions of body contraction extend laterally to the ILM obliterating the fore-gut blood plexus, while in other degrees of body contraction, the plexus is open. The nephridial system tends to be in the wall of the two large blood sinuses to either side of the rhynchocoel in the fore-gut region. Dorso-ventral muscles are absent from the intestinal region of the body.

Specimens of *L. socialis*, *L. nigricans*, *L. sanguineus*, and *L. vegetus* have been maintained for several years on a diet of *Mytilus* gonad/digestive gland. All have the same appearance at the present time. Sections of specimens of each of these species as well as of a fissiparous form from Kaikoura, New Zealand (the most frequently encountered, and most abundant nemertine beneath mid-intertidal stones there) show no differences. (Some specimens of each of these populations have had a few subepidermal glands penetrating through all of the muscle layers which I interpret to mean that the individuals are from fragments with regeneration of the anterior end.) This group does not belong to the genus *Lineus* which is characterized by paired rhynchodaeal blood vessels, subepidermal glands separated from OLM by a thick layer of connective tissue and a longitudinal muscle plate beneath the rhynchocoel. Furthermore, in the absence of any distinguishing species characters, it is my opinion that they (including the New Zealand forms) constitute a single species, *Lineus sanguineus* as Bierne *et al.*, (1993) have indicated based upon other characters. A fissiparous species which undergoes fragmentation under adverse conditions and in which fragments can form cysts could easily be transported all over the world as part of the fouling community on ships and floating debris.

Tarrhomyos gen.n.
syn. *Meckelia* Verrill 1873, *pro parte; Cerebratulus* Coe 1895, 1943, *pro parte*

Lineidae *sensu* Gibson, 1982 with autapomorphies of – three apical sensory organs; paired lateral rhynchodaeal blood lacunae; longitudinal muscles between rhynchocoel and foregut; subepithelial esophageal glands and longitudinal muscles present; proboscis with outer longitudinal, circular and inner longitudinal muscles with two muscle crosses; neur ochord cells and caudal cirrus present: with synapomorphies of – preoral diagonal muscles; subepidermal gland cells penetrate between outer-longitudinal muscle bundles; vascular plexus in foregut region; a few longitudinal muscle bundles in rhynchocoelic circular muscles anteriorly.

Type species: *Tarrhomyos luridus* (Verrill, 1873) gen.n., comb.n. (USNM # 157041)

Etymology: Gr.; *tarrhos*, interwoven; *myos*, muscle

Holotype: U.S.N.M. # 157041

Tarrhomyos luridus (Verrill, 1873) gen.n., comb.n.
syn. *Meckelia lurida* Verrill 1873:508, 630.
Cerebratulus luridus Coe 1895:497, 1943:255.
for early references see Bürger 1904:114

A large number of specimens were collected along with the holothuroid, *Molpadia oolitica* from the muddy soft-bottom community dominated in biomass by a sponge of the genus *Suberites* at 20–25 m depth in Cape Cod Bay, Massachusetts. Major characters are listed in Table 1. The subepidermal circular muscles are not concentric, but instead are cross-hatched (Fig. 6), and are inter-

Table 1. Morphological features of some North American heteronemerteans. (*Lineus longissimus* from Spain included for comparison.) (Names used are those *in lit.* and do not imply acceptance of generic placement.)

Genus and species	A	B	C	D	E	F	G	H	I	J	K	L	M
Cerebratulus													
marginatus	N	+	+	+	T	+	+	+	+	+	+	3	2
lacteus	N	+	+	+	T	+	+	+	+	+	+	3	2
melanops	N	+	+	+	T	+	+	+	+	+	+	2C	2C
montgomeryi	C			+		+						2C	
occidentalis	C			+	+	–	+	–	+	–	+		
herculeus	C		+C					–	+	–			
longiceps	C	+		+		–		–		–	–	3C	2C
signatus	C	+		+		–		–		–	+	2	2C
Tarrhomyos n.g.													
luridus	N	+	+	+	+	+	+	+	+	–	+	3	2
Parvicirrus n.g.													
dubius	N	+	+	+	+	–	+	–	+	–	+	3	2
Tenuilineus n.g.													
bicolor	N		–	–	+	–	+	–	+	–	+	tr	
Micrura													
affinis	N	+	+	–	–	–	–	–	+	–	+	2	2
alaskensis	C		+C	–	–	–	–	–	+	–	+		
pardalis	C	+		–	–	–	–	–	+	–	+		
nebulosa	C	+		–	–	–	–	–	+	–	+		
Lineus													
longissimus		+	–	+	T	–	+	–	+	–	+	2	2
flavescens	C		–	+		–	+	–	+	–	+	tr	
ruber	N	+	–	–	–	–	+	–	+	–	+	2	2
viridis	N	+	–	–	–	–	+	–	+	–	+	2	2
torquatus	C		–	–	–	–	+	–	+	–	+	2	
wilsoni	C	+	–	–	–	–	+	–	+	–	+		
socialis	N	+	–	–	+	–	+	–	–	–	+	2	
pictifrons	C		–	–	–	–	+	–	–	–	+	tr	
rubescens	C		–	–	+	–	+	–	+	–	+	tr	
albolineatus	C		–	–	+	–	+	–	+	–	+	tr	2

A Source – personally collected in New England (N); Holotype slides of Coe (C).
B Apical sensory organs present (+) or absent (–).
C Caudal cirrus present (+) or absent (–) or C according to Coe.
D Two lateral rhynchodaeal blood vessels present (+) or absent (–).
E Subepidermal glands penetrate outer longitudinal muscles (+) or are restricted (–), or restricted by a distinct CT layer (T).
F Diagonal muscles present in subepidermal gland zone (+) or absent (–).
G Connective tissue encloses brain (+) or ganglion cells ramify amongst surrounding tissues (–).
H Longitudinal and circular muscles of anterior rhynchocoel wall interwoven (+) or separate (–).
I Longitudinal muscles between rhynchocoel and foregut (+) or absent (–).
J Blood vessels in foregut parenchyme (+) or an open sinusoidal plexus (–).
K Subepithelial glands in foregut (+) or absent (–).
L Muscle layers in proboscis, outer longitudinal, circular, inner longitudinal (3), transition to heterotype (tr), circular and inner longitudinal (2), 3C 2C according to Coe.
M Number of muscle crosses . (C according to Coe).

woven with the subepidermal longitudinal muscles (Fig. 7). The amorphous intercellular substance forms a thick layer beneath the epidermis and accompanies numerous radial muscles through the OLM. Subepidermal glands penetrate between the bundles of the OLM especially in the CT associated with the radial muscles. The amorphous CT contains many small longitudinal muscle bundles as well as the subepidermal glands. Diagonal muscle fibers which appear to be circular, are abundant in the precerebral region and separate the subepidermal glands from the medullary longitudinal muscles, frontal glands, and cephalic nerves. The rhynchodaeum is attached dorsally as in the genus *Cerebratulus* (Fig. 8) so that there are two lateral cephalic vessels which fuse anteriorly, but muscles between the rhynchodaeum and body wall produce two transient dorsolateral spaces additionally which open at intervals into the large laterals. There are a few small bundles of ILM embedded initially among the circular muscle fibers of the rhynchocoel. Horizontal fibers dorsal and ventral of the lateral nerve cords cross, as 'circular' muscles dorsal and ventral to the foregut but outside the esophageal longitudinal muscles. Dorsoventral muscles are well developed between the intestinal diverticula. The CT enclosing the brain is thin and does not appear fibrous but does restrict the neurons so that they do not spread into the surrounding tissues. Neurochord cells are present in the ventral ganglia. The epithelium of the esophagus is similar to that of the buccal cavity with extensive development of subepithelial glands. The epithelium changes abruptly at the juncture of the esophagus with the intestine. Initially, the intestinal mucosa is without folds and the epithelial cells are short and of a uniform height. This portion of the intestine is surprisingly long and of uniform horizontal diameter. It does not expand dorsally around the rhynchocoel as does the esophagus. The height of the epithelium increases greatly near the origin of the first diverticula. The diverticula are lobed. The endothelium of the blood vascular plexus associated with the foregut is backed by amorphous connective tissue which becomes progressively thicker posteriorly, breaking up the plexus into large meandering endothelial lined lacunae. Toward the posterior end of the esophagus, small thick-walled blood vessels form a secondary plexus united with the lacunae. The excretory system extends through most of the esophageal region to the beginning of diverticulation of the intestine. Coe, (1895:197), in his only reference to the morphology of this species stated that 'the large lateral lacunae in the anterior esophageal region are covered' with 'nephric tubules'. In this region, they actually lie within the lacunae and only begin to spread into the surrounding CT near the origin of the intestine. The cyrtocytes project into the lacunae and do not enter the walled vessels. The nephric system is almost totally dorsal to the lateral nerve cords and lateral to the gut, and thus does not spread beneath the gut as in the genus *Cerebratulus*. A single nephric duct exits dorsal to the lateral nerve cord near the posterior end of the esophagus and a second one toward the posterior prediverticular region of the intestine on each side.

Verrill (1892) stated the *Cerebratulus medullatus* Hubrecht 1878 was 'probably not a *Cerebratulus*' and perhaps 'belonged to one of the species' which he described. In 1895 he noted (p. 144) that *C. medullatus* might be identical with '*C. luridus*'. It is possible to visualize the differences in the description of the subepidermal glands to be the result of treatment of specimens. However, the dorsal nerve of *Tarrhomyos luridus* is distinct for only a short distance behind the mouth and throughout most of the body is only evident when centripetal nerves pass from it. The dorsal nerve of *C. medullatus* impressed Hubrecht by its large size of about one-fourth the diameter of the lateral nerves.

The topography of the subepidermal glands, distribution of diagonal muscle fibers, forgut blood sinus, and wickerwork of sub-epidermal muscles (a character which is even more pronounced in the type slides of *Micrura nebulosa* Coe, 1905) differ from the characters of the genus *Cerebratulus* (see Table 1). These differences are shared with *Cerebratulus praealbescens* Cantell, 1982, which, thus, is transferred as *Tarrhomyos praealbescens* (Cantell, 1982) n.com.

Discussion

Line drawings or lithographs of preserved specimens have often been used in the descriptions of nemertines. These illustrations indicate the condition of the specimen, but are valueless in identifying species in contrast with figures of living animals. Differential contraction following fixation affects most tissues and organ systems but these are not fixation artifacts since they represent conditions which could be attained in living individuals. Morphological observations apply to specimens and thus, relative thickness of body wall layers, relative position of dorsal and ventral ganglia of the brain, closeness of brain to mouth of heteronemertines, and height of epidermal cells of different surfaces or regions of the body applies to the specimen, not the species, and should never be part of a species diagnosis. The brain, which is the most solid organ in a nemertine, is also influenced by body contractions. Differential growth described by Iwata (1972, 1985) alters the arrangement of muscles in the pre-oral region of the body so that contraction of the longitudinal musculature routinely shifts the ventral ganglia slightly posteriorly, and draws the buccal cavity into close proximity to the brain. In heavy bodied heteronemertines, the pressure of the buccal cavity against the brain usually diverts the cerebral organ nerve laterad (Fig. 13) or even anteriad, and the lateral nerve cords laterad from the ventral ganglia. The elongate mouth of carnivorous heteronemertines may expand following fixation, producing crenulated lips surrounding the open buccal cavity, or contract to form a small opening with the much folded lips drawn into the cavity.

Large stereocilia (bristles) (Fig. 2) are present at the front of the head of *Amphiporus frontalis*. Norenburg (1985) noted the common occurrence of these 'sensory bristles' on the heads of most larvae and many adult nemertines. An eversible organ consisting of support cells surrounding the pores from subepidermal mucous gland cells lies at the apex of the head of most, if not all, hoplonemertines. The morphology of the organ is reminiscent of the acoelomorph turbellarian frontal organ from which it can be derived. Apical organs (Fig. 3) of heteronemertines are not homologous with those of hoplonemertines. Those of lineids (*sensu* Gibson, 1985) are usually three in number, one dorsal and two ventrolateral. These are sensory organelles, designated as apical sensory organs by Riser (1989). Storch & Moritz (1971) described the ultrastructure of sensory tissue on the 'anterior edge of the head' of *Lineus ruber*, but did not distinguish between the apical sensory organs and the head epithelium in their text nor in the labels of their figures, however, Abb. 1a is typical of our results with the apical sensory organs of various species of *Lineus* (Abb. 2a, appears to be surface epithelium). They postulated that the cell bodies containing the nuclei of the sensory cells were near the brain or in the cerebral ganglia. The nucleate cell bodies in fact form a cup-like ganglion around the base of the inverted organ. Precerebral contraction produces a ventral trough and carries apical organelle(s) posteriorly in the mid-dorsal (also lateral in heteronemertines) wall of the trough. Extreme apical contraction in which the normally subterminal rhynchodaeal pore is retracted into the body results in the specific apical organelles coming to lie in the wall of a tube within the head.

The lateral cephalic fissures of heteronemertines terminate anteriorly slightly behind the apex of the head. Contraction of the apex may bring the fissures forward. Pre-oral contraction drawing the brain and buccal cavity into close proximity extends the fissures posterior to the brain, sometimes producing a tubular internal pocket as in Tab. 6, Fig. 9 of Bürger, 1897–1907.

The epidermis varies in height according to contraction of the muscles of the body wall and thus may be of common height in a round region of the body but of different heights on dorsal and ventral surfaces if the region is umbonate. Pedersen (1968) observed that the epidermis could be lifted off the underlying basement membrane which contained a greater concentration of fibrillae than the connective tissue (CT) continuum with which it was associated. The CT consists of amorphous intercellular substance in addition to fibrillae, as discussed by Pedersen (1991). The amorphous tissue contains cells and may form a thick layer between the epidermis and body musculature as in *Amphiporus angulatus*. The fibrillae form a neurilemma around the ganglia of the brain of most heteronemertines. The neurilemma may or may not be apparent in histological preparations, but its presence can be seen in the confinement of the ganglionic cells. Such a confinement is not present in the genus *Micrura* in which the ganglion cells ramify among the surrounding tissue.

Subepidermal gland cells characteristically occur in the CT of heronemertines. They may be restricted by fibrous elements to a circumferential zone as in the genus *Lineus*, or extend inward between the muscle bundles as in *Tenuilineus*. A similar distribution of the glands between the muscles was described in the type descriptions of *Micrura olivaris* Coe, 1905, and *M. pardalis* Coe, 1905, species for which no morphological characteristics which would allow generic placement were given. The subepidermal glands of *Lineus ruber* and *L. viridis* press against the muscle bundles of the OLM and rarely bulge between muscle bundles; however, they hypertrophy during the reproductive period and the mucous glands then extend to the ICM. Diagonal muscles with associated connective tissue which contains isolated bundles of longitudinal muscle fibers restrict the glands to a zone in the genus *Cerebratulus* (Fig. 9). In *Amphiporus angulatus*, *Cyanophthalma cordiceps* and *C. obscura* (Schultze, 1851), necks from multicellular mucous glands pass through the thick hyaline CT layer to open through the epidermis. They are most abundant anteriorly.

Hubrecht (1887) postulated a number of theories in the 'Anatomical Investigations' section in an effort to make a cohesive pattern from his observations. He expressed delight at the amount of information he was able to obtain from the Challenger collection of 'truncated' 'fragments' of specimens collected about ten years earlier. He was probably the first nemertinologist to use a microtome, the Caldwell rocking microtome, however it is doubtful that he had access to the Darwinian modification which made serial sections possible since it was developed while he was working on the collection. His student, Oudemans (1885) who was his assistant at the same time referred to sections varying in thickness 'between 1/70 and 1/100 millim', and stated that the Giesbrecht adhesive was used to attach the sections to slides; this implies that sections were obtained individually. Hubrecht prestained with Ranvier's picrocarmine, a stain which yields perfectly adequate preparations for interpretation by an experienced histologist. It was the contracted condition of the specimens that forced him to postulate in order to bring all of his observations together in a logical pattern. He noted that his assumptions could be 'open to dispute' and specially that certain conclusions 'may after all not be' final. Bürger (1895, 1897–1907) accepted the postulates without comment, and later workers simply repeated or reworded Bürgers statements, Coe, 1905, 1943; Hyman, 1951; Gontcharoff, 1961; and Gibson, 1972. This has made it difficult for beginners to comprehend anatomical observations, and has handicapped efforts to analyze the phylogeny of the phylum.

The basic nemertine body plan becomes disorganized in the precerebral region of most and in the preoral region of all groups. Stiasny-Wijnhoff (1923) concluded that the nemertines were in the process of developing a head. The differential growth which is responsible for the architectonic differences between the 'head' and body proper is clearly demonstrated in Iwata's (1972, 1985) studies on axial changes during development within the various groups of nemertines. This differential growth impacts directly upon the musculature and the nervous system.

The neurons associated with the lateral nerve cords of the palaeonemertine *Tubulanus* are primarily concentrated in the base of the epidermis, some distance from the neuropil which is in the subepidermal CT and abuts the circular muscles. Circular muscles are not carried into the head of *Procephalothrix spiralis* and the central nervous system comes to lie between the rhynchodaeal and body wall longitudinal muscles. With the advent of circular muscles beneath the subepidermal CT the lateral nerve cords become connected by a typical plexus which is in the primitive position in the CT and against the circular muscles. The neural primordium detaches from the embryonic ectoderm at the gastrula stage in the hoplonemertines studied by Iwata. Differential growth and the peculiar aggregation of cells to form organ primordia allow the muscles of the body wall to develop external to the nervous system. In adult *Amphiporus angulatus* and *Cyanophthalma cordiceps*, circular muscles continue to serve as an organizer of the nervous system. Outer fibers from the rhynchocoelic circular muscles diverge to enclose the foregut ventrally and the dorsal and ventral nerves from the nerve cords pass directly to travel along the muscle surface. A plexus has not been observed connecting the two cords, but small nerves extend outward into the body wall musculature and epidermis. Hubrecht (1887) beautifully described the hoplonemertine nervous system and the reasons for the absence of a nerve plexus connecting the lateral cords, dorsal nerve, and rhynchocoelic nerve.

The heteronemertine nervous system loses its connection (relationship) to the epidermis in development, but retains its relationship to the circular muscles when they develop. The two lateral nerve cords are connected by a plexus on the outer surface of the ICM. A longitudinal nerve tract, which when obvious is referred to as the dorsal nerve, is present mid-dorsally in the plexus and connects between circular muscle bundles with a plexus in the rhynchocoel. An enlargement in the dorsal wall of the rhynchocoelic plexus indicates the presence of a longitudinal rhynchocoelic nerve.

Hubrecht (1887:53) 'designated' the cellular layers 'outside the more or less stratified connective tissue' as 'integument' (sic). This introduced a problem with two of the species, *Cerebratulus macroren* and *C. medullatus* which he described in the first section of his manuscript, because the connective tissue in those two species did not form a restrictive layer, and the subepidermal glands penetrated between the longitudinal muscles even to the circular muscles. He was not aware that the CT extended between the muscle bundles to the nerve plexus bounding the circular muscles. He postulated that the subepidermal muscles were 'epiblastic' and the major muscles 'mesoblastic' but (p. 62) adamantly declared the fallibility of this hypothesis. The contracted condition of his material prevented him from recognizing that the 'integument' was invaded by the muscles. Iwata (1958) noted that the two muscle layers [inner circular (ICM) and inner longitudinal (ILM)] developed in *Micrura akkeshiensis* Yamaoka, 1940, at 119 hours and were the only muscles present at 39 days. At 143 hours, the longitudinal muscles passed inside the brain lobes and commissures while the circulars occurred between the epidermis and brain. Muscles of the proboscis system were 'not well differentiated' at 39 days. Further morphogenesis can only be postulated from the adult.

It is customary in heteronemertines to refer to five muscle layers, each with a specific designation indicating them to be discrete entities. There are two muscle layers, a circular and a longitudinal, each of which decussates anteriorly and posteriorly near the mouth. The migration of the blastopore which forms the juncture of the fore-gut and intestine, the development and movement of the stomodaeum, and the opening of the mouth and positioning of the blastodiscs in heteronemertines directly affect the muscular system. The basic acoelomate bilaterian body plan in which subepidermal muscles (SCM) followed by longitudinals (SLM) prevails in heteronemertines. They pass around the mouth beneath the oral epithelium which occurs between the epidermis and buccal epithelium: a position corresponding closely to the location of the original blastopore and point of proliferation of the mesoderm. A major number of circular fibers are carried dorsally (Fig. 11) as the ICM at the posterior end of the mouth, but additional fibers participate in this to the anterior end of the mouth. Some SCM fibers retain their subepithelial position along the sides of the buccal cavity leaving in the vicinity of the esophageal nerves and continuing dorsally to abut the ventral side of the horizontal muscle plate. In the description of *Perissonemertes pyrrhocephaleus* Gibson, 1990, Gibson (1990) stated that these muscles 'contribute' to the horizontal

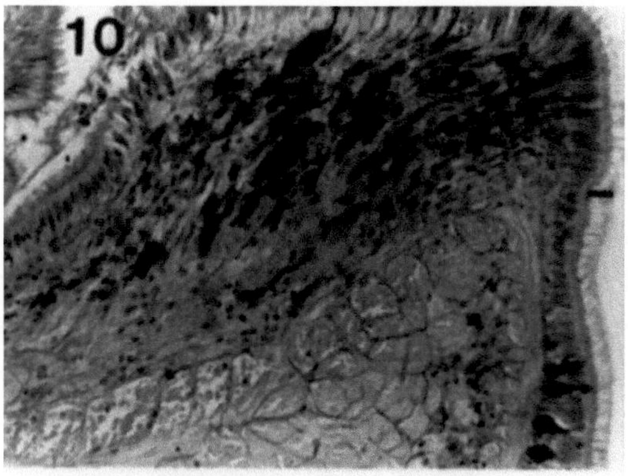

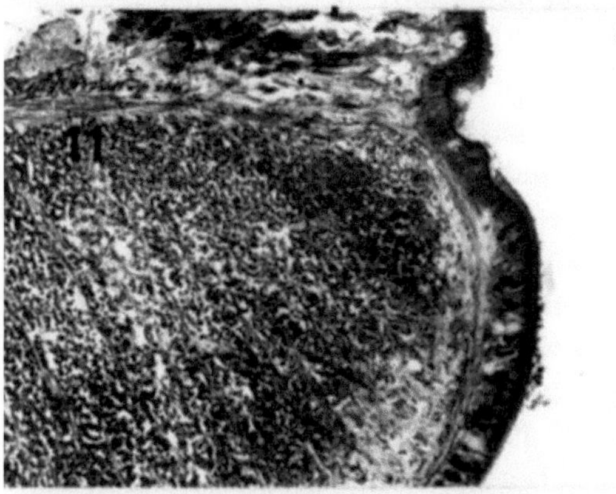

Fig. 10. Micrura affinis. Cross section through oral lip with all three longitudinal muscles uniting beneath oral epithelium.
Fig. 11. Tarrhomyos luridus. Cross section with subepidermal circular muscles continuous with the inner circular muscles internal to the esophageal nerve.
Fig. 12. Cerebratulus marginatus. Cross section behind mouth with diagonal muscles divaricating from subepidermal circular muscles.

muscle plate. This is true topographically, but the horizontal plate which is formed from radial muscles and the bow of subepithelial circular muscles which contacts its ventral surface serve different functions. Other fibers leave the point of origin around the mouth to transit the salivary glands and become incorporated in the ICM outside the esophageal nerves. Additional circular fibers radiate into the head of *Tarrhomyos* and in the genus *Cerebratulus* such fibers spread out posterior (Fig. 12) as well as anterior to the mouth. These fibers are referred to as diagonal fibers. Many of these fibers near the mouth course through the subepidermal glands, but eventually, most come to lie in the CT on which the glands abut. Circular muscles which are not connected to the oral ring are present precerebrally in some lineids. The circular cylinder which is present around the rhynchodaeal complex in some species is a forward extension of the ICM. A thick ring of subepidermal circular muscles accompany the rhynchodaeum for its entire length in a few heteronemerteans. These are assumed to be SCM carried inward with the rhynchodaeal invagination. In many species, radial fibers enter the rhynchodaeum through its connection to the surrounding tissues. There is a thin layer of subepidermal circular muscles in most of these forms primarily near the rhynchodaeal septum but the relationship of this ring to the rest of the circular musculature has not been resolved.

The longitudinal muscles pass around the mouth as a unit (Fig. 10) inside the SCM. A portion of the layer expands dorsally, retaining a position internal to the SCM and epithelium of the wall of the buccal cavity. These bundles continue forward through the brain as part of the walls of the blood lacunae and at the anterior end of the

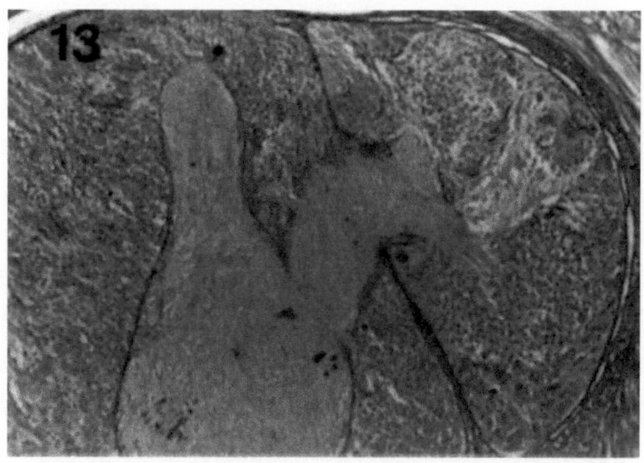

Fig. 13. Cerebratulus lacteus. Cross section through posterior part of brain of a contracted specimen in which the brain and buccal cavity were tightly appressed.

latter regain continuity with the other longitudinal muscles of the head. Posterior to the mouth, they constitute that portion of the ILM internal to the ICM. Another group of longitudinal muscle bundles continue posteriorly in the connective tissue between the ICM and SCM. The SCM is cross-hatched rather than all fibers parallel and the longitudinal muscles intermingle with the circulars (Fig. 6, 7) to form a wickerwork beneath the epidermis in *Tarrhomyos luridus* and to a greater extent in '*Micrura nebulosa*'. The proportional distribution of serous and mucous subepidermal glands, their individual degree of development, and their concentration (*i.e.* tightly appressed vs. spaces between) directly affect the disposition of the longitudinal muscles between the SCM and ICM. Norenburg (1993) records a simple arrangement of the longitudinal muscles between the SCM and ICM in the absence of subepidermal glands in the genus *Riserius*. Extensive development of the subepidermal glands isolates bundles of longitudinal muscles against the SCM in the genera (*sensu strictu*) *Lineus*, and *Micrura*. For descriptive purposes these bundles should be referred to as subepidermal longitudinal muscles and the bundles between the bases of the glands and the ICM as outer longitudinal muscles (OLM). The regionalization of gland development in *Tenuilineus bicolor* where the glands reach the ICM and are so dense anteriorly that the muscle bundles would all be referred to as subepidermal while in the esophageal region of the body subepidermal would alternate with OLM between glands and in the intestinal region (Fig. 5), all would be OLM does not obviate the usefulness of these terms. The distinction is valid in *Tarrhomyos luridus* where some glands extend between the OLM bundles some of which thus extend from the SCM to the ILM. The genus *Cerebratulus* is characterized by a dense layer of CT which contains longitudinal muscle fiber bundles as well as diagonal [correctly assigned to the circular layer by Cantell (1982)] fibers (Fig. 9) beneath the subepidermal glands. One group of ILM fibers is isolated in some genera as part of the wall of the ventral blood lacuna in the cerebral region and continues posteriorly as the subrhynchocoelic longitudinal muscles. These muscles surround the esophagus and continue posteriorly around the initial part of the intestine to the first intestinal caeca in *Notospermus geniculatus* (Delle Chiaje, 1828). They may become interwoven with the muscles of the sling formed from tangential muscles to support the rhynchocoel beneath the rhynchocoelic blood vessel. The medial walls of the ILM of the two lateral lacunae of the head, interweave initially with the rhynchocoelic circular muscles in the genus *Cerebratulus*. A few, 3–5, large longitudinal bundles continue posteriorly in the circular layer of the dorsal wall of the rhynchocoel for some distance farther than the smaller bundles in the rest of the wall in available series of sections of *Cerebratulus lacteus* and *C. marginatus*.

Some SLM fibers are carried inward with the stomodaeal invagination. In those forms in which a stomodaeal diverticulum develops in embryology, these fibers are retained as esophageal subepithelial longitudinal muscles lying immediately beneath the entire layer of epithelial cells. In other forms, they are absent. Gibson (1985) noted that 'splanchnic muscle layers' are 'almost always associated with' esophageal subepithelial glands. References to splanchinic muscles in much of the heteronemertine literature are difficult to interpret. The subrhynchocoelic longitudinal muscles press into the subepithelial glands of specimens in which the rhynchocoel is inflated in the esophageal region. References to dorsal splanchnic muscles probably refer to these. The subrhynchocoelic longitudinal muscles enclosing the esophagus have always been identified as 'splanchnic' or esophageal. The distinction between subepithelial longitudinal muscles and those associated with the subrhynchocoelic plate is of systematic as well as morphologic significance. Subepithelial circular muscles are restricted to the walls of the buccal cavity. Radial and tangential fibers insert in the epithelium of the esophagus. Some of these fibers arboresce among the subepidermal cells producing branches which extend a short distance circularly or diagonally and are most frequently encountered ventrally. They never occur between the epithelial cells and subepithelial longitudinal muscles. In some species they descend from the dorsal body wall and cross beneath the rhynchocoel or diverge in passing the rhynchocoel to continue around the esophagus. The combination of radial, tangential, and fibers forming the dorsal sling produce a thick muscular encirclement of the esophagus in *Parborlasia corrugatus* (McIntosh, 1876) and a major sphincter at the juncture of the fore-gut and intestine. These muscles have taxonomic value when they enclose the gut and appear to form a circular layer, but their origin from radial fibers should be noted and the term 'splanchic' must be avoided.

Some muscles from the inner portion of the OLM of the body wall pass across the anterior face of the brain to form the heteronemertine proboscideal longitudinal muscles. [For exceptions, see Iwata (1993)]. The two proboscideal nerves enter with these muscles. In some genera they are located beneath the epithelium while in others such as *Micrura* they lie within the muscles, a feature which Friedrich (1960) interpreted as resulting in two longitudinal layers. The proboscideal nerves are united by a plexus which is separated from the inner longitudinal muscles by circular muscles. The *de novo* appearance of the circular muscles in sections is indicative of the retraction of the proboscis away from the diaphragm. Some longitudinal muscles may continue for a short distance in the main body of the proboscis between the plexus and epithelium producing the transition to heterotype muscle arrangement as in *Tenuilineus*. The circular muscles and nerve plexus disappear at the posterior end of the proboscis and the two longitudinal layers present in *Cerebratulus*, *Parvicirrus* and *Tarrhomyos* reunite as a single layer similar to that of genera which lack outer longitudinal proboscideal muscles. The outer longitudinal muscles beneath the epithelium appear to be derived rather than primitive, but the derivation of the two muscles in *Paralineopsis* is not homologous to that of lineids. Gibson (1985) stated that the 'sequence' of 'outer and inner longitudinal layers 'separated by a circular layer' is likely to represent the primitive arrangement' of muscles in the proboscis of heteronemertines. Proboscideal spines described by Riser (1990) are permanent structures present in the proboscis epithelium of all lineid species which I have examined. They are comprised of seven microvillar plates enclosing a sensory cilium similar to receptors recorded in other phyla.

Data obtained from the holotype slides of species of *Cerebratulus*, *Lineus*, and *Micrura* described by Coe and of species in my personal collection are presented in Table 1. Features not clearly evident, are not recorded. Presence or absence of a caudal cirrus on Coe's species is from his description of the species. It is evident that the generic placement of some species must be revised '*Micrura leidyi*' a very common northeastern Atlantic species is not included because in preparing this paper it was discovered that two species which appeared to be similar to the naked eye were present and that they furthermore appear to belong to two different genera.

Acknowledgements

The conscientious criticism of this manuscript by Dr R. Gibson has hopefully prevented misinterpretations. Serendipity played a major role in the Discussion section. Evidence was already present in the literature, but it had not been brought into context. The embryology and development of *Micrura akkeshiensis* by Iwata; the morphology of *Micrura fasciolata* by Cantell, of *Perissonemertes* by Gibson, and of *Riserius* by Norenburg demonstrated that long-held concepts based upon efforts to interpret badly preserved material were erroneous. The contribution of those workers must be acknowledged. Some of the serial sections were processed by Joan Ferraris-Norenburg and Gloria M. Philips working as graduate research assistants.

References

Bierne, J., M. Tarpin & G. Vernet, 1993. A reassessment of the systematics and a proposal for the phylogeny of some cosmopolitan *Lineus* species (Nemertea). Hydrobiologia 266: 159–168.

Brunberg, L., 1964. On the nemertean fauna of Danish waters. Ophelia 1: 77–111.

Bürger, O., 1895. Die Nemertine des Golfes von Neapel und der angrenzenden Meeres-Abschnitte. Fauna und Flora des Golfes von Neapel, 22: 1–743.

Bürger, O., 1897–1907. Nemertini (Schnurwürmer). In H. G. Bronn (ed), Klassen und Ordnungen des Tierreichs. Supplement. 4: 542 pp

Bürger, O., 1904. Nemertini. Das Tierreich, 20: 1–151.

Cantell, C.-E., 1975. Anatomy, taxonomy and biology of some Scandinavian heteronemertines of the genera *Lineus*, *Micrura*, and *Cerebratulus*. Sarsia 58: 89–122.

Cantell, C.-E., 1982. The anatomy and taxonomy of *Cerebratulus praealbescens* sp. n. (Nemertini) from the west coast of Sweden and Norway. Sarsia 67: 11–20.

Coe, W. R., 1895. On the anatomy of a species of nemertean (*Cerebratulus lacteus* Verrill), with remarks on certain other species. Trans. Conn. Acad. Arts Sci. 9: 479–514.

Coe, W. R., 1901. Papers from the Harriman Alaska Expedition. XXI The nemerteans. Proc. Wash. Acad. Sci. 3: 1–110.

Coe, W. R., 1905. Nemerteans of the west and northwest coasts of America. Bull. Mus. Comp. Zool., Harvard College. 47: 1–319.

Coe, W. R., 1943. Biology of the nemerteans of the Atlantic coast of North America. Trans. Conn. Acad. Arts Sci. 35: 129–328.

Corrêa, D. D., 1961. Nemerteans from Florida and Virgin Islands. Bull. Mar. Sci. Gulf Carribb. 11: 1–44.

Friedrich, H., 1933. Morphologische Studien an Nemertinea der Kieler Bucht, I und II. Z. Wiss. Zool. 144: 496–509.

Friedrich, H., 1935. Studien zur Morphologie, Systematik und Ökologie der Nemertinen der Kieler Bucht. Arch. f. naturgesch. N.F. 4: 293–375.

Friedrich, H., 1960. Bemerkungen über die Gattung *Micrura* Ehrenberg 1831 und zur Klassifikation der Heteronemertinen nebst vorläufigen Bestimmungsschlüssel. Veröff. Inst. Meeresforsch. Bremerh. 7: 48–62.

Gibson, R., 1972. Nemerteans. Hutchinson, London, 224 pp.

Gibson, R., 1982. Nemertea. In: Parker S. P. (ed.), Synopsis and Classification of Living Organisms. McGraw-Hill, New York: 823–846.

Gibson, R., 1985. The need for a standard approach to taxonomic descriptions of nemerteans. Am. Zool. 25: 5–14.

Gibson, R., 1990. The macrobenthic nemertean fauna of Hong Kong. In Morton, B.S. (ed.) Proceedings of the Second International Marine Biological Workshop: The Marine Flora and Fauna of Hong Kong and Southern China. Hong Kong Univ. Press, Hong Kong. 1: 33–212.

Gibson, R. & F. B. Crandall, 1989. The genus *Amphiporus* Ehrenberg (Nemertea, Enopla, Monostiliferoidea) Zool. Scr. 18: 453–470.

Girard, C., 1855. Nemertidae. In W. Stimpson, Synopsis of the Marine Invertebrata of Grand Manan. Smithson. Contrib. Knowl. 6: 28.

Gontcharoff, M., 1961. Embranchment des Nemertiens (Nemertini G. Cuvier 1817; Rhynchocoela M. Schultze 1851). In P.-P. Grassé (Dir.) Traité de Zoologie. Anatomie, Systematique, Biologie 4: 785–886.

Hubrecht, A. A. W., 1887. Report on the Nemertea collected by H.M.S. 'Challenger' during the years 1873–6. Challenger Report, Zoology, 90: 1–151.

Hyman, L. H., 1951. The Invertebrates: Platyhelminthes and Rhynchocoela, The Acoelomate Bilateria. McGraw-Hill, New York. 550 pp.

Iwata, F., 1958. On the development of the nemertean *Micrura akkeshiensis*. Embryologia 4: 103–131.

Iwata, F., 1972. Axial changes in the nemertean egg and embryo during development and its phylogenetic significance. J. Zool. London 168: 521–526.

Iwata, F., 1985. Foregut formation of the nemerteans and its role in nemertean systematics. Am. Zool. 25: 23–36.

Iwata, F., 1993. *Paralineopsis taki* gen. et sp. nov., a littoral heteronemertean from Japan, provided with special proboscideal, circulatory and sensory organs of significance to nemertean systematics. Hydrobiologia 266: 185–201.

Norenburg, J. L., 1985. Structure of the nemertine integument with consideration of its ecological and phylogenetic significance. Am. Zool. 25: 37–51.

Norenburg, J. L., 1986. Redescription of a brooding nemertine, *Cyanophthalma obscura* (Schultze) gen. et comb.n., with observations on its biology and discussion of the species of *Prostomatella* and related taxa. Zool. Scr. 15: 275–296.

Norenburg, J. L., 1993. *Riserius pugetensis* gen.n., sp.n. (Nemertina: Anopla), a new mesopsammic species, and comments on phylogenetics of some anoplan characters. Hydrobiologia 266: 203–218.

Oudemans, A. C., 1885. The circulatory and nephridial apparatus of the Nemertea. Q. J. Micro. Sci. Suppl. 25: 1–80.

Pedersen, K. J. 1968. Some morphological and histochemical aspects of nemertean connective tissue. Z. Zellforsch. 90: 570–595.

Pedersen, K. J., 1991. Structure and composition of basement membranes and other basal matrix systems in selected invertebrates. Acta zool., Stockh. 72: 181–201.

Riser, N. W., 1987. Observations on the genux *Ophelia* (Polychaeta: Opheliidae) with the description of a new species. Ophelia 28: 11–29.

Riser, N. W., 1989. Speciation and time-relationship of the nemertines to the acoelomate metazoan bilateria. Bull. Mar. Sci. 45: 531–538.

Riser, N. W., 1990. New Zealand nemertines from kelp holdfasts: Heteronemertinea 1. *Adenorhagas aurantiafrons* gen.n., sp.n. N. Z. J. Zool. 17: 597–606.

Riser, N. W., 1991. New Zealand nemertines from kelp holdfasts: Heteronemertinea II. *Notospermus geniculatus* (Delle Chiaje, 1828) n.comb. N. Z. J. Zool. 18: 427–438.

Stiasny-Wynhoff, G., 1923. Die Entstehung des Kopfes bei den Nemertinen. Acta Zool. 4: 223–240.

Storch, V. & K. Moritz, 1971. Zur Feinstruktur der Sinnesorgane von *Lineus ruber* O.F. Müller (Nemertini, Heteronemertini). Z. Zellforsch. 117: 212–225.

Verrill, A. E., 1873. Report upon the invertebrate animals of Vineyard Sound and the adjacent water, with an account of the physical characters of the region. Rep. U.S. Comm. Fish & Fisheries for 1871 and 1872: 295–778.

Verrill, A. E., 1874a. Brief contributions to Zoology from the Museum of Yale College. No. XXVI. Results of recent dredging expeditions on the coast of New England. Am. J. Sci. Ser. 3 7: 38–46.

Verrill, A. E., 1874b. Brief contributions to Zoology from the Museum of Yale College. No. XXVIII. Results of recent dredging expeditions on the coast of New England. Am. J. Sci. Ser. 3 7: 405–414.

Verrill, A. E., 1879. Notice of recent additons to the marine invertebrata of the northeastern coast of America, with descriptions of new genera and species and critical remarks on others. Proc. U.S. Nat. Mus. 79: 165–206.

Verrill, A. E., 1892. The marine nemerteans of New England and adjacent waters. Trans. Conn. Acad. Arts Sci. 8: 382–456.

Verrill, A. E., 1895. Supplement to the marine nemerteans and planarians of New England. Trans. Conn. Acad. Arts Sci. 9: 523–534.

Hydrobiologia **266**: 159–168, 1993.
R. Gibson, J. Moore & P. Sundberg (eds), Advances in Nemertean Biology

A reassessment of the systematics and a proposal for the phylogeny of some cosmopolitan *Lineus* species (Nemertea)

Jacques Bierne [1], Michel Tarpin [1] & Guy Vernet [2]
[1] *Laboratoire de Biologie Cellulaire;* [2] *Laboratoire de Zoologie et des Sciences de l'Environnement; UR4R, Université de Reims Champagne-Ardenne, B.P. 347, 51062 Reims, France*

Key words: heteronemerteans, biogeography

Abstract

Morphological, behavioural and biochemical characters of 9 cosmopolitan, benthic species of the nemertean genus *Lineus* are used both to clarify their taxonomy and to discuss their phylogenetic relationships. Data analysis shows that all fissiparous species of *Lineus* collected in the world seas up to date can be placed in a single species, *L. sanguineus*, with three recognisable subspecies, *L. s. sanguineus*, *L. s. nigricans* and *L. s. pseudolacteus*. These taxa were originally described as separate species mainly according to their geographic origin under the names *L. sanguineus* (shores of European seas), *L. socialis* (Atlantic North American shores), *L. vegetus* (Pacific North American shores), *L. pseudolacteus* (shores of the English Channel) and *L. nigricans* (Mediterranean coasts of Italy and France). Examination of the character value matrix suggests a phylogenetic tree developing from two ancestral branches. The first branch includes the two related species *L. ruber* and *L. viridis*, the second branch the three species *L. longissimus*, *L. lacteus* and *L. sanguineus*. This approach to the taxonomic and phylogenetic relationships of *Lineus* nemertean species is in agreement with the intra- and interspecies histocompatibility data in nemerteans: (i) grafts transplanted from donors to recipients of the same species succeed and grafts transplanted from donors to recipients of different species fail; (ii) grafts are rejected more rapidly when the donor and recipient species diverged earlier in evolution and, *a contrario*, grafts are accepted better when the donor and recipient species have been more recently isolated.

Introduction

Although the genus *Lineus* is one of the oldest nemertean taxa, most of the species included in the genus are as yet inadequately described and their diagnosis remains uncertain, in particular because different names have been attributed to the same species by American and European zoologists. Phylogenetic relationships between the species of *Lineus* have remained entirely unknown.

In the past two decades we have had the opportunity, from observations and experiments related to developmental events, to identify and analyse a number of characters that allow us to clarify the systematics and to propose a phylogeny of some cosmopolitan species of intertidal *Lineus* collected from various locations around the world. The worms of these species were reared in the laboratory with a view to testing various biological capabilities: regeneration, reconstruction, rejuvenation and reproduction. The 27 characters investigated were obtained from different aspects of *Lineus* biology and biochemistry; they include external or internal morphology (9), development (2), patterns of reproduction (3), response to stimuli (1) and biochemical characteristics of the eggs (12).

The taxonomic reassessments resulting from data analysis, as well as the phylogenetic relationships derived from the overall examination of the character value matrix, are compared with intra- and interspecies histocompatibility rules derived from data of graft survival recorded since 1967: (i) grafts transplanted from donors to recipients of the same species succeed and grafts transplanted from donors to recipients of a different species fail; (ii) grafts are rejected more readily when the donor species and the recipient species diverged earlier in evolution and, *a contrario*, grafts are accepted better when the donor and recipient species have been more recently isolated.

Material and methods

Many specimens of nine *Lineus* species collected over 30 years from various geographic locations have been studied. The species and their geographic origins are: *Lineus ruber* (shores of the English Channel [Brittany coast at Roscoff and Tregastel; Normandy coast at Cherbourg] and the Atlantic Ocean [Asturian coast of Spain at Luanco and Banugues]); *Lineus viridis* (shores of the English Channel, Baltic Sea and northern Atlantic Ocean [Asturian coast of Spain and American coast at Nahant Massachusetts]); *Lineus sanguineus* (English Channel, Atlantic Ocean [Spanish coast at the Vigo area] and the Mediterranean Sea); *Lineus socialis* (Uruguayan and North American coasts of the Atlantic Ocean); *Lineus nigricans* (Mediterranean Sea [French coast at Villefranche-sur-Mer]); *Lineus vegetus* and *Lineus pseudolacteus* (Brittany coast of the English Channel); *Lineus longissimus* (English Channel at Roscoff and Cherbourg and the Atlantic Ocean at Vigo); and *Lineus lacteus* (Mediterranean Sea [Banyuls-sur-Mer] and the Atlantic coast of Spain [Vigo] and France [La Rochelle]).

Many of the morphological characters were examined *in vivo*, on worms either moving in sea water or immobilized by 8% $MgCl_2$ in distilled water between slide and cover glass. Others were derived from *post-mortem* studies on histological sections.

Biochemical characters were obtained by polyacrylamide gel electrophoresis of total sodium-dodecyl sulphate (SDS) extracts from new laid eggs, *i.e.*, from a suitable stage of development for significant comparisons between major molecular components of different species of *Lineus*.

The characters

The study is based on 27 characters recorded for nine *Lineus* species (Table 1) described systematically according to two states for binary coding (Table 2).

Anatomical levels (characters 1 to 3)

The body of a *Lineus* specimen consists of 10 regions distinguishable along the antero-posterior axis (Bierne, 1985; Vernet & Bierne, 1988). From these 10 anatomical levels, numbers 1 (antecerebral end), 4 (post-cerebral, pre-oral) and 7 (anterior intestinal, without gonads) are significant for diagnosis because of their presence and/or size. To describe a constant level 1 we have selected the relative size, *i.e.*, level 1 is compared with another level (cerebral level 2) of the same specimen rather than absolute size that is variable according to the developmental stage (growth, regeneration). The sizes of levels 4 and 7 are species-specific for *L. lacteus*.

Although discrete, level 4 in all fissiparous *Lineus* allows us to construct worms with either a deletion or a duplication of this level by heterotopical grafting. By contrast, such an experimental design cannot be carried out with *L. ruber* and *L. viridis* in which level 4 is completely absent.

Morphological features (characters 4 to 7)

Asconiform cells in the proboscis epithelium of various *Lineus* species have been described and compared by Moretto & Vernet (1985). The presence or absence of sexual dimorphism of genital ducts, genital pores and cutaneous glands was observed for the 9 *Lineus* species by Bierne (1970, 1983).

Table 1. List of characters and their states for taxonomic analysis of nine benthic species of *Lineus*. All character states are listed in accordance with the binary coding.

No.	Character	Character states
1	Anatomical level 1	0: short ($\leqslant 1.5 \times$ level 2) 1: long ($\geqslant 2 \times$ level 2)
2	Anatomical level 4	0: absent 1: present (discret or large)
2a	Anatomical level 4	0: short 1: long
3	Anatomical level 7	0: discret or absent 1: large
4	Proboscis asconiform cells	0: short 1: long
5	Gonoduct sexual dimorphism	0: absent 1: present
6	Female-specific cutaneous glands	0: absent 1: present
7	Male-specific cutaneous glands	0: absent 1: present
8	Number of mature oocytes per ovary	0: < 10 (1–8) 1: > 10 (20–50)
9	Fertilization	0: external 1: internal
10	Eggs laid in gelatinous mass	0: no 1: yes
11	Embryonic development	0: indirect (pilidium larva) 1: direct
11a	Embryonic development	0: Desor larva 1: Schmidt larva
11b	Electrophoretic pattern of the spawn gelatinous mass extract	0: low MW 1: high MW
12	Asexual reproduction (anterior regeneration)	0: absent 1: present
12a	Pattern of anterior regeneration	0: early conical blastema 1: late biconvex blastema
13	Behaviour to stimuli	0: coiling 1: contraction
14	Cutaneous pigmentation	0: absent or discret 1: present
15	Colour pattern of the adult antecerebral end	0: paler than the body 1: as deeply coloured as the body
16	Vitellin component of $MW \times 10^{-3}$ 145	0: absent 1: present
17	Vitellin component of $MW \times 10^{-3}$ 78	0: absent 1: present
18	Vitellin component of $MW \times 10^{-3}$ 75	0: absent 1: present
19	Vitellin component of $MW \times 10^{-3}$ 68	0: absent 1: present
20	Vitellin component of $MW \times 10^{-3}$ 64	0: absent 1: present
21	Vitellin component of $MW \times 10^{-3}$ 44	0: absent 1: present

Table 1. (Continued)

No.	Character	Character states
22	Vitellin component of MW × 10^{-3} 42	0: absent 1: present
23	Vitellin component of MW × 10^{-3} 41	0: absent 1: present
24	Vitellin component of MW × 10^{-3} 40	0: absent 1: present
25	Vitellin component of MW × 10^{-3} 39	0: absent 1: present
26	Vitellin component of MW × 10^{-3} 33	0: absent 1: present
27	Vitellin component of MW × 10^{-3} 31	0: absent 1: present

Reproductive pattern (characters 8 to 11)

The number of eggs per ovary is a taxonomic character reported by Bierne (1970, 1983). Internal *versus* external fertilization is a datum derived from works by Bierne (1970) and Langlet (1972). Direct and indirect patterns of embryonic development were reported by many authors (reviewed by Cantell, 1969, 1972): in *L. lacteus*, *L. longissimus* and the five fissiparous *Lineus* species, eggs are shed directly in the sea water to develop into pilidium larvae; in *L. ruber* and *L. viridis* eggs are laid in a gelatinous mass, the jelly being previously produced by the female-specific cutaneous glands (Bierne, 1970). The flabby gelatinous mass of *L. viridis* spawnings and the firm gelatinous mass of *L. ruber* spawnings allow an easy distinction between spawnings from the two species. Biochemical studies by polyacrylamide gel electrophoresis of their SDS extracts provide two species-specific patterns of components (Fig. 1A). Gelatinous mass consistency is dependent on component molecular weights, low for the flabby mass of *L. viridis* and high for the firm mass of *L. ruber*.

Asexual reproduction (anterior regeneration, character 12)

The capability to reproduce asexually (clonability) is shared between all the nemerteans of the five fissiparous species in which anterior regeneration occurs after sectioning at any body level. Early, conical blastemata occur in *L. sanguineus* (Oxner, 1909) and *L. socialis* (Coe, 1929a, b, 1930). Late, biconvex blastemata take place in

Table 2. Character values matrix for phylogeny analysis of 9 benthic *Lineus* species. All character states are numbered in accordance with the binary coding.

Taxon	Character number																														
	1	2	3	4	5	6	7	8	9	10	11	12	13	14	15	16	17	18	19	20	21	22	23	24	25	26	27	2a	11a	11b	12a
L. ruber	0	0	0	1	1	1	1	0	1	1	1	0	1	1	1	0	0	0	1	0	1	1	0	0	1	0	0		1	1	
L. viridis	0	0	0	1	1	1	1	0	1	1	1	0	1	1	1	0	0	0	1	0	0	0	1	0	0	1	0		0	0	
L. sanguineus	1	1	0	0	0	0	0	1	0	0	0	1	0	1	1	1	1	0	0	0	0	0	0	1	0	0	0	0			0
L. socialis	1	1	0	0	0	0	0	1	0	0	0	1	0	1	1	1	1	0	0	0	0	0	0	1	0	0	0	0			0
L. vegetus	1	1	0	0	0	0	0	1	0	0	0	1	0	1	0	1	1	0	0	0	0	0	0	1	0	0	0	0			1
L. nigricans	1	1	0	0	0	0	0	1	0	0	0	1	0	1	0	1	1	0	0	0	0	0	0	1	0	0	0	0			1
L. pseudolacteus	1	1	0	0	0	0	0	1	0	0	0	1	0	0	1													0			1
L. longissimus	1	1	0	0	0	0	0	1	0	0	0	0	0	1	1	0	0	0	0	1	0	0	1	0	1	0	1	0			
L. lacteus	1	1	1	0	0	0	0	1	0	0	0	0	0	0	1	0	0	1	0	0	0	0	1	1	0	0	0	1			

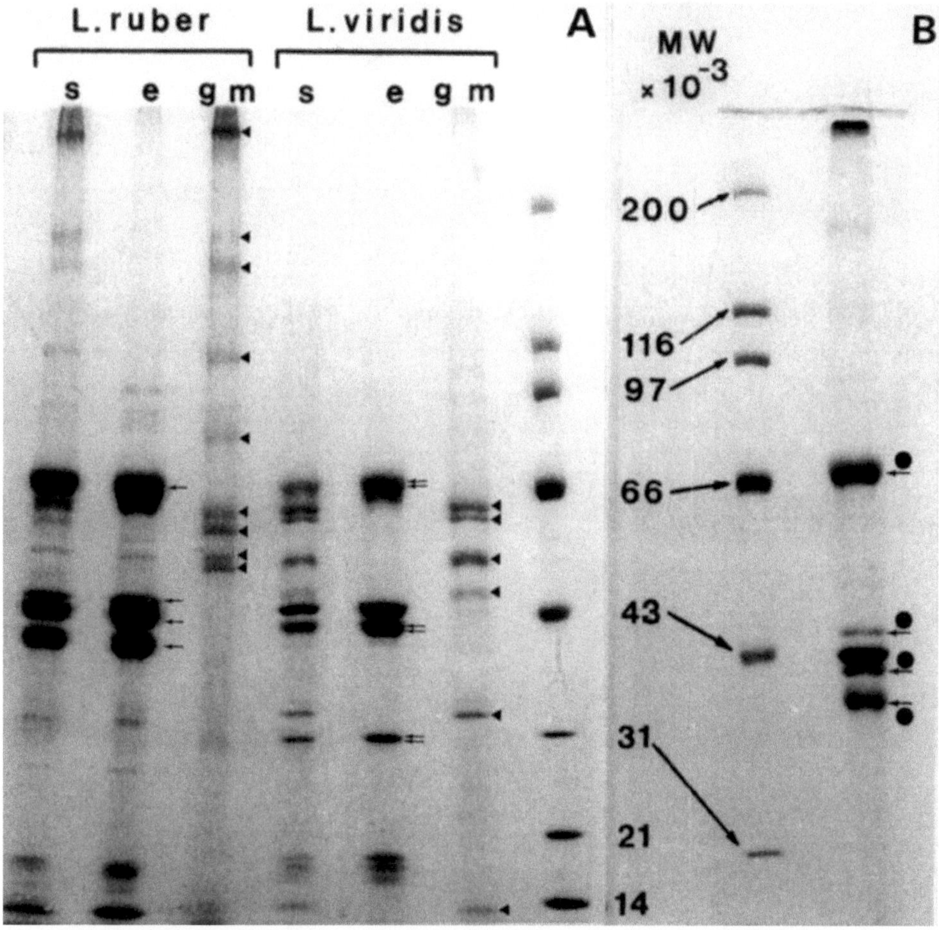

Fig. 1. (A) Polyacrylamide gel electrophoresis of SDS extracts from *L. ruber* and *L. viridis* spawnings (s), eggs (e) and spawning gelatinous masses (gm). *L. ruber* vitellin components are indicated by arrows. *L. viridis* vitellin components are indicated by double arrows. Height and low molecular weight protein patterns from gelatinous mass extracts of respectively *L. ruber* and *L. viridis* are indicated by arrow heads. (B) Polyacrylamide gel electrophoresis of SDS extracts from *L. ruber* eggs performed in a homogeneous gel to show the 4 vitellin components (arrows and discs) of MW $\times 10^{-3}$68, 44 (species-specific), 42 (species-specific) and 39. The arrows show the one-to-one mapping of molecular weights from Fig. A to Fig. B.

L. nigricans (Monastero, 1928) and *L. vegetus* (Tucker, 1959). Very late, biconvex blastemata occur in *L. pseudolacteus*. Up to 12 °C, spontaneous fragmentation of the body, rapid regeneration from pieces and mediocre health of asexually-derived progeny are characteristics of most specimens of *L. socialis* and *L. sanguineus* reared clonally in the laboratory. By contrast, at the same temperature, rare fragmentation, slow regeneration, resistance to parasites and thriving specimens are characteristics of *L. vegetus* and *L. nigricans*.

Behaviour in stress (character 13)

Responses to stimuli such as mechanical irritation are both classical and easily attainable features for diagnosis: *L. ruber* and *L. viridis* contract without coiling whilst all fissiparous *Lineus*, *L. lacteus* and *L. longissimus* react by coiling.

Cutaneous pigmentation and colour (characters 14 and 15)

The pigment is totally absent from *L. lacteus*, discretely located in *L. pseudolacteus* although a few specimens may show rare minute but very pigmented blotches, and abundantly distributed in *L. ruber*, *L. viridis*, *L. sanguineus*, *L. socialis*, *L. nigricans* and *L. vegetus* (Vernet & Bierne, 1983).

The cephalic end is paler than the body in *L. vegetus* and *L. nigricans*. On the other hand, it is at least as deeply coloured as the body in *L. socialis* and *L. sanguineus* (Fig. 2).

Vitellin components (characters 16 to 27)
Separation of yolk proteins by polyacrylamide gel electrophoresis of egg SDS extracts indicate a species-specific pattern of their components (Fig. 1). These vitellin components, characterized by their molecular weight, are significant enough to identify a *Lineus* species from its eggs. The vitellin component pattern of *L. pseudolacteus* is at present unknown because no female specimen has so far been collected.

Systematics

The five fissiparous species of *Lineus* show little or no difference between the character states (Tables 1 and 2). When minor state differences

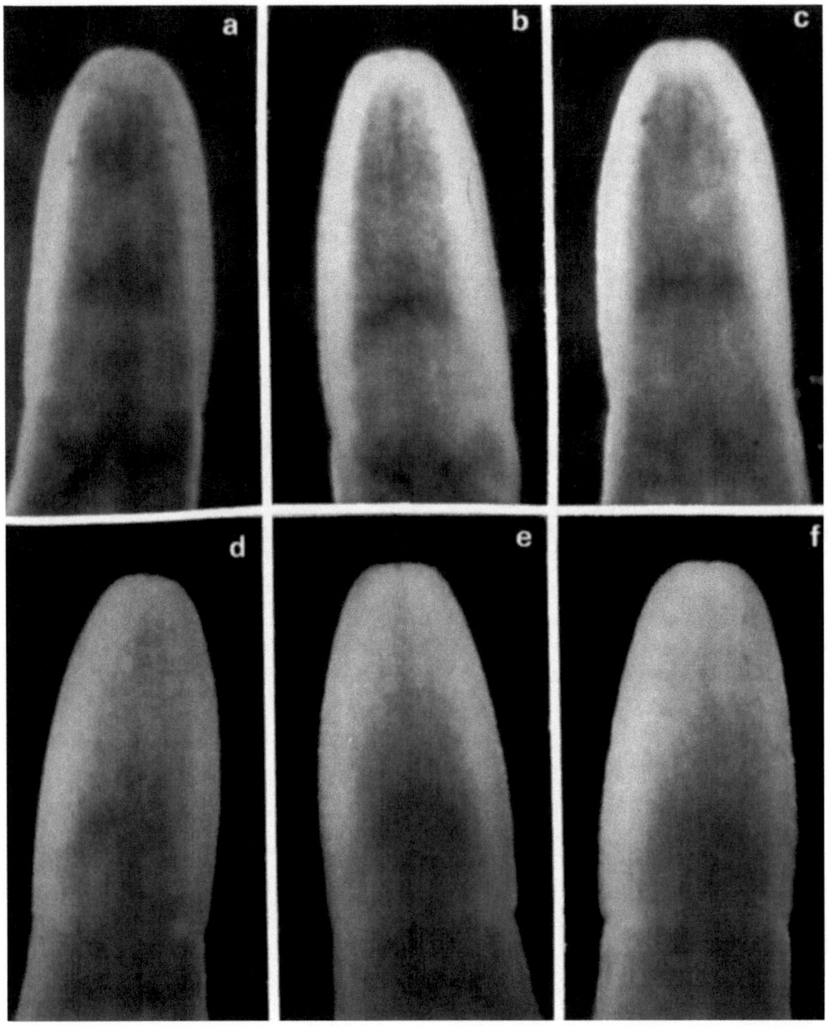

Fig. 2. Cephalic ends from fissiparous *Lineus*. (a) *l. sanguineus* from Roscoff (Brittany shores of the English Channel); (b) *L. socialis* from Nahant (American shores of the Atlantic Ocean); (c) *L. sanguineus* from Banyuls-sur-Mer (French shores of the Mediterranean Sea); (d) *L. nigricans* from Villefranche-sur-Mer (French shores of the Mediterranean Sea); (e) *L. nigricans* from Banyuls-sur-Mer; (f) *L. vegetus* from Roscoff.

are observed, they are easily attributable to intraspecific variation because the absence of intermediate states very likely results from reproductive isolation by asexual multiplication. Consequently, there is no reason to maintain the five fissiparous species as separate. It is accordingly proposed that all the fissiparous *Lineus* be included in a single species, *L. sanguineus*, on the basis of the priority rule. However, subtle, stable features transmissible by asexual reproduction allow us to suggest a division of the species *L. sanguineus* into three sub-species (Table 3), defined phenotypically by their pattern of regeneration and pigmentation. The first sub-species *L. s. sanguineus* (s. sp. nov.) contains *L. sanguineus* and *L. socialis*. The specimens of this sub-species fragment spontaneously at 12 °C into many pieces that rapidly develop small worms both from conical, cephalic blastemata and from posterior regenerates. The antecerebral end of worms is as pigmented as the body (Fig. 2). The second sub-species, *L. s. nigricans* (s. sp. nov.), includes *L. nigricans* and *L. vegetus*. The specimens of this sub-species seldom fragment at 12 °C. When transections cause worm regeneration from body fragments, the late cephalic blastemata are biconvex. The antecerebral end always remains paler than the body (Fig. 2). Nevertheless, asexual reproduction is efficient because of resistance of the worms to parasites (Gregarina, Orthonectida). The third sub-species, *L. s. pseudolacteus* (s. sp. nov.) is probably derived from *L. s. sanguineus* by mutation of a pigment gene. The pigment dots that can occur in the skin of rare specimens suggest an unstable mutation. In one clone of 60 worms, we have counted 4 worms with one blotch and 2 with 2 blotches. These worms were old, regenerated for more than two years. Such worms cut across a pigmented blotch result in two fragments that respectively regenerate anteriorly and posteriorly without pigment dots. These observations may suggest a genetic transposition mechanism similar to the one observed in maize by McClintock (1951).

Data from Riser (1993) are in agreement with our suggestion of keeping only one species designating all the fissiparous *Lineus*.

By contrast, some biological as well as biochemical criteria allow us clearly to define *L. ruber*, *L. viridis*, *L. longissimus* and *L. lacteus* as distinct species. Highly reproducible vitellin component patterns with specific components – vt 145 and vt 78 (fissiparous *Lineus*); vt 75 (*L. lacteus*); vt 64 (*L. longissimus*); vt 44 and vt 42 (*L. ruber*) – provide taxonomy with an interesting complement to the usual classical approaches. Such a taxonomic investigation by biochemical methods will probably be able to improve our knowledge of other nemerteans in future.

Phylogeny

The phylogeny that we put forward in the present study results both from an analysis of character

Table 3.

Current taxonomy	Proposed taxonomy	
Species	Sub-species	Species
L. longissimus (Gunnerus, 1770)		*L. longissimus*
L. ruber (Müller, 1771)		*L. ruber*
L. viridis (Müller, 1774)		*L. viridis*
L. sanguineus (Rathke, 1799) L. socialis (Leidy, 1855)	*L. s. sanguineus*	
L. nigricans (Burger, 1892) *L. vegetus* (Coe, 1931)	*L. s. nigricans*	*L. sanguineus*
L. pseudolacteus (Gontcharoff, 1951)	*L. s. pseudolacteus*	
L. lacteus (Rathke, 1843)		*L. lacteus*

states (Table 1) and from the overall examination of a character value matrix constructed with the aim of performing a subsequent cladistic analysis (Table 2).

The anatomical levels show three species-group-specific states: *L. ruber* type (*L. ruber*, *L. viridis*), *L. sanguineus* type (*L. sanguineus*, *L. longissimus*) and *L. lacteus* type. The morphological features, the reproductive pattern and the behaviour in stress clearly make two separate species-group-specific types stand out: *L. ruber* type (L. ruber, *L. viridis*) and *L. longissimus* type (*L. longissimus*, *L. sanguineus*, *L. lacteus*). By contrast, pigmentation and cutaneous colour, although species-specific (in animals, skin colour often is only individually specific), do not make species groupings stand out. The same is true of vitellins. The pattern of protein components present in eggs does not let us propose species reassembly. However, the high reproducibility of the pattern revealed in all fissiparous *Lineus* is a convincing argument to suggest that they belong to a single species. Subsequent molecular analysis of vitellins will probably be significant in the

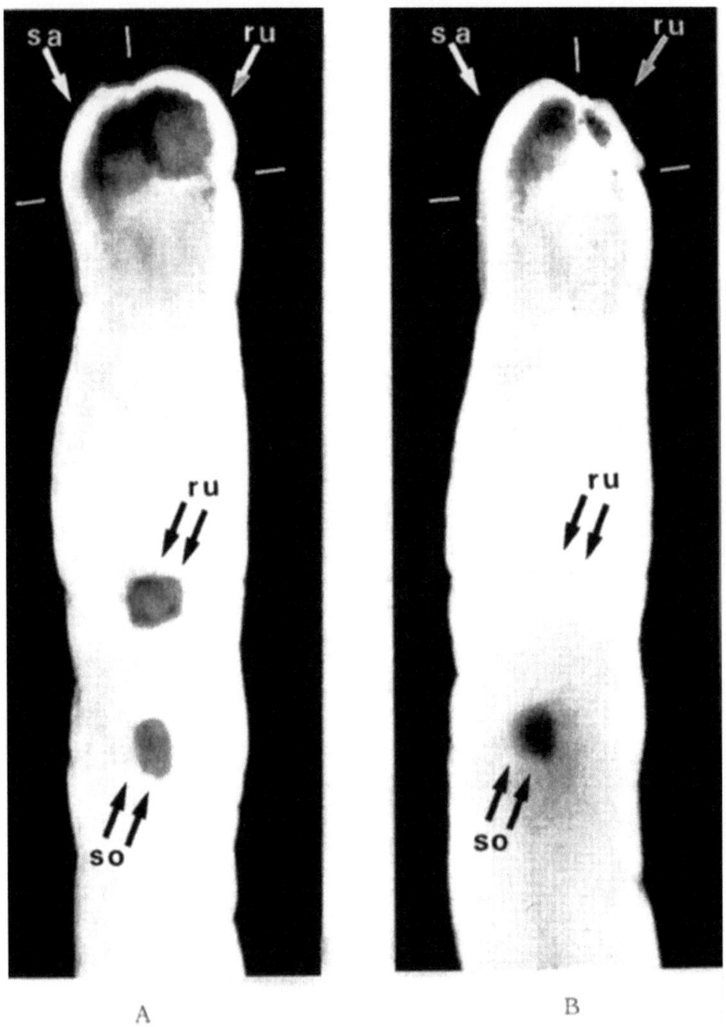

Fig. 3. Antecerebral (arrows) and cutaneous (double arrows) grafts from donors *L. ruber* (ru), *L. sanguineus* (sa) and *L. socialis* (so) transplanted on to a recipient *L. pseudo- lacteus.* (A) No sign of rejection 80 days after transplantation of *L. sanguineus* and *L. socialis* grafts, and 10 days after transplantation of *L. ruber* grafts. (B) Success (survival) of grafts from donors *L. sanguineus* and *L. socialis* and failure (rejection) of grafts from donors *L. ruber* viewed one month later.

search for phylogenetic affinities due to similarity between their amino-acid sequences. The ability to reproduce asexually (anterior regeneration) is obviously a character state shared only by fissiparous *Lineus*.

Eventually, grouping of species possessing the same character states among the 15 biological features show a clear dichotomy – 11 antagonistic states *versus* 4 differently shared states – between the two related species *L. ruber* and *L. viridis* on the one hand and the three species *L. longissimus*, *L. lacteus* and *L. sanguineus* on the other. In terms of evolution, this means that two lines of *Lineus* must have arisen early: the *L. ruber* type line and the *L. longissimus* type line.

Conclusion

Analysis of the character states clearly shows that all the fissiparous *Lineus* belong to a single species: *L. sanguineus* (Rathke, 1799) and examination of the character value matrix concerning finally five *Lineus* species suggests a phylogenetic tree developing from two ancestral branches. The first branch includes the two related species *L. ruber* and *L. viridis*, the second branch includes the three species *L. longissimus*, *L. lacteus* and *L. sanguineus*.

This taxonomic and phylogenetic approach to some cosmopolitan, benthic *Lineus* species is in total agreement with previous histocompatibility tests (Bierne, 1967, 1970; Bierne & Langlet, 1974; Langlet & Bierne, 1973, 1977, 1979, 1982a, b, 1983). Some facts and/or rules attest that *in vivo* confrontation of cells from any two *Lineus* by grafting allows us to appraise affinities between them.

1. Grafts from one fissiparous *Lineus* to another survive indefinitely following the example of allogeneic grafts performed in other *Lineus* species (Fig. 3). By contrast, grafts from non-fissiparous *Lineus* are rejected subacutely (*L. longissimus*, *L. lacteus*) or acutely (*L. ruber*; *L. viridis*) by any fissiparous *Lineus* (Fig. 3). Acceptance of grafts exchanged between all clonable *Lineus* reveals, without doubt, their very close taxonomic relationships. On the other hand, rejection of grafts exchanged between *Lineus* worms reveals distant taxa.

2. Acute or subacute rejection of grafts occurs when donors and recipients belong to two different groups of species; one group contains *L. ruber* and *L. viridis* and the other *L. longissimus*, *L. sanguineus* and *L. lacteus*. This strong histoincompatibility most probably implies an early phylogenetic divergence.

3. By contrast, late rejection of grafts exchanged between *L. ruber* and *L. viridis* suggests that the two species are rather closely related.

4. It is the same for rejection of grafts exchanged between any two among the three species *L. sanguineus*, *L. lacteus* and *L. longissimus*. Long survival of all transplants suggests also that these species are closely related.

Based on these data and facts the following phylogeny is proposed for the five *Lineus* species. Two ancestral lineages of species can be recognized from the criteria of strong histoincompatibility and antagonistic states of most biological characters. The first lineage includes the two species *L. ruber* and *L. viridis*, the second lineage includes the three species *L. longissimus*, *L. lacteus* and *L. sanguineus*.

Acknowledgements

We should like to thank Professor N. W. Riser for *Lineus socialis* and *L. viridis* collected on the American shores of the northern Atlantic Ocean. We are grateful to Professor R. Gibson for reading and improving the manuscript. We thank A. Carlier and J. Leleu for technical assistance.

References

Bierne, J., 1967. Régénération postérieure des chimères interspécifiques chez les Lineidae. Bull. Soc. zool. Fr. 92: 351–359.

Bierne, J., 1970. Recherches sur la différenciation sexuelle au cours de l'ontogenèse et de la régénération chez le Némertien *Lineus ruber* (Müller). Annls Sci. nat. Zool., Paris 12: 181–298.

Bierne, J., 1983. Nemertina. In K. G. Adiyodi & R. G. Adiyodi (eds), Reproductive Biology of Invertebrates. Vol. 1: Oogenesis, Oviposition, and Oosorptiont, John Wiley & Sons, Chichester: 147–167.

Bierne, J., 1985. Histocompatibility in Nemertines: Fates of multiparental *Lineus* constructed by grafting of pieces from many donors. Amer Zool. 25: 135–144.

Bierne, J. & C. Langlet, 1974. Recherches sur l'immunité de greffe chez les Némertiens du genre *Lineus*. Etude de la réponse primaire à la transplantation hétérospécifique. C.r. hebd. Séanc. Acad. Sci., Paris 278: 1445–1447.

Cantell, C. E., 1969. Morphology, development and biology of the pilidium larvae (Nemertini) from the swedish west coast. Zool. Bidr. Upps. 38: 61–112.

Cantell, C. E., 1972. Studies on the morphology, taxonomy and larval development of heteronemertines (Nemertina). Acta Univ. Upps. 218: 1–7.

Coe, W. R., 1929a. Transplantation in nemerteans. Anat. Rec. 44: 223–224.

Coe, W. R., 1929b. Regeneration in nemerteans. J. exp. Zool. 54: 411–459.

Coe, W. R., 1930. Asexual reproduction in nemerteans. Physiol. Zool. 3: 297–308.

Langlet, C., 1972. Contribution à l'étude cytologique et histologique du développement abortif de certains embryons de *Lineus ruber* Müller. Thèse de 3ème cycle, Université de Reims Champagne-Ardenne.

Langlet, C. & J. Bierne, 1973. Recherches sur l'immunité de greffes chez les Némertiens du genre *Lineus*. Evolution de transplants homospécifiques et hétérospécifiques. C. r. hebd. Séanc. Acad. Sci., Paris 276: 2485–2488.

Langlet, C. & J. Bierne, 1977. The immune response to xenografts in nemertines of the genus *Lineus*. In J. B. Solomon & J. D. Horton (eds), Developmental immunobiology, Elsevier/North Holland Biomedical Press, Amsterdam: 17–26.

Langlet, C. & J. Bierne, 1979. Intervention d'immunocytes circulants dans le rejet des greffes chez les Némertiens du genre *Lineus* (Invertébrés, Acoelomates). C.r. hebd. Séanc. Acad. Sci., Paris 288: 1003–1006.

Langlet, C. & J. Bierne, 1982a. Immune character nemerteans of the genus *Lineus*. Eur. J. Immunol. 12: 705–

Langlet, C. & J. Bierne, 1982b. Immunocompetent cells are responsible for rejection of incompatible xenogeneic grafts in *Lineus* (Invertebrata, nemertea). Transplantation 34: 8–12.

Langlet, C. & J. Bierne, 1983. Experimental evidence for cell-mediated immune responses to incompatible grafts in *Lineus* (Nemertea). Develop. Comp. Immunol. 7: 617–620.

McClintock, B., 1951. Chromosome organization and genic expression. Cold Spring Harb. Symp. Quant. Biol. 16: 1–53.

Monastero, S., 1928. Esperienze sulla rigenerazione dei Nemertini (*Lineus nigricans* Bürger 1892 e *Prostoma melanocephalum melan* Johnst. 1827). Boll. Ist. Zool. R. Univ. Palermo 21: 1–16.

Moretto, H. J. A. & G. Vernet, 1985. Les cellules asconiformes de la trompe des hétéronémertes du genre *Lineus*: Aspects morphologiques et fonctionnels. Bull. Soc. zool. Fr. 110: 129–138.

Oxner, M., 1909. Sur deux modes différents de régénération chez *Lineus ruber*. C. r. hebd. Séanc. Acad. Sci., Paris 148: 1424–1426.

Riser N. W., 1993. Observations on the morphology of some North American nemertines with consequent taxonomic changes and a reassessment of the architectonics of the phylum. Hydrobiologia 266: 141–157.

Tucker, M., 1959. Inhibitory control of regeneration in nemertean worms. J. Morph. 105: 569–593.

Vernet, G. & J. Bierne, 1983. Etude de la couleur du corps de trois némertes: *Lineus sanguineus*, *L. pseudolacteus* et *L. lacteus*. Application au marquage des tissus des chimères interspécifiques. Bull. Soc. zool. Fr. 108: 257–266.

Vernet, G. & J. Bierne, 1988. Neuroendocrine control of gonadogenesis in regenerating *Lineus lacteus* (Heteronemertea). Hydrobiologia. 156: 53–60.

Hydrobiologia **266**: 169–174, 1993.
R. Gibson, J. Moore & P. Sundberg (eds), Advances in Nemertean Biology

A proposal for a check-list of characteristics to be used in the descriptions of palaeonemertean species

Richard Hylbom
Department of Zoology, University of Gothenburg, Medicinaregatan 18, S-413 90 Göteborg, Sweden

Key words: nemertea, morphological characters

Abstract

The need for a standardized description of nemerteans is developed by suggesting a check-list of characters to be used in descriptions. Some problematic characters are discussed and illustrated.

Introduction

Nemertean taxonomy is, and has been, almost entirely based on morphological characters. The first descriptions included only a few external characters, like the colour and general shape of the animal, and maybe also some note on the habitats where the worm was found. With an increasing awareness of the high number of species actually existing, the descriptions also increased in detail. In spite of the more detailed descriptions today, there are still inconsistencies in the way they are written and Gibson (1985) urged for a standardized description of nemerteans. This paper is a proposal for such a standardized description for paleonemerteans; *i.e.*, standardized in the sense that *at least* these characters should be reported. There will of course be characters unique for a species, or a group of species, that will not fit in my scheme. But, the 'form' will not only be added by this kind of characters, it should also be improved and developed over the years as our knowledge increases. Similar lists of characters to be used as a basis when describing hoplonemerteans can be found in e.g. Sundberg (1989a, b).

The characters I propose to be used in future descriptions of paleonemertean species are listed below. Certain characters demand an explanation concerning difficulties in examination or definition of the characters, and such remarks follow immediately after the characters.

Collection data

Locality, substrate, depth, number of specimens, etc.

Observations of living individuals

a) Body length and width.
   Measurements while gliding.
b) Colour and pattern.
c) Shapes attained by head.
d) Head-furrows, eyes and other visible external features of living specimens visible to the naked eye, or with the aid of dissecting or compound microscope.

Epidermis

a) Thickness of epidermis.
   Measurements in relation to cross-section.

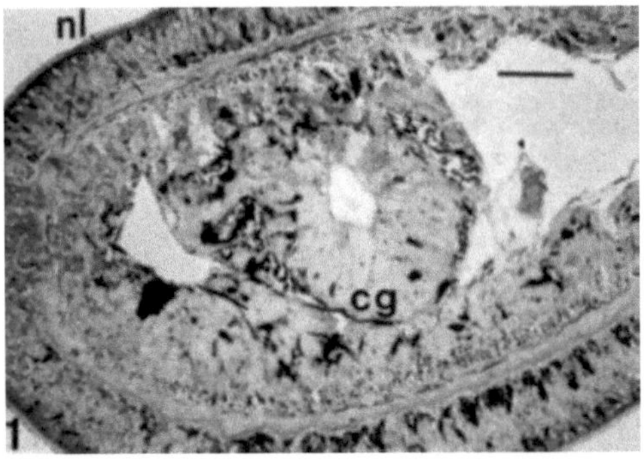

Fig. 1. Tubulanus annulatus. Cross-section showing cephalic glands, *cg*, in rhynchodeum and cephalic tissue. Nervous layer, *nl*, beneath basement membrane. Scale = 100 microns.

b) Glands, characters and distribution.
Type of secretion (acidophile, basiophile, granular, etc.).
Frequency of different gland-types in different regions of body, as well as position in epidermis (close to surface or deep).

c) Cephalic glands.
There are two different types of so called cephalic glands, *i.e.* glands in the head region in front of the brain. One type is the cells with well developed nucleus typical for many *Tubulanus* species, situated both epidermal and subepidermal, and in the epithelium of the rhynchodaeum (Fig. 1). The other type is the granular cells found in Cephalotrichidae (Hylbom, 1957: plate 3, Fig. 16).

Basement membrane or subepidermal connective tissue

a) Thickness in different regions.
Measurements of thickness in brain-mouth and intestinal regions. Even if the thickness may be influenced by the fixation, measurements are better than subjective characters like 'thick' or 'thin'.

b) Processes into epidermis and/or mesh-like structure of the inner layer of the basement membrane.
See Figs 2, 3.

Body musculature

a) Arrangement of muscle layers, relative thickness and extent of distribution in body.
The thickness of muscle layers can be greatly influenced by fixation, but measurements can still be valuable.

b) Presence of extra muscle layers in relation to alimentary canal and proboscis sheath.
Most commonly a longitudinal muscle plate between proboscis sheath and foregut (Fig. 4).

c) Muscle crosses between circular muscle layers.

d) Conspicuous radial fibres in relation to mouth and foregut-region.
Radial muscle fibres are shown in Fig. 4.

Proboscis apparatus

Rhynchodaeum

a) Variation of height of epithelium in different regions.
This character may vary with fixation, but notes on the variation of the height of epithelium along rhynchodaeum can be of interest compared with other species.

b) Presence of nerves in epithelium.
See Fig. 5.

c) Presence of glands in epithelium, with description.

Figs 2–7. Fig. 2a–4. Tubulanus annulatus. 2a. Cross-section showing basement membrane with processes, *p*, into epidermis and mesh-like inner layer, *m*, of the membrane. *2b.* Lateral vessels in a lacunar, *lvl*, system around oesophagus. *3.* Upper dorsal nerve, *udn*, and lower dorsal nerve, *ldn*. Mesh-like structure, *m*, of basement membrane. *4.* Mouth with radial muscle fibres, *r*, and longitudinal muscle plate, *lmp*, between proboscis sheath and mouth. *Fig. 5. Carinina coei.* Cross-section through rhynchodeum in anterior mouth-region. In rhynchodeum, nerves, *n*, in epithelium, circular muscles, *cm*, and horizontal muscles, *hm*. Radial muscle

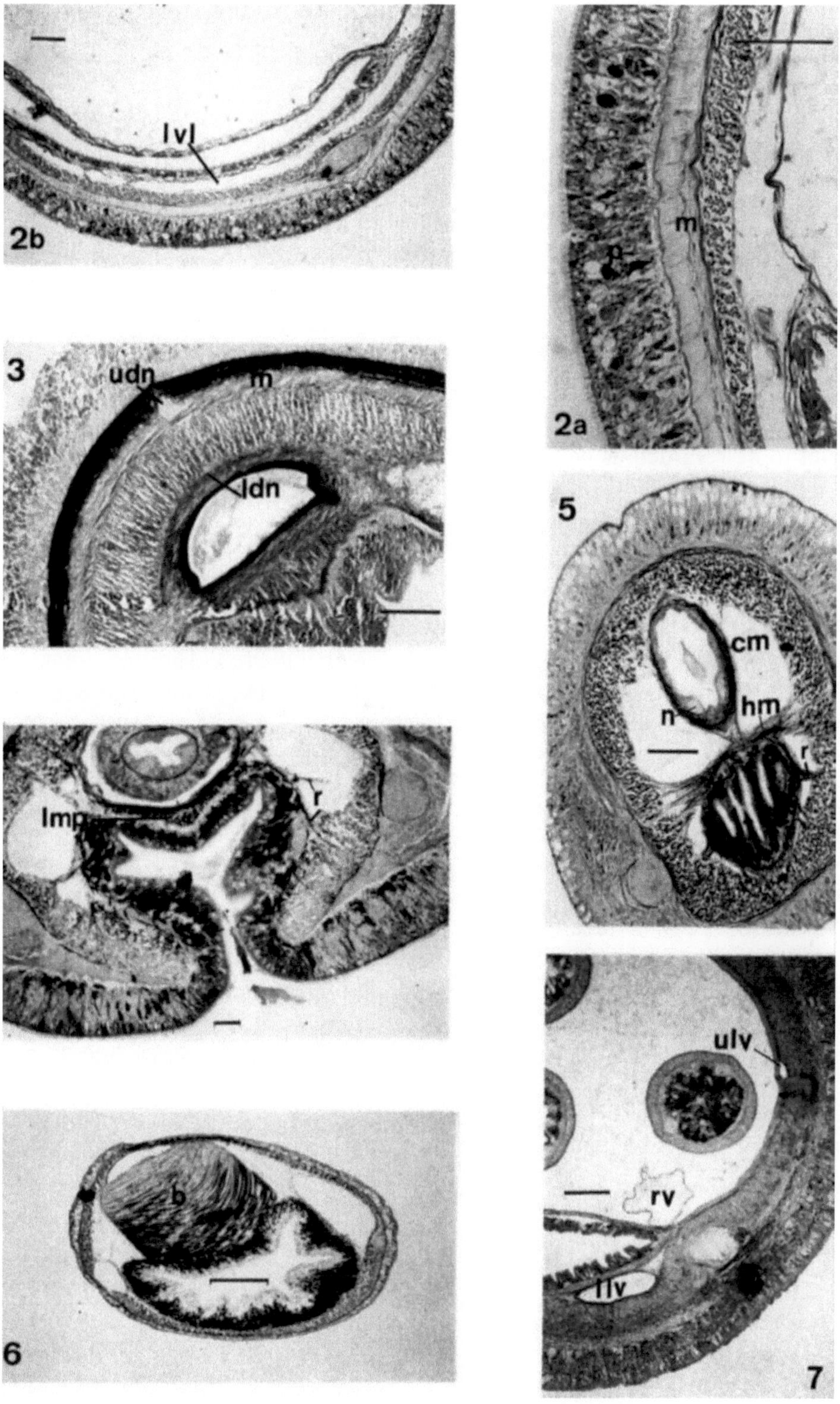

fibres, *r*, in connection to mouth can also be seen. *Fig. 6. Callinera bürgeri.* The muscle bulb, *b*, in rear end of proboscis sheath. *Fig. 7. Carinoma armandi.* Upper and lower lateral vessels, *ulv* and *llv*, and rhynchocoel vessel, *rv*. Scale in Fig. 2a = 50 microns, in 2b–7 = 100 microns.

d) Presence and kind of rhynchodaeum musculature.
In the posterior region, the rhynchodaeum is often surrounded by a circular muscle forming a kind of sphincter (Fig. 5). Sometimes a horizontal muscle plate occur ventrally to the rhynchodaeum (Fig. 5).
e) Presence of caecum.

Proboscis sheath

a) Muscle layers.
The longitudinal muscle plate discussed above is not to be regarded as belonging to the proboscis sheath.
b) Remarkably strong developed circular layer.
In *Carinoma* and *Callinera*, the circular muscles of the proboscis sheath can be extremely well developed in the posterior part (Fig. 6).
c) The extension of sheath in the body.
To be able to observe this character, it is important to section the posterior region of the body.
d) Presence of caeca.
Caeca on rhynchodaeum and proboscis sheath can be artefacts due to conservation contractions. Investigate more than one specimen whenever possible. See Sundberg (1979) for an account of intraspecific variation in nemerteans.

Proboscis

a) Muscle layers.
b) Position of proboscis insertion in relation to brain.
c) Regional differentiation.
d) Nerves.
e) Glands in epithelium, with characters.

Alimentary canal

a) Location of mouth in relation to other parts of body.
b) Presence of caeca.
c) Description of different regions, epithelium, glands etc.

Blood vascular system

a) Lateral vessels.
Number and position in relation to muscular layers. In *Carinoma* the upper pair of lateral vessels, close to, but outside the proboscis sheath have been called 'lateral rhynchocoelomic vessels' (Fig. 7). Wijnhoff (1924) wrote in her description of *Carinoma caraibica* 'The lateral rhynchocoelomic vessels and the rhynchocoelomic vessels themselves'.
b) Presence of rhynchocoel vessels (two lateral and/or one median).
Rhynchocoel vessels may be paired (Fig. 8) or unpaired. The mid-dorsal vessel is situated between the proboscis sheath and the intestine and/or partly inside the ventral wall of the proboscis sheath. This vessel is equivalent to the unpaired rhynchocoel vessel.
c) Presence of rhynchocoel villi.
The rhynchocoelic villi can be difficult to separate from rhynchocoel vessels. The position is about the same, but the villi are glandlike processes without conspicuous lumen. Extremely well developed villi are shown in Fig. 10.
d) Lateral vessels surrounding oesophagus in a lacunar system.
See Fig. 2.
e) A cephalic lacuna developed or not.

Excretory organs

a) Present or absent.
Many short excretory organs are present in Cephalotrichidae, which may be difficult to observe, especially in small specimens.
b) Numbers of nephridial canals.
Can be observed on cross-sections on each side.

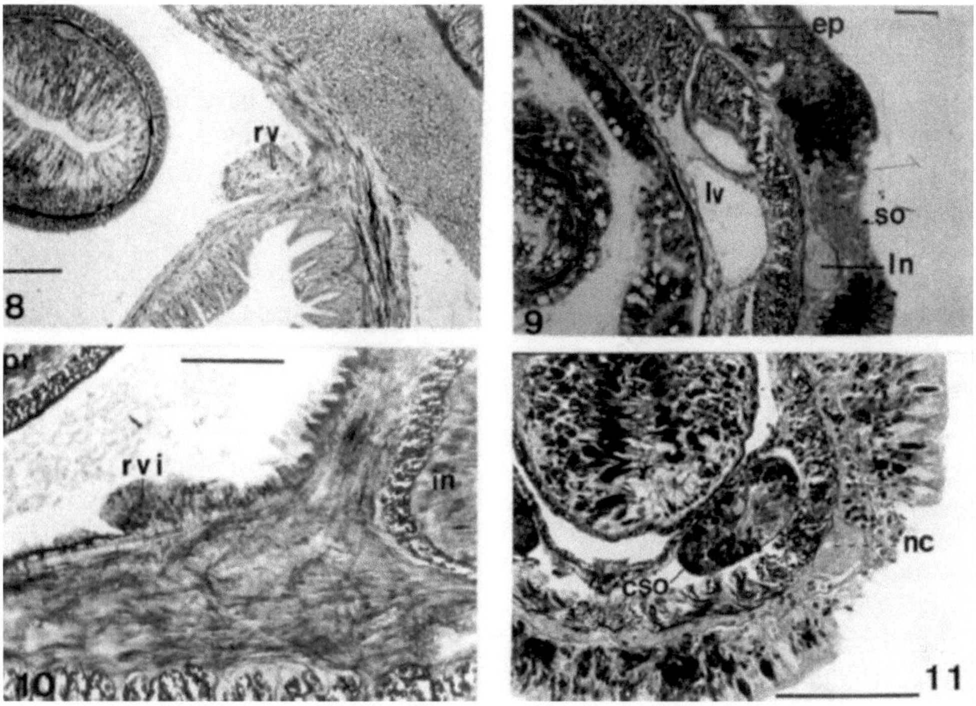

Figs 8–11. Fig. 8. Tubulanus superbus. Rhynchocoel vessel, *rv*, in proboscis sheath. *Fig. 9. Tubulanus théeli.* Side-organ, *so*, excretory pore, *ep*, lateral vessel, *lv*, and lateral nerve, *ln*. *Fig. 10. Tubulanus* sp.. Rhynchocoelic villi, *rvi*, in proboscis sheath; intestine, *in*, proboscis, *pr*. *Fig. 11. Hubrechtella dubia.* Cerebral sense organ, *cso*, in lateral blood vessel; neurochord, *nc*. Scales in Fig. 8–11 = 100 microns.

c) Position of nephridial canals in relation to muscle layers.
d) Presence/absence of glandular expansion of the anterior part of the nephridial canals into lateral blood vessels.
e) Excretory pores, numbers and position.

Nervous system

a) Position of lateral nerves in relation to layers of body wall.
b) Upper and lower dorsal nerve – distinct or not apparent.
   The lower dorsal nerve can sometimes be very difficult to detect and generally does not reach far backwards. The so-called rhynchocoel nerve is synonymous to the lower dorsal nerve (cf. Fig. 3).
c) Brain commissures – position, numbers, and thickness.
d) Nervous layer in the head region.
   A layer of nerves situated outside or inside the basement membrane in head region (Figs 1, 3).
e) Presence/absence of four big nerves in the head region.
   The four big nerves in head region is typical for Cephalotrichidae (cf. Gerner, 1969: Abb. 1). The four nerves are often associated with the cephalic glands.
f) Presence of neurochords or neurochord cells.
g) Presence of buccal nerves, paired or unpaired.
   In Cephalotrichidae, the unpaired nerve soon divides and becomes paired.

Sense organs

a) Cerebral sense organs, description.
   The cerebral sense organs are generally simple ciliated ducts which do not penetrate the

basement membrane. In *Hubrechtella*, however, they are large and situated inside the lateral vessels (Fig. 11). The same type organs is described from the genera *Hubrechtia* and *Coeia*.

b) Eyes, presence, number of.
c) Side organs.
   The side-organs are situated close to the excretory pores outside the lateral nerves (Fig. 9). The side-organs must not be mixed up with the cerebral sense organs. In german literature the cerebral sense organs are sometimes called 'Seitenorgane'. For example, Bürger (1892) when describing *Carinina grata* stated: 'Die Seitenorgane bilden epitheliale Einschnitte'. Later Bürger (1895) in an account of the same species stated 'Seitenorgane sind nicht vorhanden'. Bürger (1904) made clear the difference between 'Seitenorgane und Cerebralorgan' in his description of *Carinina antarctica*.
d) Sensory pits in the head region.
   This organ is typical for the genus *Carinoma*, but has also been described from the genus *Statolithonemertes*.
e) Presence of statolithes.
   A character easily overlooked.

Gonads

a) Separate sexes or hermaphrodite.
   The only known hermaphroditic paleonemertean is *Procephalothrix hermaphroditicus* (Gibson *et al.*, 1990).
b) Arrangement of the gonads in a row on each side of the body or clustered above each other.

Discussion

The generic diagnoses for the paleonemertean genera do not include the same characters in a consequent way and it is difficult to find out what is unique for one genus compared with another. Many characters said in a diagnosis to be unique for a genus are in fact be found in a number of species from other genera (Hylbom, pers. obs.). A reason for this may be that species have not been described in a standardized way, *i.e.* the same characters have not been included in the descriptions. It is my hope that the proposal presented here will remedy this to at least some degree.

Acknowledgements

Thanks are due to Drs P. Sundberg and N. Riser for comments and advice, and to Inger Holmqvist for photographic assistance. Financial support for taking part in the Bangor Symposium 1991 was given by Wilhelm and Martina Lundström's foundation.

References

Bürger, O., 1892. Zur Systematik der Nemertinenfauna des Golfes von Neapel. Nachr. Königl. Ges. Wissensch., Göttingen 5: 137–178.

Bürger, O., 1895. Die Nemertinen des Golfes von Neapel und der Angrenzenden Meeres-Abschnitte. Fauna Flora Golf. Neapel 22: 1–743.

Bürger, O., 1904. Nemertinen. Expéd. Antarct. Belge, Résult. Voyage S.Y. 'Belgica' 1897–1899, Zool., 1–10.

Gerner, L., 1969. Nemertinen der Gattungen *Cephalothrix* und *Ototyphlonemertes* aus dem marinen Mesopsammal. Helgoländer wiss. Meeresunters. 19: 68–100.

Gibson, R., 1985. The need for a standard approach to taxonomic descriptions of nemerteans. Am. Zool. 25: 5–14.

Gibson, R., M. Sánchez & M. Méndez, 1990. A new species of *Procephalothrix* (Nemertea, Anopla, Archinemertea). J. Nat. Hist. 24: 277–287.

Hylbom, R., 1957. Studies on palaeonemerteans of the Gullmar Fjord area (West coast of Sweden). Arkiv f. Zool. 10: 539–582.

Stiasny-Wijnhoff, G., 1924. On a collection of nemerteans from Curacao. Bijdr. Dierk. 24: 97–120.

Sundberg, P., 1979. Statistical analysis of variation in characters in *Tetrastemma laminariae* (Nemertini), with a redescription of the species. J. Zool., Lond. 189: 39–56.

Sundberg, P., 1989a. Classification of family Plectonemertidae (Nemertea): a phenetic comparison. Zool. J. Linn. Soc. 97: 57–68.

Sundberg, P., 1989b. Phylogeny and cladistic classification of the paramonostiliferous family Plectonemertidae (phylum Nemertea). Cladistics, 5: 87–100.

Hydrobiologia **266**: 175–184, 1993.
R. Gibson, J. Moore & P. Sundberg (eds), Advances in Nemertean Biology

The genus *Valdivianemertes* Stiasny-Wijnhoff 1923 (Nemertea, Enopla, Hoplonemertea): nomenclatural status and proper systematic position

Frank B. Crandall
Turkey Run Research Institute, 900 Turkey Run Road, McLean, Virginia 22101-1700, USA

Key words: Akrostomum, Acrostomum, Cratenemertidae

Abstract

The genus *Akrostomum* was erected by Grube in 1840 for a single species *A. stannii* which remained a *nomen dubium* until redescribed by Bürger in 1895 as *Amphiporus stanniusi.* Bürger (1909) described *Drepanophorus valdiviae* which shared many characters with *A. stannii.* Stiasny-Wijnhoff (1923) noted that these two forms belonged neither to *Amphiporus* nor *Drepanophorus* and united them in a new genus *Valdivianemertes* but did not mention the earlier name *Akrostomum* nor a family affiliation. Several major characters are examined. The nomenclatural status is analyzed, with retention of the junior synonym *Valdivianemertes*, and the type species is fixed as *V. stannii* by subsequent designation. Family affiliation is discussed, and the genus is transferred to the Cratenemertidae.

Introduction

Through the first decade of this century the Hoplonemertea consisted of only a few families, broadly and loosely defined on the basis of very few characters. The next sixty years saw great advances in an understanding of nemertean morphology and its relationship to the systematics of the Hoplonemertea. Two of the early genera typifying such families, *Amphiporus* Ehrenberg and *Drepanophorus* Hubrecht, accumulated species of widely disparate morphology due to the poorly understood significance of many of the observed characters. Later, when clearly-related species were drawn from these heterogeneous assemblages to be united in smaller, more precise groupings, the question of family relationships and placement often was not addressed and new generic names were sometimes assigned without regard to nomenclatural priorities. *Valdivianemertes* Stiasny-Wijnhoff (= *Akrostomum* Grube) is such a case. This study examines the nomenclatural status of the genus and its systematic position.

History of the constituent species

Grube (1840) described a new genus and species, *Akrostomum stannii*, from a single, fragmented alcoholic specimen giving as generic diagnosis: mouth and anus at ends of body directly opposite one another; body not wormlike, like *Nemertes* but rather shorter, broad, and flattened; head demarcated from body by a distinct groove; no eyes. His specific description noted that: the animal was 37 mm long and 7 mm wide, not segmented, with the ends abruptly tapering to a point; head, at its base, narrower than forward part of body; dissection revealed, above the gut, a retractile proboscidiform organ which protruded about one-fourth of the body length out through the opening in the tip of the triangular head and also appeared in the canal of the mouth; pale flesh colored; skin entirely smooth. Grube placed his new genus in Ehrenberg's (1828–1831) family Gyratricina into which Ehrenberg had grouped *Prostoma* Duges and his new genera *Orthostoma*, *Gyratrix*, *Tetrastemma*, *Hemicyclia*, *Ommatoplea* and *Amphiporus*. Only the genera *Prostoma*, *Tetrastemma* and *Amphiporus* survive, albeit quite differently defined, and Ehrenberg's higher taxonomy was abandoned long ago.

For all of its shortcomings, Grube's description correctly noted the key nemertean features of proboscis apparatus and separate mouth and anus at opposite ends of the body. His further observations of the partially-everted proboscis lying in what was clearly the rhynchodaeum (notwithstanding that he referred to it as though it were the esophagus) and then passing through the single anterior aperture establish the animal as an enoplan and most likely monostiliferous. Diesing (1850, 1862) listed the species as *Acrostomum stannii* and Carus (1885) (doubtfully) mentioned *Akrostomum stannii*; neither added new information.

Hubrecht (1879), in speaking of *Amphiporus hastatus*, commented, 'There is a certain probability that Grube established his genus *Akrostomum* on a spirit-specimen of this species in which case his generic name might be re-established and more closely defined. However, provisionally I propose to leave them with the Amphipori.' Hubrecht subsequently (1883) listed *Akrostomum hastatus* with *Amphiporus hastatus* as synonym. How Hubrecht reached his conclusions is unclear since the descriptions of *A. hastatus* and *A. stannii* were inadequate to identify a specimen in hand.

Vaillant (1890) first listed *Acrostomum stannii* as a *species indeterminata* but in the same article noted Hubrecht's suggested synonymy and gave a somewhat confused account of the species' probable status. Joubin (1890) listed *Akrostomum stannii* as a synonym for specimens he identified as *Amphiporus hastatus*, adding the following comments (author's translation from French), 'It differs from the description of that author [McIntosh, 1873–74] in that the eyes are large and very obvious although deeply situated beneath the integument. The cephalic grooves are very distinct. The proboscis opening is very close to the mouth, very much as noted by Hubrecht. I do not believe, however, that it may be used to re-establish the old genus *Akrostomum* of Grube for a species which, because of all its characters, is clearly an *Amphiporus*.' Joubin later (1894) doubtfully repeated this synonymy. It is clear that the species Joubin examined, with its large and obvious eyes, was not *A. stannii*, which Grube described as eyeless, and that his synonymy was unjustified. Moreover, Gibson & Crandall (1989) concluded that the various fragmentary and confused accounts of *Amphiporus hastatus* probably referred to different forms and they regarded it as a *nomen dubium*.

Grube's description is no more than a *nomen dubium* and his species might so remain but for the fact that a preserved collection from Naples was sent to Bürger for identification while he was in the process of examining the nemertean holdings of the Berlin Natural History Museum. He was able to compare specimens from Naples with the fragments of Grube's original specimen contained in the Berlin collections and satisfy himself that they belonged to the same species. Later, he published a much-amplified description (Bürger, 1895a) under the name *Amphiporus stanniusi*, listing *Acrostomum stannii* (Grube, 1840) as a synonym. Bürger (1895b) repeated much of his redescription, again as *Amphiporus stanniusi*, but with *Akrostomum stannii* (Grube, 1840) as the synonym. His change in the spelling of the specific name was later (Bürger, 1904) corrected to *Amphiporus stannii* with the other spellings listed as synonyms.

Bürger (1909) described *Drepanophorus valdiviae* from a specimen taken at 155 m off the southern tip of Africa by the steamer *Valdivia* during the German Deep Sea Expedition (1898–99). The specimen was accompanied by notes and sketches from life. His description noted (author's translation from German), 'This nemertean is like a *Drepanophorus* externally and brings to mind especially *D. crassus*. Unfortunately, the proboscis is not present and the organization exhibits some features pointing more toward *Amphiporus* than toward *Drepanophorus*, yet this specimen is placed in the genus *Drepanophorus* since the rhynchocoel has, in the most anterior sections, a few outpouchings arising laterally opposite one another, known until now only from *Drepanophorus*.' Further on he stated, 'It was previously emphasized that the rhynchocoel of *D. valdiviae* differs very significantly from the previously known littoral species of *Drepanophorus* in that its pockets are limited to the anterior section of the body (*i.e.*, essentially to the nephridial region). These pockets originate at regular intervals opposite one another and represent moderately deep sacs or, a better way of expressing it, protrusions, which possess a very heavy wall (Taf. VIII: Figs 5, 6).' Elsewhere his description noted, 'The wall of the rhynchocoel has an unusually thick muscle layer which, as in the case of other littoral drepanophorans, exhibits the characteristic that its constituent circular and longitudinal muscle fibers do not form separate layers but, on the contrary, intertwine themselves with one another, indeed in such a manner that the circular predominate toward the outer side and the longitudinal toward the inner side.' He also noted that, 'The gut agrees in a surprising manner with that of *D. crassus*: mouth and proboscis openings are separate.'

Bürger was an excellent histologist, a careful observer, and a highly skilled illustrator whose figures often added important information to the description given in the text (notably for such features as diagonal muscle layers and esophageal-rhynchodaeal relationships). Although the text of his account (1895a) of *A. stanniusi* does not mention the interwoven rhynchocoel musculature, this feature is clearly illustrated (Taf. 17, Figs 5, 13–14). Since the same plate shows with equal clarity bilayered rhynchocoel musculature for other figured species, this character, as illustrated, must be given full credence. Another figure (Bürger, 1895b: Taf. III, Fig. 10) clearly shows the interwoven rhynchocoel musculature, this time in sagittal section. Why his description of *D. valdiviae* failed to compare the rhynchocoel musculature and diverticula to *A. stannii* remains unclear.

It must be noted that Bürger's (1895a) description of *A. stanniusi* stated 'Proboscis and mouth openings fall together. The common opening lies strikingly far, namely 1.5–2.0 mm, behind the tip of the head', and his later (1895b) account contained an almost identical statement. However, in his transverse view of the precerebral region (1895a, Taf. 17, Fig. 13) they appear somewhat separated and his later sagittal view (1895b, Taf. III, Fig. 10) shows the rhynchodaeal and esophageal canals as fully separate almost to the aperture in the body wall. Indeed, some of Friedrich's and Stiasny-Wijnhoff's species, which are described as 'proboscis and mouth opening into a common atrium', are not appreciably different from Bürger's illustration, including the continuation of epidermal epithelium right up to the oesophageal opening. Were the head of the animal figured by Bürger somewhat contracted (or an atrium somewhat everted), the mouth and proboscis pore might well appear as separate openings. Unfortunately, Bürger's (1909) account of *D. valdiviae* does not provide equally explicit detail about esophageal-rhynchodaeal relationships, but this seeming difference between the two species may have negligible significance.

Gering (1913), in the course of discussing *Amphiporus bergendali*, noted the dorsoventral flattening of the posterior portion of *D. valdiviae* figured by Bürger (1909: Taf. VIII), but added no new information. Gering (1912), in his original description of *A. bergendali*, had noted its swimming behavior and the similarity of this feature to that described by McIntosh (1873–74) for *A. pulcher* and also referred to Coe (1905) who described similar morphology and presumed behavior for *A. punctatulus*. I can confirm a close resemblance of this morphological feature of *D. valdiviae* and *A. punctatulus* based on studies of the original specimen series used by Coe in describing the latter. Friedrich (1955) transferred *A. bergendali* and *A. punctatulus* to *Cratenemertes* and then (1968) *Nipponnemertes*. Berg (1985) regarded *N. bergendali* as synonymous with *Nipponnemertes pulcher* and made additional observations on the posterior dorsoventral flattening and swimming behavior in a large number of live specimens. However, this synonymy may be unjustified, since Berg (1972), in his redescription of *N. pulcher*, observed no diagonal muscle layer in the body wall, while Gering (1912, 1913) stated that such a layer was present in *A. bergendali* though difficult to see. Berg (1985) regarded Gering's diagonal fibers as a contraction phenomenon, but I have found diagonal fibers of either fasciated or non-fasciated type in all members of the Cratenemertidae thus far examined. It is true that the non-fasciated type is very difficult to see in contrast to the fasciated type which was noted, but not fully described, by Coe (1905) for *N. punctatulus*. Furthermore, biomechanical analysis of circular-diagonal-longitudinal muscle layer relationships, whether the diagonal layer is fasciated or non-fasciated, makes it extremely difficult, if not impossible, to accept the contraction phenomenon theory. Moreover, Berg (1985) did not mention the short, forwardly directed diverticulum arising from the ventral wall of the rhynchocoel which Gering (1912) found in both of his specimens, but his earlier (Berg, 1972) paper redescribing *N. pulcher* regarded such pouches as, once again, contraction artifacts.

Brinkmann (1914–15), in discussing his new genus *Uniporus*, commented (author's translation from German), 'It is most striking that in the case of these species, which show a whole series of primitive enoplan characters, we find the strongest specialization of the rhynchocoel diverticula. Structures which in the higher forms of the family [Drepanophoridae] have become gradually reduced, at first their branching, then their muscle layer loss and finally even their reduction in number, so that in the case of *Drepanophorus valdiviae* Bürger they only occur in the anterior region of the body, and in the case of the pelagic nemerteans (which belongs to an investigation of all of the *Drepanophorus* groups to be published later) finally disappear entirely.'

Brinkmann (1917) said 'Within the genus *Drepanophorus* we find one species, *Drepanophorus valdiviae* Bürger, which indeed for the genus possesses remarkable characteristics, it has, no doubt through reduction phenomena, significantly approached the pelagic nemerteans. Foremost, this is true of the rhynchocoelic diverticula; these are, of course, large and branched in the case of *Uniporus* and their wall is strongly muscular; in the case of *Drepanophorus*, with the exception of *D. lankesteri* Hubrecht, the diverticula are unbranched; the musculature diminshes in thickness within the genus ultimately to disappear entirely, and in the case of *D. valdiviae* we find short diverticula which no longer occur throughout the entire body but, rather, sparsely, and indeed are found only in the anterior region of the body. The cerebral organs are known in the case of the drepanophorans to be very large and the cerebral canal provided with a broad characteristic sac. In the case of *D. valdiviae* these organs have become not only strikingly small but also the sac has completely disappeared. However, even if we also find here an approach to the pelagic nemerteans, the difference is still quite significant.' It seems evident from the foregoing that Brinkman fully accepted Bürger's placement of *D. valdiviae* as a valid drepanophoran and to a considerable degree saw polystiliferans as monophyletic with the various character states simply representing evolutionary continua, a view perhaps not widely held today.

The genus *Valdivianemertes*

Stiasny-Wijnhoff (1923) recognized the close correspondence between *A. stannii* and *D. valdiviae* and that they could be properly placed in neither *Amphiporus* nor *Drepanophorus*. After noting the characteristic presence of paired rhynchocoelic diverticula in the Reptantia and their very unusual occurrence in *A. stannii*, she commented, 'However these diverticula are quite other structures, or at least much more primitive; they are present in the nephridial region alone and are very small though wide. The musculature of the rhynchocoelomic cavity widens out at certain places. These are the sacs that have a wide lumen and open with a wide mouth into the rhynchocoelomic cavity. The muscular walls of these sacs are the regular continuations of, and just as thick as, the rhynchocoelomic wall, and a difference of structure seems not to exist. The wall of these diverticula in Polystilifera is as a rule much thinner, even when contracted, and we know that the mouth is provided with a sphincter, that is absent in *A. stannii*.' She compared the similarities of the two species and, contrasting them with the Polystilifera, observed, 'In *Drepanophorus valdiviae* Bürger, which species has exactly the same rhynchocoelic diverticula as *Amphiporus stannii* ... in both species these structures are restricted to the nephridial region, and in other characteristics they are very much alike too: they have no eyes, both possess small cerebral organs without a sac, behind or partly behind the brains, both have lateral nerve cords (not ventral as in *Drepanophorus*), both have a layer of glands in the head that fails in all Polystilifera and is present in *Amphiporus*, both have brains that are quite different from all Polystilifera, with very small dorsal and larger, perfectly separated ventral ganglia, in both the vascular system differs from that of the Polystilifera by the presence of a dorsal loop over the brain, etc, etc.' On these grounds Stiasny-Wijnhoff transferred the two species to a new genus which she named *Valdivianemertes*. With reference to the proboscis she noted, 'The presence of the only [single] stylet in *Valdivianemertes stannii* (Grube) makes it certain that both *V. stannii* and *V. valdiviae* belong to the monostilifera. Through this conclusion we have excluded *Drepanophorus valdiviae* from the Polystilifera, in which it might cause much trouble by the different structure of almost every organ.'

Discussion of important characters

Although Stiasny-Wijnhoff (1923) had not mentioned the interwoven rhynchocoel musculature of *Valdivianemertes*, she later (1936) gave a lengthy analysis of the rhynchocoel and its musculature in the Polystilifera, noting the universal presence of a meshwork structure in the Reptantia and the varying situations among the Pelagica, in which she noted (p. 120), 'Besides the Pelagica we find the mesh structure also in the case of *Amphiporus stannii* and *Drepanophorus valdiviae* which also have other characters in common with the Polystilifera, for example, the rhynchocoelomic pouches. Whatever the relationships of these pouches, they are different from all similar structures in the Reptantia. They are short and broad and have a wide opening into the rhynchocoel whose musculature simply continues in the outpouching. Moreover, the pouches occur in *A. stannii* and *Valdivianemertes valdiviae* only in the beginning of the rhynchocoel.'

Stiasny-Wijnhoff (1936: 127–128) discussed esophageal-rhynchodaeal relationships in the Reptantia, noting that an atrium is approached in varying degrees by most of the Inaequifurcata and *Drepanophorina* and commenting, 'The mouth has a very variable position, not only that it can lie together with the proboscis opening in a sunken area in the tip of the head which we call the atrium or lie directly on the surface of the head, but in this latter case also changes its position considerably' and 'The epithelium of the atrium is differentiated a little from the surrounding epidermis, a condition which arises from its reaction to stains. The same condition is exhibited by the epithelial band between mouth and proboscis opening in some species where I can perceive no indentation, e.g., in the case of *Paradrepanophorus nisidensis*, whereas *P. corallinicola* and *P. obiensis* have a distinct atrium.' She (1926) erected *Paradrepanophorus* to which she transferred *Drepanophorus crassus crassus sensu* Bürger as *P. corallinicola*, *D. crassus nisidensis sensu* Bürger as *P. crassus*, and *D. igneus* Bürger as *P. nisidensis*, with *P. obiensis* being added in 1936.

Stiasny-Wijnhoff did not refer to the mouth-rhynchodaeal relationship of *Valdivianemertes* in 1923 and did not refer to *Valdivianemertes* at all in 1926, but in the 1936 paper, where she discussed other features of monostiliferous species with the cerebral organs located beside and behind the brain, she noted, '... *A. stannii* and *marmoratus* with separate mouth and proboscis opening ...' Thus, we find descriptions of the mouth-proboscis pore relationship in *A. stannii* ranging from a common opening to separate. However, the observations Stiasny-Wijnhoff set forth are quite clear and straightforward and, when Bürger's statement about the similarity of *D. valdiviae* and *D. crassus* is reconsidered in this context, we may reasonably conclude that the mouth and proboscis opening of *A. stannii* and *D. valdiviae* are rather similar, *i.e.*, what would now be called an atrium, and that any actual differences would be significant only at the specific but not at the generic level. Gibson (1988), in the course of discussing a possible new approach to enoplan phylogeny, mentions *Communoporus*, *Cratenemertes variabilis*, *Duosnemertes*, *Korotkevitschia* and *Valdivianemertes* as having some degree of separation of the mouth and proboscis pore. Finally, it should be noted that the position of the mouth-proboscis opening as far back on the ventral surface of the head (1.5–2.0 mm) as described by Bürger for *A. stannii* would be quite remarkable for a monostiliferan and is more characteristic of the mouth position in the Reptantia.

Friedrich (1935) stated (author's translation), 'Stiasny-Wijnhoff (1923) made it likely that *Amphip. stannii* together with the specimen described by Bürger as *Drepanophorus valdiviae* form a separate genus, for which she proposed the name *Valdivianemertes*.' Friedrich (1955: 146), in his paper on the genera of the Monostilifera, retained *Valdivianemertes* as a separate genus, but made no mention of the interwoven rhynchocoel musculature nor of a possible relationship to his newly-named (p. 145) genus *Cratenemertes* which was based on its possession of a 'flecktwerk' rhynchocoel wall structure. He likewise made no mention of possible relationships with his new genus *Duosnemertes* (p. 148), based on separate mouth and proboscis openings, in which he placed *Amphiporus marmoratus sensu* Bürger 1895a, nor his newly erected *Communoporus* (p. 149) to which he transferred three *Amphiporus* species with an atrium. In the key forming the final section of this article, Friedrich listed *Valdivianemertes* as possessing neither separate nor atrial mouth and proboscis openings and, further, placed it among the genera with bilayered rhynchocoel musculature despite Bürger's very clear text and figures and Stiasny-Wijnhoff's equally clear statements showing the rhynchocoel wall to be composed of interwoven circular and longitudinal fibers.

Friedrich (1957) noted that rhynchocoelic diverticula had been found in *Proneurotes*, *Gurjanovella* and *Amphiporella*, that he had observed them in a few instances in *Amphiporus*, the subgenus *Amphiporus* (*Intestinonemertes*), *Emplectonema* and in some heteronemerteans, but failed to mention *Valdivianemertes*. Kirsteuer (1963), after describing the curious lateral diverticula he found on the rhynchocoel in *Tetrastemma longissimum* Bürger, discusses the other monostiliferans for which rhynchocoelic diverticula have been described. He noted, in addition to *Valdivianemertes* Stiasny-Wijnhoff 1923, *Proneurotes multioculatus* Montgomery 1897 with five and *P. baltica* Friedrich 1940 with four ventral unpaired pouches, *Gurjanovella littoralis* Uschakow 1926 with a single dorsal pouch, *Amphiporella baltica* Friedrich 1940 and *Cratenemertes bergendali* (Gering 1912) with single ventral pouches. Both Friedrich and Kirsteuer noted Coe's (1943: 289) comments regarding Montgomery's *Proneurotes multioculatus* that 'An examination of the sections which he prepared from the type specimen, however, indicates that these [diverticula] were merely artifacts caused by irregular contraction at the time of preservation.' However, Kirsteuer made the compelling observation that careful tracing of muscle fibers, particularly the circular fibers as he had done in the case of *Tetrastemma longissimum*, could resolve many uncertainties as to whether an apparent diverticulum is natural or artifactual. I can confirm the efficacy of this method in establishing whether appendages and diverticula of the foregut are indeed valid morphological characters. It would appear that rigorous re-examination, with these principles in mind, of those genera and species for which appendages or diverticula have been claimed could finally resolve many of the questions that have been raised regarding these features. Suffice it to say that, in the case of *Valdivianemertes*, the pouches form a consistent and acceptable generic character.

Nomenclatural considerations

Grube's (1840) *Akrostomum stannii* was a *nomen dubium* and his generic diagnosis was so vague that it could have applied to almost any eyeless monostiliferan. However, *Akrostomum* was validly constituted (*cf* Art. 68, International Code of Zoological Nomenclature, 3rd Ed., 1985). The last known, but very doubtful, use of *Akrostomum stannii* except as a synonym was by Vaillant (1890). Hubrecht (1883) regarded *Amphiporus hastatus* as a synonym of *Akrostomum stannii* while Joubin (1890, 1894) did the reverse. As previously shown, neither synonymy was justified and both have been consistently rejected by all workers since.

When Bürger (1895a, b) redescribed *Akrostomum stannii* he placed it in the older genus *Amphiporus* Ehrenberg, 1831. This generic placement was consistent with *Amphiporus* as understood at that time and it is *sensu* Bürger that the species has since been understood.

Aside from the citations of *Akrostomum* already mentioned, there is only one other instance of the definite application of this generic name to another species. Diesing (1862) reported *Polia canescens* (Leuckart, 1849) as *Acrostomum canescens*, but Bürger (1904) transferred it to the genus *Amphiporus* and Gibson & Crandall (1989) regarded *A. canescens* as a *nomen dubium*. Joubin (1894), in a confused account of *Amphiporus pugnax*, suggested that it and *Amphiporus hastatus* might belong in a separate genus, '? *Acrostomum* (Grube).' But this suggestion has been universally rejected and Gibson & Crandall regarded both as *nomina dubia*. After 1890 *Akrostomum* as a generic name existed only as a junior synonym of the older *Amphiporus* until Stiasny-Wijnhoff (1923) united *A. stannii* and *D. valdiviae* in a separate genus, *Valdivianemertes*, for which she did not designate a type species.

There can be a legitimate difference of opinion as to whether *Akrostomum* or *Valdivianemertes* should be retained as the generic name. One school of thought would hold that Bürger's redescription, based in part on the original specimen, rescued the taxon from *nomen dubium* status and when Stiasny-Wijnhoff withdrew *A. stannii* from *Amphiporus* it automatically reverted to *Akrostomum stannii* and that her name *Valdivianemertes* was never more than a junior synonym (*cf* Art. 23, ICZN) and should be abandoned.

The other school of thought would hold that *Akrostomum* has not been used, except in synonymy, for more than 100 years, that the redescription of Bürger (1895a, b) is the characterization by which the taxon *stannii* has been understood for more than 95 years, that *Valdivianemertes* is the only name by which the genus has been understood for more than 60 years and, therefore, the stability of the nomenclature would be threatened by now reverting to *Akrostomum*. ICZN Article 79(c) provides that the International Commission on Zoological Nomenclature may use its plenary powers to suppress an otherwise valid name if the senior name has not been used as a valid name during the immediately preceding fifty years and the junior name has been applied to a taxon, as its presumably valid name, by at least five different authors in at least 10 publications during the same period. The name *Valdivianemertes* meets the above criteria.

It is my opinion, shared by others in the nemertean taxonomic community, that the rule of priority nothwithstanding, the stability of the nomenclature and the recognition of this taxon, as understood for the past 96 years, would be jeopardized by a reversion to the name *Akrostomum*. Accordingly, an application has been submitted to the Commission for conservation of the name *Valdivianemertes* and suppression of the names *Akrostomum* and *Acrostomum*. *V. stannii* is here recognized as the type species of *Valdivianemertes* by subsequent designation (*cf* Art. 69 [a] ICZN).

Family affinities

The genus *Amphiporus* Ehrenberg 1831 formed the basis for the family Amphiporidae McIntosh 1873 and similarly *Drepanophorus* Hubrecht 1879 for the family Drepanophoridae Verrill 1892. Bürger initially (1895a) included both *Amphiporus* and *Drepanophorus* in the Amphiporidae, but subsequently (1904, 1897–1907) listed *Amphiporus* and *Proneurotes* Montgomery 1897 as comprising the Amphiporidae and the Drepanophoridae as a separate family. Brinkmann (1914–15) added *Uniporus* as the second genus of the Drepano-

Table 1. Differential diagnosis of the genera of the Cratenemertidae.

	Cratenemertes[1]	*Korotkevitschia*	*Nipponnemertes*	*Valdivianemertes*
Cerebral organs	small	large	large	small-medium
Rhynchocoelic diverticula	0	0	0,[2]	paired-lateral[3]
Mouth-proboscis opening	common	separate-atrial	common[4]	separate-atrial
Eyes	+	0	+,[4]	0
Diagonal muscle layer	+	?	most sup5	+
Nephridia	+	0	+	extensive[6]
Cephalic glands	+	0	+	0, +
Intestinal caecum	medium	0	medium	short, nearly 0

[1] *C. variabilis* Korotkevich, 1983, has atrial mouth-proboscis opening and moderately large cerebral organs. Probably belongs in a different genus, possibly *Nipponnemertes*.

[2] *N. bergendali* is described as having a ventral diverticulum on the rhynchocoel.

[3] A few pairs limited to the nephridial region.

[4] *N. scoresbyi* has probable atrium, no eyes and fairly small cerebral organs. Probably belongs in a different genus.

[5] Berg reports absence in *N. pulcher*. However, fasciated or non-fasciated diagonal layer found in all examined by author, including a non-fasciated layer in *N. pulcher*.

[6] Nephridia extend from posterior end of brain to anteriormost intestinal region and are richly branched.

phoridae and later (1917), in his subdivision of the Hoplonemertea into Monostilifera and Polystilifera and the latter into Reptantia and Pelagica, designated the Drepanophoridae as the sole family of the Reptantia. It is ironic that Stiasny-Wijnhoff (1923) took much care with the proper placement of *Siboganemertes weberi*, erecting the Archireptantia for this new taxon, but when erecting *Valdivianemertes* in an extended footnote to the same article, did not mention a family placement, although this latter genus was created by uniting species from two different families. Friedrich (1955), in his review of the genera of the Monostilifera, created several new genera for species separated from *Amphiporus* and discussed *Valdivianemertes* at the same point in the paper. However, he explicitly declined to discuss family relationships in this article.

Friedrich (1968) separated *Korotkevitschia*, *Nipponnemertes* and *Sagaminemertes* from *Cratenemertes* and established the family Cratenemertidae to contain the four genera based on a diagnosis of monostiliferans with interwoven rhynchocoel musculature and cerebral organs beside and behind the brain, but did not mention *Valdivianemertes*. Later, Kirsteuer (1973) concluded that the monotypic *Sagaminemertes* probably was a reptantic polystiliferan, and more recently Iwata (1988) gave a comprehensive redescription of *S. nagaiensis* (Iwata, 1957) transferring it to the Archireptantia as the second known genus and species of that group.

Subsequently Gibson (1972, 1982), following Friedrich's (1955) arrangement and key, listed *Valdivianemertes* as a genus of the Amphiporidae with bilayered rhynchocoel musculature. Gibson & Crandall (1989), in a review of the genus *Amphiporus*, noted only the earlier misspellings of the specific epithet by Bürger, and that *Amphiporus stannii* had been transferred to *Valdivianemertes* by Stiasny-Wijnhoff.

Conclusions

The generic name *Valdivianemertes* Stiasny-Wijnhoff, 1923, has been retained here in the interest of stability of the nomenclature with *V. stannii* as the type species by subsequent designation. The International Commission on Zoological Nomenclature (Case 2888) has been requested to conserve the name *Valdivianemertes* and to suppress the names *Akrostomum* and *Acrostomum*.

Valdivianemertes is monostiliferous, has a rhynchocoel with interwoven circular and longitudinal musculature, and has cerebral organs beside and partly behind the brain, *i.e.*, precisely the characters diagnostic for the family Cratenemertidae. Accordingly the genus *Valdivianemertes* is transferred to the Cratenemertidae.

The distinguishing features of the four genera now comprising the Cratenemertidae are shown in Table 1. However, it should be noted that one species currently placed in *Cratenemertes* and possibly several in *Nipponnemertes* are valid cratenemertids but may properly belong in other genera.

Acknowledgements

I am indebted to the Division of Worms, U.S. National Museum of Natural History, for the use of specimen collections and laboratory facilities as a visiting investigator, to the Smithsonian Institution Oceanographic Sorting Center for access to recently collected specimens, and to the Registry of Tumors in Lower Animals, USNMNH, for the use of excellent histology facilities. I am also indebted to the British Museum (Natural History) for access to its collections and to Dr Alex Muir of its Department of Zoology for his kind hospitality and helpful discussions.

I am grateful to Dr Jon L. Norenburg and Prof. Ray Gibson for commenting on an earlier draft of this paper, and to Drs Frederick M. Bayer and Raymond B. Manning, USNMNH, for discussions of nomenclatural issues, although their views may not be fully reflected in the present version.

References

Berg, G., 1972. Studies on *Nipponnemertes* Friedrich, 1968 (Nemertini, Hoplonemertini). I. Redescription of *Nipponnemertes pulcher* (Johnston, 1837) with special reference to intraspecific variation of the taxonomical characters. Zool. Scr. 1: 211–225.

Berg, G., 1985. Studies on *Nipponnemertes* Friedrich (Nemertini, Hoplonemertini). II. Taxonomy of *Nipponnemertes pulcher* (Johnston) and some other species. Zool. Scr. 14: 239–246.

Brinkmann, A., 1914–15. *Uniporus*, ein neues Genus der Familie Drepanophoridae Verrill. Bergens Mus. Aarbok No. 6: 1–29.

Brinkmann, A., 1917. Die pelagischen Nemertinen. Bergens Mus. Skr. ny Raek 3: 1–194.

Bürger, O., 1895a. Die Nemertinen des Golfes von Neapel uned der angrenzenden Meeres-Abschnitte. Fauna Flora Golf. Neapel 22: 1–743.

Bürger, O., 1895b. Beiträge zur Anatomie, Systematik und geographischen Verbreitung der Nemertinen. Z. wiss. Zool. 61: 16–37.

Bürger, O., 1897–1907. Nemertini. In: H. G. Bronn (ed.) Klassen und Ordnungen des Toerreichs 4 Suppl., Leipzig.

Bürger, O., 1904. Nemertini. Das Tierreich 20: 1–151.

Bürger, O., 1909. Die Nemertinen. Dt. Tiefsee Exped. 1898–99 16: 171–221.

Carus, J. V., 1885. Prodromus Faunae Mediterraneae sive Descriptio Animalium Maris Mediterranei Incolarum quam Comparata silva rerum quatenus innotuit adiectis locis et nominibus vulgaribus eorumque auctoribus in commodum Zoologorum. Vol. I. Coelenterata, Echinodermata, Vermes, Arthropoda. Stuttgart, E. Schweizerbart'sche Verlagshandlung (E. Koch), 524 pp.

Coe, W. R., 1905. Nemerteans of the west and northwest coasts of America. Bull. Mus. comp. Zool. Harv. 47: 1–318.

Coe, W. R., 1943. Biology of the nemerteans of the Atlantic coast of North America. Trans. Conn. Acad. Arts Sci. 35: 129–328.

Crandall, F. B., 1992. Major characters and enoplan systematics. Hydrobiologia 266: 115–140.

Diesing, K. M., 1850. Systema Helminthum. Vol. 1: 1–679. Braumuller, Vindobonae.

Diesing, K. M., 1862. Revision der Turbellarien. Sber. Akad. Wiss. Wien Math.-nat. Kl. 45: 198–307.

Ehrenberg, C. G., 1828–1831. Phytozoa turbellaria Africana et Asiatica in Phytozoorum Tabule IV et V delineata. (Pp 53–67 in: Animalia evertebrata exclusis insectis, 126 pp. Plates published in 1828, text in 1831). In: P. C. Hemprich & C. G. Ehrenberg (eds), Symbolae Physicae, seu icones et descriptiones Corporum Naturalium novorum, Pars Zoologica. Officina Academica, Berolina, 1828–1845.

Friedrich, H., 1935. Studien zur Morphologie, Systematik und Ökologie der Nemertinen der Kieler Bucht. Arch. Naturgesch. 4: 293–375.

Friedrich, H., 1955. Beiträge zu einer Synopsis der Gattungen der Nemertini monostilifera nebst Bestimmungsschlüssel. Z. wiss. Zool. 158: 133–192.

Friedrich, H., 1957. Beiträge zur kenntnis der Arktischen Hoplonemertinen. Vidensk. Meddr dansk naturh. Foren. 119: 129–154.

Friedrich, H., 1968. *Sagaminemertes*, eine bemerkenswerte neue Gattung der Hoplonemertinen und ihre systematische Stellung. Zool. Anz. 180: 33–36.

Gering, G., 1912. Neue Nemertinen der schwedischen Westküste. Zool. Anz. 39: 520–523.

Gering, G., 1913. Neue Nemertinen der schwedischen Westküste. Zool. Jb. Abt. Syst. 34: 187–202.

Gibson, R., 1972. Nemerteans. London, Hutchinson, 224 pp.

Gibson, R., 1982. Nemertea. In: S. P. Parker (ed.), Synopsis and Classification of Living Organisms 1: 823–846. New York, McGraw-Hill.

Gibson, R., 1988. Evolutionary relationships between mono- and polystiliferous hoplonemerteans: *Nipponnemertes* (Cratenemertidae), a 'missing link' genus? Hydrobiologia 156: 61–74.

Gibson, R. & F. B. Crandall, 1989. The genus *Amphiporus* Ehrenberg (Nemertea, Enopla, Monostiliferoidea). Zool. Scr. 18: 453–470.

Gibson, R. & F. B. Crandall, 1991. *Amphiporus* Ehrenberg (Nemertea): Proposed designation of *Planaria lactiflorea* Johnston, 1828, as the type species. Bull. zool. Nomenclat. 48: 22–24.

Grube, A. E., 1840. Actinien, Echinodermen und Würmer des Adriatischen- und Mittelmeers. Konigsberg, J. H. Bon, 92 pp.

Hubrecht, A. A. W., 1879. The genera of European nemerteans critically revised, with description of several new species. Notes Leyden Mus. 1: 193–232.

Hubrecht, A. A. W., 1883. On the ancestral form of the Chordata. Q. Jl microsc. Sci. 23: 349–368.

International Commission on Zoological Nomenclature, 1985. International Code of Zoological Nomenclature, 3rd ed. International Trust for Zoological Nomenclature, British Museum (Natural History), London.

Iwata. F.. 1988. On the hoplonemertean *Sagaminemertes nagaiensis* (Iwata, 1957), with consideration of its systematics. Hydrobiologia 156: 115–124.

Joubin, L., 1890. Recherches sur les Turbellaries des cotes de France (Nemertes). Archs Zool. exp. gen., Ser. 2, 8: 461–602.

Joubin, L., 1894. Les Nemertiens. Fauna Française, R. Blanchard & J. de Guerne (eds). Paris, Société d'Editions Scientifiques, 235 pp.

Kirsteuer, E., 1963. Beitrag zur Kenntnis der Systematik und Anatomie der adriatischen Nemertinen (Genera *Tetrastemma, Oerstedia, Oerstediella*). Zool. Jb. Abt. Anat. 80: 555–616.

Kirsteuer, E., 1973. A new polystiliferous hoplonemertean, *Curranemertes natans* gen. et sp. n., from the Caribbean Sea (Nemertina, Polystilifera Reptantia). Zool. Scr. 2: 125–140.

McIntosh, W. C., 1873–74. A Monograph of the British Annelids. Part 1, Nemerteans. London, Ray Society.

Stiasny-Wijnhoff, G., 1923. On Brinkmann's system of the Nemertea Enopla and *Siboganemertes weberi* n. gen. n. sp. Q. Jl microsc. Sci. 67: 627–669.

Stiasny-Wijnhoff, G., 1926. The Nemertea Polystilifera of Naples. Pubbl. Staz. zool. Napoli 7: 119–168.

Stiasny-Wijnhoff, G., 1936. Die Polystilifera der Siboga Expedition. Siboga-Exped. 22: 1–214.

Vaillant, L., 1890. Lombriciniens, Hirudiniens, Bdellomorphes, Teretulariens et Planariens. Hist. nat. Anim. Marins et d'eau douce, Paris 3, Part 2: 341–768.

Hydrobiologia **266**: 185–201, 1993.
R. Gibson, J. Moore & P. Sundberg (eds), Advances in Nemertean Biology

Paralineopsis taki gen. et sp. nov., a littoral heteronemertean from Japan, provided with special proboscideal, circulatory and sensory organs of significance to nemertean systematics

Fumio Iwata
Laboratory of Biology, Kushiro Public University of Economics, Kushiro 085, Japan

Key words: proboscideal diaphragm, cephalic blood lacuna, cerebral organs

Abstract

Some morphological features with major systematic significance are recorded in the heteronemertean *Paralineopsis taki* gen. et sp. nov. from Onomichi, Japan as follows: horizontal band of specialized epithelium extends from near apex to the opening of the cerebral organ canal on either side of the head; precerebral region filled with gelatinous (hyaline) connective tissue in which longitudinal muscles are absent; body wall muscles do not accompany rhynchodaeal invagination; rhynchodaeum initially only epidermal; inner longitudinal muscles of ventral wall of cephalic blood lacuna become intimately associated with rhynchodaeum forming a dorsal saddle over it; cerebral organs do not penetrate inner longitudinal muscles, and do not contact blood vascular system; proboscideal diaphragm post-cerebral; outer longitudinal muscles absent throughout body; longitudinal muscles of proboscis derived from inner longitudinal musculature. The systematic relationship of *P. taki* and *Paralineus elisabethae* (Schütz, 1911) from Villefranche, France is discussed.

Introduction

One of four nemertean species reported from Onomichi, Japan, the heteronemertean *Zygeupolia littoralis* (Thompson, 1900) was identified from external features of the body and no description of the internal anatomy was given (Iwata, 1951). However, four unusual systematic features were noted: (1) lateral cephalic furrows absent, epidermal depressions of specialized cells extend from anterior end to the opening of the cerebral canals; (2) outer longitudinal muscles absent; (3) longitudinal muscles absent from precerebral medulla except for the inner longitudinals around the precerebral blood lacuna; (4) caudal cirrus absent. These characters distinguish it from *Zygeupolia* and warrant the erection of *Paralineopsis taki* (gen. et sp. nov.) for this species. The new species *Paralineopsis taki* has been compared at the generic level with the related heteronemertean *Paralineus elisabethae* Schütz 1911 from Villefranche.

Material and methods

The single specimen forming the basis of the present paper was collected under a rock on sand near low water mark at Mukaijima, Onomichi, Hiroshima prefecture, near the Onomichi Marine Biological Station of Hiroshima University on 10 April, 1949. A coloured sketch of the body in the living state was made and after anaesthetizing by adding drops of 70% alcohol, the material was fixed in Bouin's fluid: it was sectioned at

10 μm and stained with Delafield's haematoxylin and eosin. An Olympus video camera with Ikegami monitor television and Hitachi printer, in addition to an Olympus BH-2 microscope, were used for observation of the material.

Description of the species

Genus Paralineopsis *gen. nov.*

Type species

Paralineopsis taki *sp. nov.*

Etymology

The generic name indicates the resemblance between this taxon and the genus *Paralineus* by the addition of the Greek suffix-*opsis* (denoting a likeness of form). The species is named in honour of Emeritus professor Iwao Taki, former director of the Onomichi Marine Biological Station of the Hiroshima University as a tribute to his eminent contributions to the description of Japanese molluscs.

Diagnosis

Heteronemertean without horizontal lateral cephalic furrows, but with a single pair of horizontal lateral shallow epidermal depressions; mouth far behind cerebral ganglia; outer longitudinal muscles absent; rhynchodaeal longitudinal muscles entirely derived from ventral inner longitudinal muscles; rhynchodaeum long and vast in dimension, extending far behind the cerebral ganglia and cerebral organs; rhynchocoel wall longitudinal muscle layer connected with body wall inner longitudinal musculature by a pair of muscular septa for the proboscideal organ or the proboscideal diaphragm at dorsolateral sides of rhynchodaeum; proboscis unbranched, containing two muscle layers (outer circular, inner longitudinal) and no muscle crosses; rhynchocoel wall circular muscles not interwoven with adjacent body wall inner longitudinal fibres; well-developed dermal connective tissue present under subepidermal glands; dorsal fibrous core of cerebral ganglia not forked into upper and lower branches; nervous system with neither neurochords nor neurochord cells; proboscideal nerves derived from ventral brain lobes extremely long, entering proboscis at the posterior end of the rhynchodaeum; foregut nerves arise from ventral brain lobes; foregut without circular or longitudinal somatic muscle layer and with subepithelial glands; mouth wall rich in salivary glands; blood system with single, well-developed cephalic lacuna running in the precerebral, cerebral and post-cerebral regions with structural differences; post-cerebral lacuna long and vast in dimension, running under the rhynchodaeum behind the cerebral organs and under the rhynchocoel up to the foregut; rhynchocoel vessel from post-cerebral lacuna enters rhynchocoel wall immediately behind the rhynchodaeum; foregut provided with dorsal and two dorso-lateral vessels only and for most of the intestinal region three longitudinal vessels which are not joined by pseudometameric transverse connectives; excretory system not well-developed and confined to anterior foregut region; eyes absent; special sensory organs composed of three small chambers; frontal glands not well-developed but confined to the middle portion of the head; caudal cirrus absent; sexes separate.

Paralineopsis taki *sp. nov.*

Zygeupolia littoralis*: Iwata, 1951**.

Type specimen

Holotype female, partial series of transverse sections in the anterior portion of the body, including the head, mouth and anterior foregut region

---

* For identification, the absence of the horizontal lateral cephalic furrows and the presence of the caudal cirrus were used, but it was thought that the caudal cirrus was damaged as described in the paper.

(four slides), part of the intestine (three slides) and posterior end (three slides). The intestinal region of the body remains unsectioned in alcohol. The material used will be deposited at the Zoological Institute, Faculty of Science, Hokkaido University.

Type locality

Under a rock on sand, lower shore, Mukaijima, Onomichi, Japan.

External features

The single specimen (Fig. 1a) was about 20 cm long and 1 mm at maximum width. The head has no horizontal lateral cephalic furrows, but instead bears a single pair of horizontal lateral cephalic shallow epidermal depressions seen in transverse sections. It does not possess eyes. At the rear of the head a small mouth is situated ventrally (Fig. 1b). The head is slender and tapers to a point when extended. The body in the foregut region is rounded, while in the intestinal region it is flattened in life, but rounded after preservation. Throughout most of its length the body is similar in width; it tapers to end in a bluntly pointed tip which does not bear a caudal cirrus.

The colour is blueish white in the head, pale yellow in the foregut region, and milky white in the intestinal region.

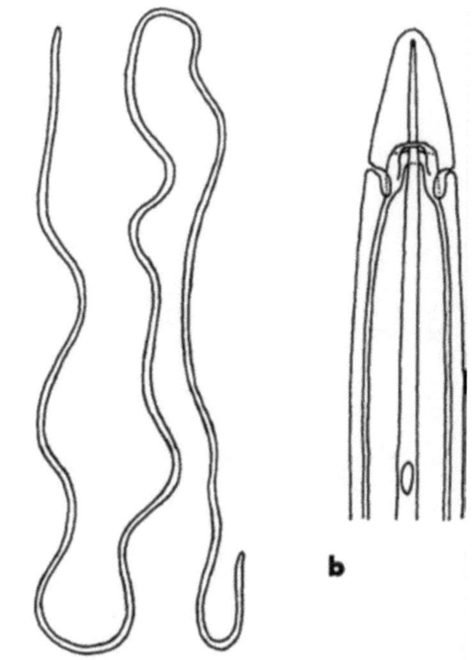

Fig. 1. Paralineopsis taki. External features in dorsal view of whole specimen based on a coloured sketch in 1949 (a) and diagram of anterior portion of body showing the brain, cerebral organs and mouth (b) (after Iwata, 1951).

Body wall, musculature and parenchyma

The head in transverse sections is divided into the precerebral region, extending from the anterior end of the head to the brain (0.39 mm long), a cerebral region containing the brain lobes and cerebral organs (0.16 mm long), and a postcerebral region ranging from the rear of the cerebral organs up to the anterior end of the foregut (1.22 mm long) (Fig. 2a, d).

The epidermis, 30–40 μm thick, is characterized by a row of thickly arranged rhabditous glands in the cephalic and foregut regions and by oval glands sparsely embedded in the intestinal region; the glands are acidophilic, but most of them are colourless because they have scanty contents. The epidermis in the intestinal region thins to half the thickness of that in the foregut region (Fig. 3f, g).

The basement layer of hyaline connective tissue, 5–10 μm thick, is a conspicuous feature of the cephalic region, though it becomes membraneous in the intestinal region (Figs 3g, 4a–d).

In the cephalic region, the circular muscles surrounding the body under the basement layer, 5 μm thick, form a distinct layer, the same as that of the body wall middle circular muscles (Fig. 4a–d). This layer becomes membraneous in the intestinal region.

Intestinal region

The body wall outer longitudinal muscle layer is four times the thickness of the middle circular and

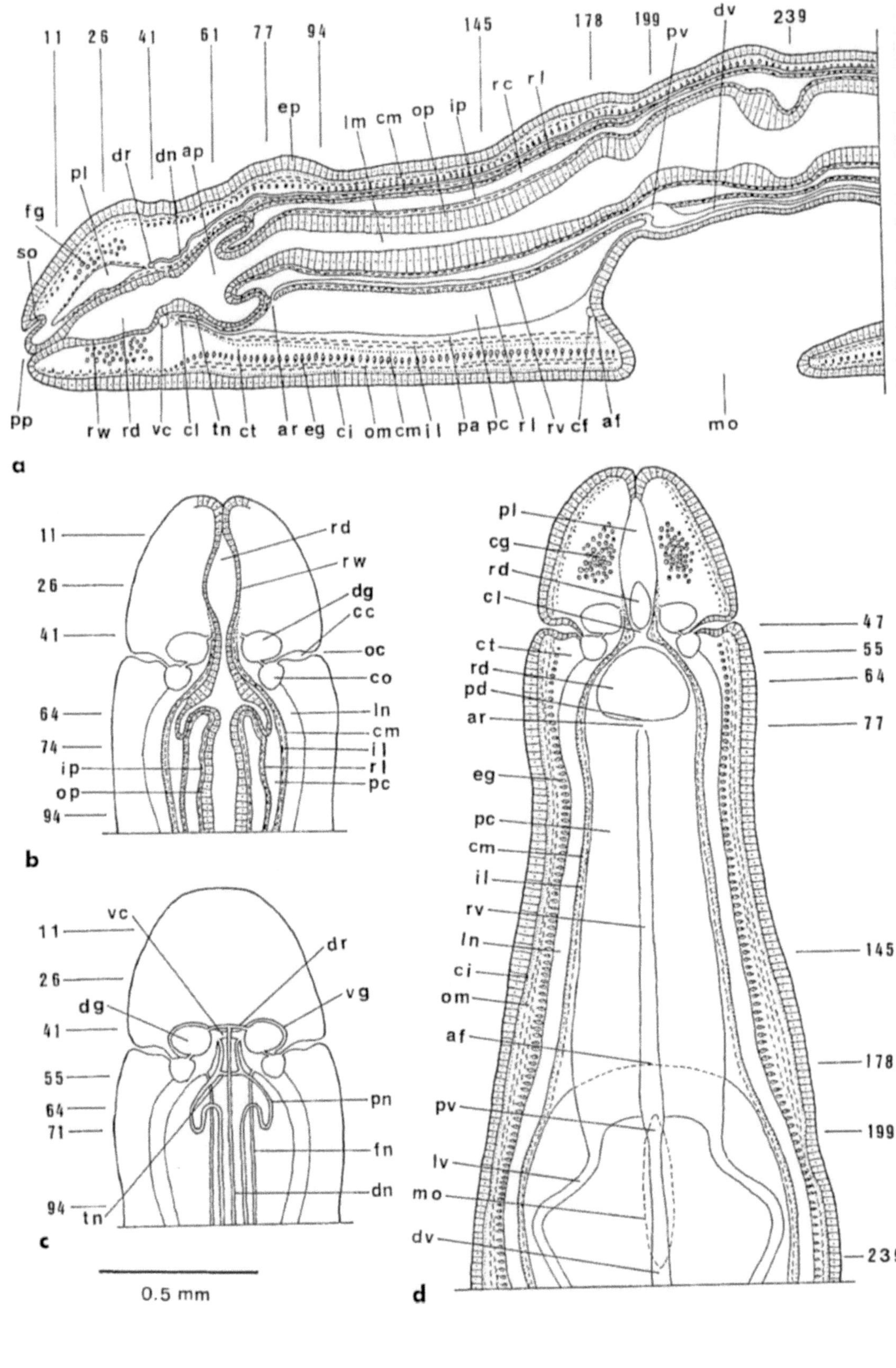
11
26
41
61
77
94
145
178
199
dv
239
pv
rc
rl
ep
lm
cm
op
ip
dr
dn
ap
pl
fg
so
pp
rw
rd
vc
cl
tn
ct
ar
eg
ci
om
cm
il
pa
pc
rl
rv
cf
af
mo
a
11
26
41
64
74
ip
op
94
rd
rw
dg
cc
oc
co
ln
cm
il
rl
pc
b
vc
11
26
dg
41
55
64
71
94
tn
dr
vg
pn
fn
dn
c
0.5 mm
pl
cg
rd
cl
ct
rd
pd
ar
eg
pc
cm
il
rv
ln
ci
om
af
pv
lv
mo
dv
47
55
64
77
145
178
199
239
d

inner longitudinal muscle layers at the mid-ventral side of the body in the post-cerebral region. Posteriorly it is reduced to twice the thickness of these layers (Fig. 4b, d).

The body wall middle circular and inner longitudinal muscle layers surrounding the foregut at the portion behind the mouth are well developed and nearly the same thickness as the body wall outer longitudinal muscle layer (Fig. 4f).

The body wall outer longitudinal muscle layer consists of three components: the outer longitudinal muscle layer, the dermal compound or subepidermal glands, and the dermal connective tissue (hyaline), in which no outer longitudinal muscles are present (Fig. 4b). There is no dermal connective tissue between the outer longitudinal muscle layer and the glands. In the head, in front of the cerebral ganglia, the outer longitudinal musculature is deficient in development, whereas the connective tissue is widely expanded and contains some cephalic nerves and the frontal glands derived from the dermal glands (Fig. 5c, d). These layers show regional differences in thickness. In the cerebral region, the layers are 135 μm thick on the mid-ventral side of the body and have a thickness ratio of 3:2:9 respectively. The connective tissue, in which the brain lobes are located, is the thickest among them (Fig. 5h). In the anterior post-cerebral region, the layers together are 100 μm thick and have a ratio of 4:10:5, showing that the glands are well developed (Fig. 4b). In the middle portion of the post-cerebral region, they are 70 μm thick, with a ratio of 4:4:1, showing that the connective tissue is much reduced in development (Fig. 4d). In the foregut region, they are 70 μm thick, but the glands and connective tissue are inconspicuous as layers (Figs 3f, 4f). In the intestinal region, tissue layers are 20 μm in total thickness and have a ratio of 5:0:3. This shows that the glands are sparsely embedded in the outer longitudinal musculature (Fig. 3g). The subepidermal glands abut the neural plexus (Fig. 6g) adjoining the inner circular muscles. A pair of symmetrically placed subepidermal glands penetrate into the inner longitudinal muscles (Fig. 6h) a short distance behind the brain of the holotype or immediately behind the rhynchodaeum.

The body wall middle circular muscle layer is not well developed, but in the foregut region is 25 μm thick, 10 times the thickness of the layer in the anterior post-cerebral region (Figs 3f, 4b, f). In the post-cerebral region, this layer is about half the thickness of the circular muscles surrounding the body under the basement membrane (Fig. 4a–d).

The body wall inner longitudinal muscle layer is nearly constant in thickness at the mid-ventral side of the body in the post-cerebral and foregut regions, measuring 30 μm (Fig. 4b, d, f). In the intestinal region, it is reduced to a thickness of 10 μm (Fig. 3g).

The body wall inner longitudinal muscle layer in the post-cerebral region is irregular in thickness in the latero-dorsal parts of the body. On the lateral sides, where the layer immediately contacts the post-cerebral lacuna, it is reduced to 20 μm thickness (Fig. 3a, b). In the dorso-lateral

Fig. 2. Paralineopsis taki. Schematic drawings of anterior portion of body showing the proboscideal, circulatory and digestive systems (a), the rhynchodaeum (b), the nervous system (c) and the circulatory system (d). Scale = 0.5 mm. 11–239: number of sections from anterior end of body. af = anterior end of foregut; ap = anterior end of proboscis; ar = anterior end of rhynchocoel vessel; cc = canal of cerebral organ; cf = commissure of foregut nerves; ci = circular muscles under basement membrane; cl = cerebral lacuna; cm = body wall circular muscle layer; co = cerebral organ; ct = connective tissue of the body wall outer longitudinal muscle layer or dermal connective tissue; dg = dorsal ganglion; dn = dorsal nerve; dr = dorsal brain commissure; dv = dorsal vessel; eg = epidermal or dermal compound glands; ep = epidermis; fg = frontal glands; fn = foregut nerve; il = body wall inner longitudinal muscle layer; ip = inner muscular wall of proboscis; ln = lateral nerve; lp = lumen of proboscis; lv = lateral vessel; mo = mouth; oc = opening of canal of cerebral organ; om = outer longitudinal muscle layer; op = outer epithelium of proboscis; pa = parenchyma; pc = post-cerebral lacuna; pd = posterior end of rhynchodaeum; pl = cerebral lacuna; pn = proboscideal nerve; pp = proboscis pore; pv = posterior end of rhynchocoel vessel; rc = rhynchocoel; rd = rhynchodaeum; rl = rhynchocoel wall; rv = rhynchocoel vessel; rw = rhynchodaeal wall; so = special sensory organ; tn = transverse cord of proboscideal nerves; vc = ventral brain commissure; vg = ventral ganglion.

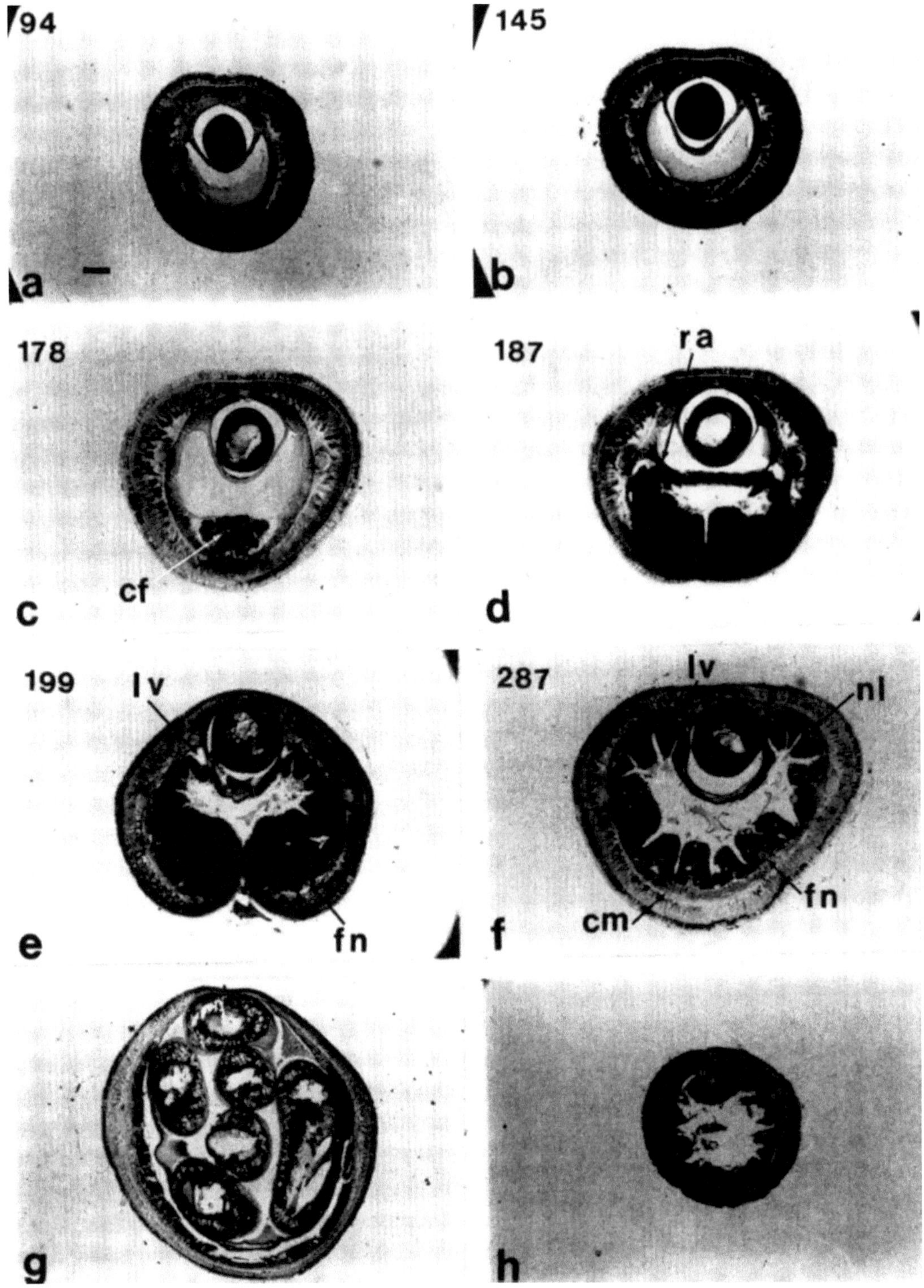

Fig. 3. Paralineopsis taki. Transverse sections from the anterior portion of the body showing the post-cerebral lacuna (a–d), mouth (e), oesophagus (f) and intestine (g and h). Scale = 100 μm. 94–287; number of sections from anterior end of body. nl = nephridial canals; ra = radial muscular bundle.

region it thickens to 50 μm and makes tight contact with the rhynchocoel wall (Fig. 3c). In the mid-dorsal region, it is deficient so that the rhynchocoel wall makes immediate contact with the body wall middle circular muscle layer (Fig. 4a, c).

The inner circular and longitudinal muscles derived from the body wall middle circular and inner longitudinal muscle layer suround the rhynchodaeum and cerebral lacuna inside the cerebral ganglia (Fig. 5h). They form the proboscideal insertion (Figs 4a, 6g) and thus the rhynchocoelic diaphragm is not fixed in location within the body as in most other heteronemerteans in which the insertion is from the outer longitudinals passing over the anterior wall of the margins of the brain. On the mid-ventral side of the body, the body wall in the post-cerebral region before the foregut is about 160 μm thick, whereas in the intestinal region it is reduced to a thickness of less than 50 μm (Fig. 3b, g).

A pair of radial muscular bundles running from the inner longitudinal muscles connect to the anterior wall of the buccal cavity (Fig. 3d). Dorsoventral muscles are absent in the intestinal region. Parenchymatous connective tissues are exceedingly sparse; only small amounts are buried in a space between the body wall inner longitudinal muscle layer and the ventral wall of the post-cerebral lacuna (Fig. 4d).

Proboscis apparatus

The proboscis pore opens ventroterminally, just behind the tip of the head. The apex of the head of the holotype is retracted into the body (Fig. 5a) so that the proboscis pore can be discerned initially in section 12, some 110 μm back in the body. The rhynchodaeum immediately behind the proboscis pore flattens and is closed; its wall consists of acidophilic cells and is surrounded by a large number of minute glandular cells (Fig. 5a). The rhynchodaeal atrium is short and is lined with a ciliated columnar epithelium lacking glands. It forms a U as the result of the retraction of the head, and opens into the rhynchodaeum proper which is pressed against the end of the retracted apical epidermis. It is not accompanied by muscles of the body wall and the rhynchodaeum, thus, consists only of a thin walled epithelial tube. The row of longitudinal muscle fibres associated with the ventral wall of the cephalic blood lacuna becomes intimately associated with the dorsal side of the rhynchodaeum (Fig. 5c, d) while the dorsal and lateral fibres exist as the internal longitudinal muscles of the head. The differential contraction resulting from the saddle-like formation of longitudinal muscles over the rhynchodaeum with the ventral half of the organ existing as an epithelial tube bound to the amorphous connective tissue (Fig. 5d, e) prevents aligning the dorsal and ventral walls. This makes it difficult to determine with certainty the sequence of morphological events.

In front of the brain lobes, the rhynchodaeum gradually widens, first into a circular then a flask-shaped space in transverse sections (Fig. 5c, d). The walls of the cephalic blood lacuna are collapsed as the complex passes through the brain (Fig. 5e). The inner longitudinal muscles are displaced beneath the dorsal nerve and are replaced in mid-line by the dorsal rhynchodaeal muscles (Fig. 5f). The proboscideal nerves pass directly from the ventral brain lobes at the ventral commissure (Figs 2c, 5g) and insert between the rhynchodaeal endothelium and ventro-lateral rhynchodaeal longitudinal muscles. The latter (as well as the inner longitudinal) then close ventrally so that the organ is totally encompassed by longitudinal muscles (Fig. 5h).

The rhynchodaeum is unusually long and more spacious than that of heteronemerteans hitherto recorded, extending posteriorly through the brain lobes up to a position a short distance behind the cerebral organs (Fig. 2b). Its length in the precerebral, cerebral, and post-cerebral regions is 390 μm, 160 μm and 220 μm respectively. Its length in the first region is nearly equal to that of the other two combined (Fig. 2a, b). The rhynchodaeum at the ventral brain commissure, the posterior end of the cerebral organs, and the posterior portion of the rhynchodaeum has a width of 75 μm, 160 μm and 300 μm respectively; the rhynchodaeum at the latter site is four times wider than that at the first site (Fig. 2b).

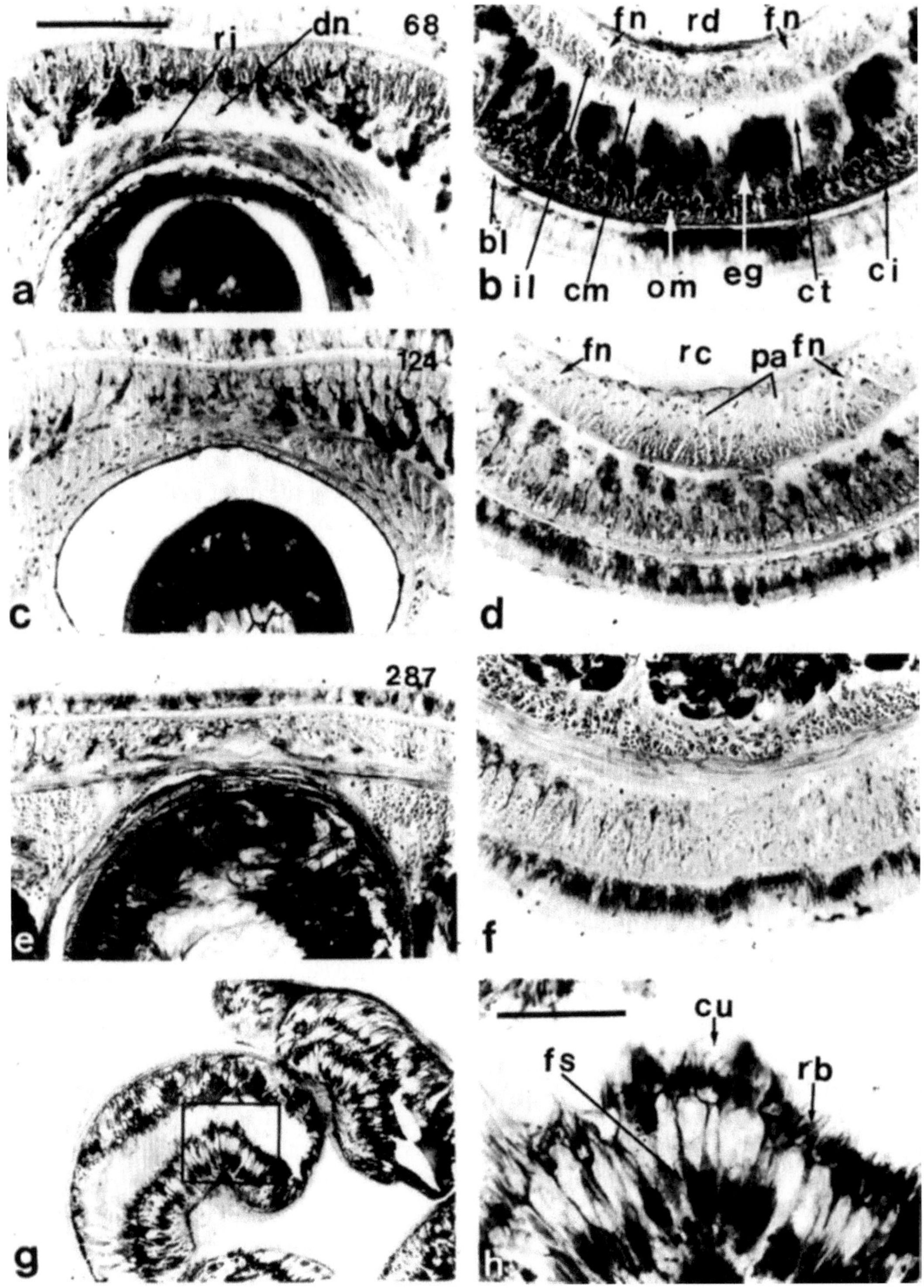

Fig. 4. Paralineopsis taki. Transverse sections of the anterior portion of body showing the body wall on dorsal (a) and ventral (b) sides in the posterior region of the rhynchodaeum, on dorsal (c) and ventral (d) sides in the middle portion of the post-cerebral lacuna, and on dorsal (e) and ventral (f) sides of the foregut region; transverse sections of the proboscis showing the proboscideal rhabditoids (g and h). Scale = 100 μm (a–g), 50 μm (h). 67–287: number of sections from anterior end of boy. bl = basement layer

The rhynchodaeum is elliptical in transverse section where it passes between the brain lobes, becoming more rounded in the post-cerebral region (Figs 5f, g, 6d). It is covered entirely by a thin circular and a thick longitudinal muscle layer originating from the body wall middle circular and inner longitudinal muscles.

Behind the cerebral organs, the lateroventral sides of the rhynchodaeum are covered by the cephalic blood lacuna, while its dorsolateral sides tightly contact the inner longitudinal muscles except mid-dorsally where the longitudinal muscles are deficient and the circular muscle layer immediately contacts the rhynchodaeal wall as in the rhynchocoel wall (Fig. 6a–f).

At the posterior end of the rhynchodaeum, fibres from the body wall inner longitudinal muscles form a pair of muscular septa for the proboscideal organ or the proboscideal diaphragm at the dorsolateral parts of the rhynchodaeum, entering immediately into the longitudinal muscle layer of the proboscis and rhynchocoel wall (Figs 2a, b, 4a, 6g). The fibres from the body wall middle circular musculature are interwoven with the rhynchocoel wall circular muscles at the mid-dorsal part of the rhynchocoel (Fig. 6g), beginning at a successive portion of the muscular septa. Their entrance into the proboscideal wall is not a recognised character. The proboscis behind the muscular septa for the proboscideal organ protrudes anteriorly into the rhynchodaeum; its outer side contacts the rhynchodaeal wall (Figs 2a, b, 6d–f).

Transverse muscles between the anterior ends of the cerebral organs and the lateral nerve cords pass through the internal longitudinal muscles and produce a thin ring of circular fibres between the proboscideal nerves and the endothelial outer wall of the rhynchodaeum (Fig. 6b).

The basophilic rhynchodaeal wall has no glandular nor ciliated columnar cells and is membraneous in appearance.

The rhynchocoel wall consists of thin circular and thick longitudinal muscle layers. The rhynchocoel extends almost to the posterior end of the body (Fig. 3h). It is narrow and cylindrical in transverse sections in the post-cerebral and foregut region. In the intestinal region it becomes more spacious and contains the folded proboscis (Fig. 3g). The fibres of the circular muscle layer of the rhynchocoel are not interwoven with the body wall inner longitudinal muscle layer.

The proboscis is composed of an outer circular and an inner longitudinal muscle layer without muscle crosses. The proboscideal nerves, situated between the thick outer epithelium and the outer circular muscle layer are clearly visible in the proboscis immediately behind the rhynchodaeum (Fig. 6e–h). The proboscis in the post-cerebral region anterior to the foregut is cylindrical, 160 μm in diameter, and occupies almost the whole rhynchocoel (Figs 2a, 3b, c).

Immediately behind the rhynchodaeum the proboscis is composed of an inner basophiilic flattened epithelium 10–30 μm thick containing glandular cells, a neural sheath, 2.5 μm thick, with a pair of distinct nerves about 35 μm long and 22 μm wide in diameter, an inner longitudinal muscle stratum 10–17 μm thick, and a basophilic lining endothelium 2–4 μm wide (Fig. 6h). No outer circular muscle layer is discernible. The lumen of the proboscis in this region is filled with basophilic substance.

The proboscis in the intestinal region, about 250 μm in diameter, is folded inside the rhynchocoel, comprising an inner epithelium 80 μm or more thick. There is also an outer circular muscle layer about 4 μm thick, an inner longitudinal muscle coat about 10 μm across, and a flattened endothelium 1 μm wide.

The epithelium of the proboscis is dominated by mucus and rhabditoid producing cells. In addition, there are widely scattered gland cells containing large acidophilic globules. The rhabditoids

of hyaline connective tissue; cu = cup-like packet for proboscideal rhabditoids; fs = flattened sheath for proboscideal rhabditoids; rb = rhabditoid; ri = rhynchocoel longitudinal muscles from the body wall inner longitudinal muscles, forming a septum for the proboscideal organ.

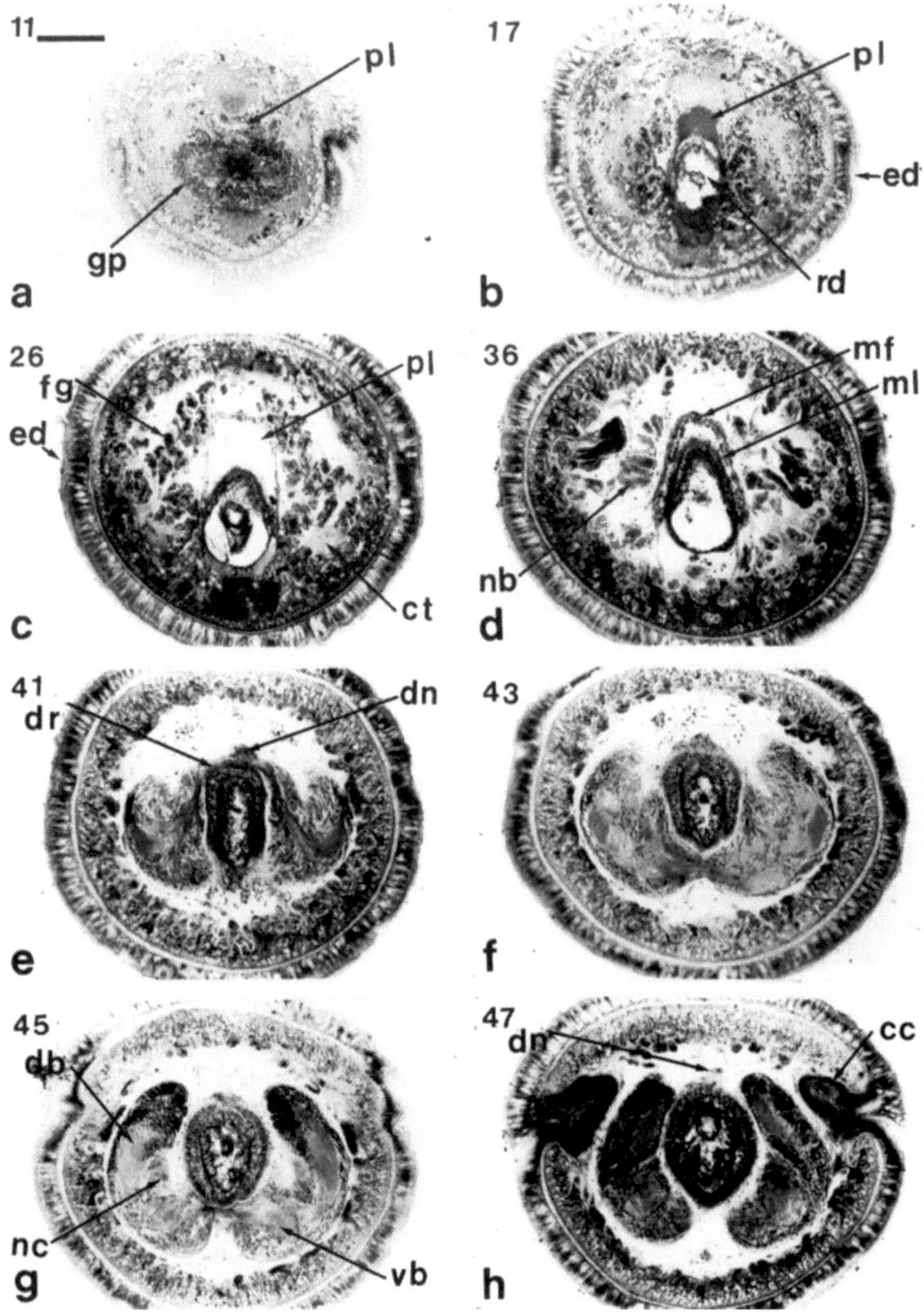

Fig. 5. Paralineopsis taki. Transverse sections of cephalic region of body showing the rhynchodaeum, precerebral and cerebral lacuna and horizontal lateral cephalic epidermal depressions (a–g). Scale = 100 μm. 11–47: number of sections from anterior end of body. db = dorsal brain lobe; ed = horizontal lateral shallow epidermal depression; gp = glandular cells around proboscis pore; mf = muscle fibres from the body wall circular and inner longitudinal muscle layers; ml = muscle layer on rhynchodaeum; nb = nerve cord from brain lobes; nc = nerve cell of the third type; vb = ventral brain lobe.

which are approximately 16 μm long project from the cells (Fig. 4g, h) in small packets. The sheath of spent rhabditoids frequently remained attached to the cells which produced them. Many cells containing stored and/or developing rhabditoids are present.

Alimentary canal

The small mouth opens as a narrow slit with the lips deeply invaginated, measuring 0.5 mm long in transverse sections (Figs 1b, 2a, 3e). The distance between the tip of the head and mouth, 1.9 mm long, is about 5 times the distance from the tip of head to the brain lobes, 0.39 mm in length (Fig. 2a, d). The mouth is surrounded by a subepithelial salivary cell layer associated with the foregut (Fig. 3d, e). There is no delicate layer of circular somatic muscles nor a somatic muscle plate between this part of the gut and the rhynchocoel.

The foregut extends to a point anterior to the mouth, with its anterior end bathed in the post-cerebral lacuna. The distance between the tip of the head and foregut is about 4.5 times the length between the tip of the head and the brain lobes (Fig. 2a, d). The foregut anterior to the mouth slit is horizontal on its dorsal side under the rhynchocoel (Fig. 3d). In the foregut region it elongates dorsally along the lateral sides of the rhynchocoel (Fig. 3f). Subepithelial gland cell zone is present in the foregut (Fig. 3f).

The intestine flattens dorsoventrally under the spacious rhynchocoel, its lateral ends reaching nearly to the dorsal surface of the body (Fig. 3g). The lateral diverticula of the intestine and the dorso-ventral muscle bundles between the diverticula are absent. The anus opens terminally and no caudal cirrus is present.

Blood system

The cephalic supply consists of a thin-walled lacuna running through the precerebral, cerebral and post-cerebral regions with structural differences in each region.

The precerebral lacuna found on the dorsal side of rhynchodaeum is straight, extending from the tip of the head to the front of the brain lobes (Fig. 2a, d). Just behind the proboscis pore it forms a narrow flattened crevice surrounded by the dermal connective tissue (Fig. 5a). It then becomes more spacious toward the posterior in the middle portion of the precerebral region, forming a rectangle in transverse sections on its dorsolateral sides (Fig. 5b, c). At this region, the rhynchodaeum protrudes dorsally on the ventral side of the lacuna.

Near the cerebral ganglia, the precerebral lacuna narrows dorsoventrally. It is crescent shaped due to the fact that the rhynchodaeum becomes more spacious dorsally (Fig. 5d). It is covered by outer circular and inner longitudinal muscle fibres derived from the body wall middle circular and inner longitudinal muscle layers, though their features are membraneous in the anterior portion of the head. On the ventral side of the precerebral lacuna, the muscle fibres become thicker as a layer, are not distinguishable as separate outer circular and inner longitudinal musculature, and cover the dorsal side of the rhynchodaeum forming a dorsal saddle over it.

The cerebral lacuna follows the precerebral lacuna and runs inside the brain lobes and the cerebral organs. It flattens more dorsoventrally with the well-developed muscle layers and becomes indistinguishable with an arrangement of the basophilic nuclei remaining (Fig. 5e). Immediately behind the dorsal brain commissure, it is divided into a pair of lateral lacunae by the muscle layers fused at the mid-dorsal part of the rhynchodaeum (Fig. 5f, g). Immediately behind the ventral brain commissure, the muscle layers covering the lacuna fuse again in the mid-ventral part of the rhynchodaeum. Thus, the body wall middle circular and inner longitudinal muscle layers encircle the cerebral lacuna and rhynchodaeum (Fig. 5h). In the middle portion of the brain lobes, the cerebral lacuna with a space for the blood flow is found at the mid-ventral side of the rhynchodaeum (Fig. 5h) and a flattened space representing it is clearly recognizable at the anterior end of the cerebral organs, immediately behind

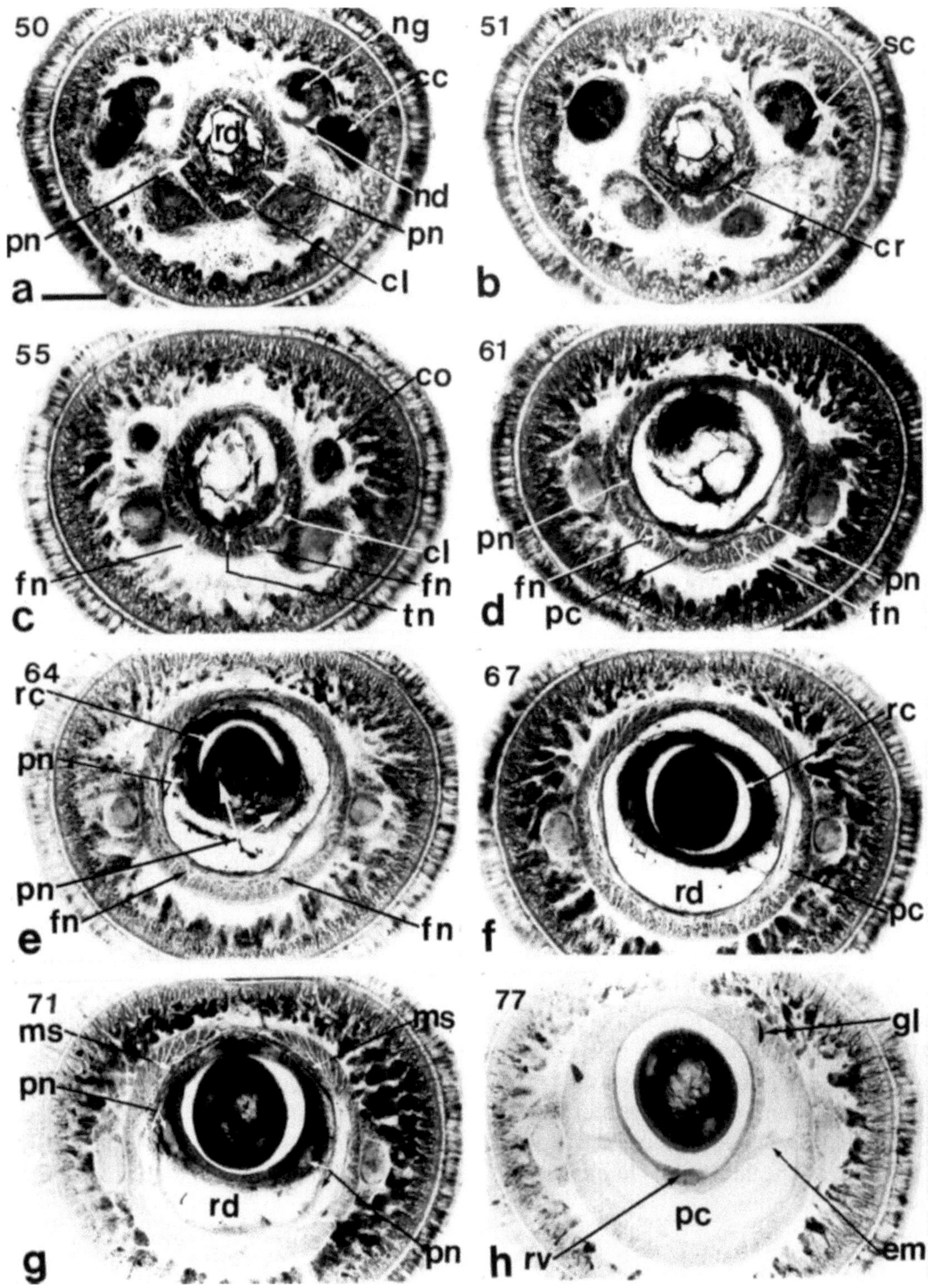

Fig. 6. Paralineopsis taki. Transverse sections of anterior portion of body showing the cerebral organs and the post-cerebral lacuna (a–c), the posterior region of the rhynchodaeum (d–g), and the opening of the post-cerebral lacuna in the rhynchocoel vessel behind the rhynchodaeum (h). Scale = 100 μm. 50–77; number of sections from anterior end of body. cr = circular muscles running into rhynchodaeal wall; em = endothelial membrane; gl = glandular cell in the inner longitudinal muscle layer of the body

the cerebral ganglia (Fig. 6a). At the posterior end of the cerebral organs the cerebral lacuna widens toward the dorsal sides of the body. At the latero-ventral parts of the rhynchodaeum, it is divided into three spaces by contacts between the rhynchodaeal wall and body wall inner longitudinal muscle layer (Fig. 6c). Behind the cerebral organs, the post-cerebral lacuna is divided into three spaces in transverse section (Fig. 6d). It becomes semicircular in space below the rhynchodaeal wall due to the loss of the contacts between the muscle layer and rhynchodaeal wall in the region near the posterior end of the rhynchodaeum (Figs 2d, 6e, f).

Immediately behind the rhynchodaeum, the post-cerebral lacuna becomes suddenly spacious, surrounding the latero-ventral sides of the rhynchocoel (Fig. 6g, h). At this location, the rhynchocoel vessel running along the inside floor of the rhynchocoel wall branches off from the post-cerebral lacuna (Fig. 6h) and two endothelial membranes running transversely at the lateral sides of the rhynchocoel divide it into a pair of dorsal spaces and broad ventral space, their antero-posterior length, 250 μm, ranging between the sections 75 and 100.

The post-cerebral lacuna between the posterior end of the rhynchodaeum and the anterior tip of the foregut, 1 mm long, is about 3.5 times the length between the posterior ends of the brain lobes and the rhynchodaeum (Fig. 2a, d, 3a–c).

The single spacious post-cerebral lacuna is posteriorly divided by the appearance of the foregut into two parts, which become the lateral vessels (Fig. 3d–f). At the anterior end of the mouth, the rhynchocoel vessel running inside the rhynchocoel wall comes downward in the connective tissue between the rhynchocoel wall and foregut and becomes the dorsal vessel (Figs 2a, d, 3e). The lateral vessels in the foregut region are not developed into a vascular plexus and consist of two laterodorsal vessels only (Fig. 3f). There are no pseudometameric transverse connectives between the mid-dorsal and ventrolateral vessels in the intestinal portion of the body.

Nervous system

The brain lobes are small and simple in structure. They are suspended in a wide space of dermal connective tissue in the body wall outer longitudinal muscle layer of the head. On the ventral side of the head immediately behind the ventral brain commissure, the body wall outer longitudinal muscle layer is present in a 3:2:9 ratio to the three components of the wall. The outer longitudinal muscle layer is 30 μm thick, the glands 20 μm thick, and the connective tissue 85 μm thick respectively (Fig. 5h). The brain lobes located near the anterior end of the body are 100 μm long. In comparison to the antero-posterior length of the precerebral region, 390 μm, cerebral organs, 60 μm, and the post-cerebral region to the posterior end of the rhynchodaeum, 220 μm, this shows the small size of the brain lobes (Fig. 2c).

The small brain lobes are transversely joined by a dorsal brain commissure, 15 μm thick, which is anterior to the strong (50 μm in diameter) ventral commissure (Figs 2c, 5e, f). The brain is enclosed by a delicate outer neurilemma, but the inner neurilemma is indistinct and only clearly evident in the lateral nerve cords. Amongst the ganglionic tissue of each cerebral lobe, on the rhynchodaeal side of their fibrous cores, there is no single large neurochord cell. Neurochords are not distinguished in the lateral nerves. The dorsal brain lobes appear first in the transverse sections and soon send out a dorsal brain commissure. The ventral brain lobes extend posteriorly from the root of the dorsal commissure and are connected by a short thick ventral commissure. The dorsal brain lobes immediately behind the ventral commissure are columnar in shape and protrude dorsally, their dorsal tips above the level

wall; ms = muscular septum for the proboscideal organ; nd = nerve cord from the dorsal brain lobe; ng = nerve cells of the cerebral organ; sc = sensory canal of the cerebral organ.

of the dorsal side of the rhynchodaeum (Fig. 5g, h). The ventral brain lobes are elliptical in shape, arranged horizontally at a level below the ventral side of the rhynchodaeum. The dorsal and ventral lobes thus form an acute angle, and a wide section of connective tissue occurs between the brain lobes and rhynchodaeum (Fig. 5g). The dorsal brain lobes soon end, while the ventral lobes remain voluminous at a level of the latero-ventral side of the rhynchodaeum, extending posteriorly beyond the dorsal lobes (Figs 2c, 6a). The ventral brain lobes suddenly narrow and become the lateral nerves at the posterior end of the cerebral organs (Fig. 6b, c). The lateral nerves, thus formed, enlarge and run dorsally toward the lateral sides of the rhynchodaeum (Fig. 6d). In each dorsal lobe the fibrous core is not forked into upper and lower points at its rear, but immediately sends out a pair of thick nerves to the cerebral organs (Figs 2c, 6a).

The brain lobes contain four types of nerve cells. The first, 2.5 μm in diameter, are arranged thickly around the central core of the dorsal ganglia. The second type, 5 μm in diameter and about twice as large as the first, surround the central core of the ventral ganglia. The third type is composed of several large nerve cells located in the middle portion of the dorsal and ventral brain lobes outside the rhynchodaeum. One of these cells is 10 μm wide and 15 μm long (Fig. 5g). The fourth type, 15 μm wide and 28 μm long, is composed of three cells at the margin of the ventral lobe on either side in sections 49 and 50, a fibrous membrane around each cell is not clearly observed.

The two proboscideal nerves branch upward from the ventral brain lobes at the ventral commissure (Figs 2c, 5g) and are situated outside the rhynchodaeal wall (Fig. 5h). These nerves run posteriorly to the hind end of the rhynchodaeum and enter the proboscis as a pair of the proboscideal nerves (Figs 2c, 6a–g). A remarkably thick transverse cord connects the proboscideal nerves in the region near the posterior end of the cerebral organs (Fig. 6c).

The foregut nerves arising from the ventral brain lobes near the lateral cords (left nerve in section 53 and right one in section 54) extend to the anterior end of the foregut, where the nerves are connected by a thick transverse cord (Figs 2c, 4b, d, 6c–h). On the latero-ventral sides of the mouth and oesophagus they run posteriorly without transverse connection (Fig. 3e, f).

The dorsal nerve from the dorsal brain commissure extends posteriorly in the dermal connective tissue without immediate contact with the peripheral nerve plexus running outside the body wall middle circular muscle layer (Figs 2a, 4a, 5e–h, 6a–g). In the post-cerebral region, the dorsal nerve projects downward and immediately contacts the body wall middle circular muscle layer (Figs 2a, 4c, e). Several nerve cords branch off from the anterior end of the cerebral ganglia and enter the broad dermal connective tissue (Fig. 5d).

Special sensory organs and frontal glands

The special sensory organs contain three short, small canals arranged in a triangle above the shallow transverse slit at the opening of the rhynchodaeum.

The frontal glands, which are derived from the dermal glands, are not well developed and are confined in the broad central portion of the dermal connective tissue (Fig. 5c).

Sensory organs

The cerebral organs, 120 μm high, 85 μm wide, and 80 μm in length, are small and ovoid. They are located a short distance behind the dorsal ganglia (Fig. 6a–c). These organs float in the broad and thick dermal connective tissue of the body wall outer longitudinal muscle layer, and are embedded obliquely in their arrangement in the head. They do not penetrate the inner longitudinal muscles so that they do not contact the blood vascular system. The dorsal ganglia send out a pair of thick nerve cords toward the anterior ends of the organs without immediate contact between them; many nerve cells are located around the cord (Fig. 6a). The slender sensory canal of the organ runs posteriorly in the dorsal side of the organ to its bottom, where the glandular cells are filled (Fig. 6b, c). The short and broad sac-like canals of the organs, with thickly arranged cilia, open widely at the mid-lateral surface of the head (Fig. 5h).

There are no horizontal lateral cephalic furrows, but in their place a pair of shallow epidermal depressions are clearly recognizable in transverse sections, extending from the anterior end to the opening of the cerebral canals (Fig. 5b–e). Just anterior to the opening of the canal of the cerebral organ the horizontal lateral cephalic depression widens dorsoventrally (Fig. 5f, g). At the tip of the head, these depressions are deeply invaginated with thickly arranged eosinophilic cells (Fig. 5a). In the epidermis of the depression, the glandular cells are absent, but cilia are thickly arranged.

Excretory and reproductive systems

The nephridial canals found in the parenchyma are arranged on the dorsolateral sides of the foregut behind the mouth slit (Fig. 3f). In the material sectioned, the efferent ducts were not observed.

The immature oogonia are found in the parenchyma, located on the dorsolateral sides of the body above the lateral nerves.

Systematic discussion

In 1911, Schütz reported the heteronemertean *Paralineus elisabethae* (nov. gen. et sp.) from Villefranche and pointed out four characteristics of the body for comparison with the genus *Parapolia* (Coe, 1895) as follows: horizontal lateral cephalic furrows, absent in both genera; muscle layers of the proboscis, three in *Parapolia* and two in *Paralineus*; cerebral organs, plate-like in the former and sac-like in the latter; caudal cirrus, present in the former and absent in the latter.

In later work, Friedrich (1960) listed 11 genera of the Heteronemertea lacking horizontal lateral cephalic furrows and arranged into five sections as follows: oblique cephalic grooves present (*Baseodiscus*); proboscis opening on the ventral side of the head between the anterior tips of the head and the brain lobes (*Valencinura*, *Oxypolella*, *Valencinia*); cephalic grooves present on the dorsal and ventral sides of the head (*Poliopsis*); cerebral organs separated from the dorsal ganglia (*Oxypolia*, *Parapolia*); cerebral organs attached immediately to the dorsal ganglia (*Planolineus*, *Cephalomastax*, *Paralineus*, *Zygeupolia*). The last section, however, was subdivided into *Cephalomastax* with the mouth under the brain and the remainder with the mouth behind the brain, and *Paralineus* was separated from *Zygeupolia* in not having the caudal cirrus.

Thus, the two genera *Parapolia* and *Paralineus* used for comparison by Schütz (1911, 1912) were far separated by Friedrich's taxonomy.

Gibson (1985, 1989) reassessed all the genera of the Heteronemertea hitherto recorded by eight stand-points mainly based on the structure of the proboscideal wall. He proposed that all the genera of the Heteronemertea could be divided into 7 families and 2 unplaced genera.

The 11 genera listed by Friedrich (1960) were separated into three families Cerebratulidae (*Valencinura*, *Oxypolella*, *Oxypolia*, *Parapolia*), Lineidae (*Planolineus*, *Paralineus*, *Zygeupolia*), and Valencinidae (*Baseodiscus*, *Valencinura*, *Poliopsis*, *Cephalomastax**).

Paralineopsis belongs to the family Lineidae as does *Paralineus* in Gibson's taxon. *Paralineus* and *Paralineopsis* have little in common with other genera lumped in this taxon other than the arrangement of the muscles in the proboscis and even here there is a marked distinction in that the longitudinal muscles of the proboscis are not derived from the outer longitudinal muscles in these two genera. Furthermore, the two genera lack outer longitudinal muscles in the body wall and

* *Cephalomastax* belongs to the family Lineidae, because its proboscideal wall consists of the outer circular and inner longitudinal muscle layer (Iwata, 1957). Iwata's description followed an old rule, but not Coe's definition (Coe, 1943), as in the cases of *Uchidana* and *Hinumanemertes* (Iwata, 1967, 1970).

thus there are no medullary precerebral muscles; muscles do not accompany the rhynchodaeum from the pore, and thus the ventral wall of the rhynchodaeum is strictly epithelial for much of its length: lateral cephalic fissures are absent and the cerebral organs do not penetrate the inner circular muscles of the body nor do they project into the blood vascular system: the cephalic lacuna extends far behind the brain lobes. These morphological features clearly separate the two genera from the others listed by Gibson (1985). These fundamental differences including the origin of the proboscideal longitudinal muscles warrant the removal of these two genera from the Lineidae *sensu* Gibson 1985, and their placement in a new family. The absence of outer longitudinal muscles in the body wall of members of this family necessitates a revision of the definition of the order Heteronemertea which at present includes the presence of these muscles as a fundamental character.

The presence of 2 muscle crosses in *Paralineus* was cited in Gibson's description (Gibson, 1989), but Schütz (1911, 1912) merely noted that 'Die innere Ringfaserschicht gibt einige Fasern von sich, die zur Peripherie gelangen'.

Schütz (1912) described the epidermis of *Paralineus elisabethae* in great detail including the depression at the entrance of the cerebral organ canals. He stressed that there was no indication of anything comparable to the cephalic fissures of lineids. Thompson (1901) described specialized epidermal lateral grooves near the origin of the intestine of *Zygeupolia litoralis*, but only specialized 'ciliated pits' at the origin of the cerebral canals in the head region. Norenburg (1993) described similar pits without modified epidermal tracts leading to them in *Riserius pugetensis* (Norenburg, 1993), a species lacking subepidermal glands, with cerebral organs not enclosed in cephalic blood lacunae, but with a typical proboscis insertion from the body wall musculature. Schütz (1912) indicated in his figures that the proboscis insertion of *P. elisabethae* was in the vicinity of the cerebral commissures but he did not describe the proboscis insertion nor any attachments that would fix the diaphragm at that point. The absence of restraints at the diaphragm in the absence of outer longitudinal muscles forming the proboscis insertion in *Paralineopsis taki* would allow the diaphragm to be moved back and forth as the result of contraction and extension of the head as well as by peristaltic waves in the rhynchocoel wall.

In the absence of information on the relationship of the cerebral organs to the cephalic vascular system, on possible specialized lateral epidermal depressions extending posteriorly along the head to the cerebral canals, and on the nature of the proboscis insertion, it is not appropriate to designate the Onomichi specimen as a second species of *Paralineus* differing in living colour, circular muscles under the basement membrane, radial muscular bundles and subepithelial glands of the foregut, endothelial membranes of the postcerebral lacuna, and intestinal diverticula. It is concluded that the present species must now be placed in a separate genus.

Acknowledgements

I thank Dr Nathan W. Riser of the Northeastern University for comments and suggestions which substantially improved the manuscript. The Kushiro Public University Research Committee provided partial travel funding.

References

Coe, W. R., 1943. Biology of the nemerteans of the Atlantic coast of North America. Trans. Conn. Acad. Arts Sci. 35: 129–328.

Friedrich, H., 1960. Bemerkungen uber die Gattung Micrura Ehrenberg 1831 und zur Klassifikation der Heteronemertinen nebst vorlaufigem Bestimmungsschlussel. Veroff. Inst. Meeresforsch. Bremerh. 7: 48–62.

Gibson, R., 1985. The need for a standard approach to taxonomic descriptions of nemerteans. Am. Zool. 25: 5–14.

Gibson, R., 1989. The macrobenthic nemertean fauna of Hong Kong. In: Proc. Sec. Inter. Work. Mar. Flo. Fau. Hong Kong and S. China 1. Hong Kong Univ. Pres.: 33–212.

Iwata, F., 1951. Nemerteans in the vicinity of Onomichi. J. Fac. Sci. Hokkaido Univ. Ser. 6. Zool. 10: 125–138.

Iwata, F., 1957. Nemerteans from Sagami Bay. Publs. Akkeshi mar. biol. Stn, 7: 1–31.

Iwata, F., 1967. *Uchidana parasita* nov., gen. et nov. sp., a new parasitic nemertean from Japan with peculiar morphological characters. Zool. Anz. 178: 122–136.

Iwata, F., 1970. On the brackish water nemerteans from Japan, provided with special circulatory and nephridial organs useful for osmoregulation. Zool. Anz. 184: 133–154.

Norenburg, J. L., 1993. *Riserius pugetensis* gen. n., sp. n. (Nemertina: Anopla), a new mesopsammic species, and comments on phylogenetics of some anoplan characters. Hydrobiologia 266: 203–218.

Schütz, V., 1911. *Paralineus elisabethae* (nov. gen. et sp.). Zool. Anz. 37: 449–457.

Schütz, V., 1912. *Paralineus elisabethae* (nov. gen. et sp.). Z. Wiss. Zool. 102: 111–135.

Thompson, C. B., 1900. Preliminary description of *Zygeupolia litoralia*, a new genus and new species of heteronemertean. Zool. Anz. 23: 151–153.

Thompson, C. B., 1901. *Zygeupolia litoralis*, a new heteronemertean. Proc. Acad. nat. Sci. Philad. 53: 657–739.

Hydrobiologia **266**: 203–205, 1993.
R. Gibson, J. Moore & P. Sundberg (eds), Advances in Nemertean Biology

Riserius pugetensis gen. n., sp. n. (Nemertina: Anopla), a new mesopsammic species, and comments on phylogenetics of some anoplan characters

Jon L. Norenburg
Department of Invertebrate Zoology, National Museum of Natural History, Washington, DC 20560, USA

Key words: nemertine, systematics, phylogeny, meiofauna

Abstract

Riserius pugetensis gen. n., sp. n. is described from the northwest coast of Washington, U.S.A. It is probably a heteronemertine and, as such, would be the first known mesopsammic member of that order; it lives in the interstices of medium to coarse, moderately clean sands. Its morphology presents some attributes considered characteristic of mesopsammic fauna. The effect of some of these attributes is a generalized appearance of anatomical 'simplicity' and, as with other mesopsammic metazoans, this presents difficulties in distinguishing between primitiveness and reduction. In the new species such problematic features include the lack of subepidermal glandular cells and connective tissue, reduced proboscideal musculature, simple blood-vascular system, and the presence of a 'secondary', outer circular musculature in the body wall. The general appearance of this new species is very similar to the so-called palaeonemertine *Hubrechtella* and characteristics of both suggest relationship with the baseodiscid heteronemertines. These relationships are explored in light of a modified interpretation of proboscideal musculature, a traditional cornerstone of heteronemertinean taxonomy.

Introduction

Riselius pugetensis gen. n., sp.n. fits a basic structural definition of the Heteronemertina but resembles palaeonemertines in its simplicity. The orders Heteronemertina and Palaeonemertina are usually defined on the basis of body wall organization, arrangement of the dermis, and position of the nervous system (e.g, Hylbom, 1957; Gibson, 1982). Heteronemertina are defined as having a body wall constructed as follows: an outer zone (dermis) containing connective tissue, subepidermal glands and sometimes circular muscle fibers; an outer longitudinal musculature; the central nervous system between this longitudinal musculature and a deeper circular musculature (Gibson, 1982). Palaeonemertina (*sensu* Iwata, 1960) are Anopla having the nervous system at the base of the epidermis and lacking outer longitudinal musculature and subepidermal gland cells. Although not explicit on the issue, Bürger (1895) effectively recognized the palaeonemertines as paraphyletic, but the heteronemertines as monophyletic. In contrast, Hylbom's (1957) phylogenetic scheme, which echoed the opinions of several other principal nemertinologists, results in both orders being at least paraphyletic. A 'cutis' comprising the subepidermal cell bodies of glandular cells and often an associated zone of connective tissue, is characteristic of heteronemertines. Potential homologues of these cutis glandular cells are known from several palaeonemertinean genera, e.g., *Hubrechtella* Bergendal, 1902, in which submerged cell bodies form a region of pseudostratified epidermis (Gibson, 1979a; Norenburg, 1985). *Riserius pugetensis* has neither pseudostratified epidermis nor components of a cutis, a condition known among other heteronemertines only from the lineid *Colemanniella albulus* (Gibson, 1981).

Effectively, the order Heteronemertina can be separated from the order Palaeonemertina (*sensu* Iwata, 1960) only by the presence of the outer longitudinal musculature lying between the epidermis and the central nervous system; *i.e.*, this is a synapomorphy for Heteronemertina. However, less discrete longitudinal musculature may also be found in the dermis or among the epidermal cells of Palaeo- and Hoplonemertina (Friedrich, 1936, 1979). Similarly, some species of palaeo- and heteronemertines have a more or less distinct dermal or epidermal outer circular musculature, as seen in *R. pugetensis*.

Riserius pugetensis has been alluded to previously as a mesopsammic heteronemertine (Norenburg, 1987, 1988a, b). However, one may legitimately ask, is it in fact a heteronemertine and, if so, is it a relict species of a largely extinct clade or is its simplicity a derived function of its specialized habitat?

Material and methods

Six specimens of *Riserius pugetensis* gen. n., sp. n. were collected in August, 1979, from sand at 0 m (near the bottom of a moderate slope), 200 m south of an oil terminal and depot at Richmond Beach along Puget Sound, Washington (47° 40′ N, 122° 25′ W). Three more specimens were obtained from sand samples collected by divers in July, 1981, from Minnesota Reef near Friday Harbor, San Juan Island, Washington (48°31′48″N, 122°58′55″), at about 10 to 15 m depth. Sand samples were collected with a trowel to a depth of 5–10 cm, placed in buckets or plastic Whirl-Paks and brought back to the laboratory. Specimens were separated from the sand by swirling a subsample of sand in a bucket with seawater and quickly decanting the water through a 63-μm mesh sieve. The worms were examined while alive with a stereomicroscope and, while lightly squeezed under a coverslip, with a compound microscope utilizing bright-field or Nomarski optics. For preservation and histology specimens were anaesthetized briefly (5–6 min) in 7.5% $MgCl_2$, fixed in Hollande's cupri-picri-formal-acetic fixative, and embedded in polyester wax (Steedman, 1960; Norenburg & Barrett, 1987). Transverse and longitudinal serial sections were cut at 5–7 μm and stained with Heidenhain's azan or Heidenhain's iron hematoxylin methods.

Systematic account

Riserius pugetensis *gen.n., sp.n.*

Generic diagnosis
Four body-wall muscle strata, outer circular, outer longitudinal, middle circular, and inner longitudinal; cutis (subepidermal glandular cells and connective tissue) lacking. Central nervous system and plexus between outer longitudinal and middle circular musculature; dorsal ganglia not bifurcated posteriorly; without neurochord elements; middorsal nerve from dorsal commissure. Proboscis with inner longitudinal musculature; inner and outer circular musculatures reduced or lacking. Rhynchocoel short; with outer circular and inner longitudinal musculature. Cerebral organs not fused to cerebral ganglia; not projecting into vascular lacunae; open via lateral pits. Transverse cephalic furrow encircles body in front of mouth. Rhynchodaeal aperture terminal; mouth ventral, posterior to cerebral organs. Blood vascular system with single cephalic lacuna; two lateral vessels and one middorsal; supra-anal anastomosis. Cyrtocytes single and clustered, projecting into lateral vascular lacunae; one pair of collecting ducts and nephric pores.

Riserius pugetensis *gen.n., sp.n.*

Specific diagnosis
Characters of the genus. Cephalic glands and frontal organ lacking. Inner circular musculature reduced and longitudinal muscle plate lacking.

Blood vessels simple, lacking cross-connections, middorsal vessel enters rhynchocoel, precerebral lacuna a single, medial sac. Cerebral organs large; openings deep lateral pits. Rhynchodaeum thin-walled. Rhynchocoel approximately one-quarter of body length. Stomach and intestine histologically distinct, without constrictions or diverticula. Gonochoric.

Type specimens
Holotype (USNM 149925), and single paratype (USNM 149926) deposited in the Division of Worms, National Museum of Natural History (USNM), Smithsonian Institution, Washington, DC; both specimens were collected 17 July 1981, anaesthetized in magnesium chloride, fixed in Hollande's cupri-picri-formal-acetic fixative and subsequently transferred to 70% ethyl alcohol.

Additional specimens
Two sets of serial sections of specimens, from the Richmond Beach site, in the author's possession will be deposited in the USNM at a later date. One is a set of three slides bearing transverse serial sections (stained with Heidenhain's iron hematoxylin) of the anterior one-fifth of a specimen. The other consists of eight slides bearing transverse serial sections of two specimens side by side (stained with Heidenhain's azan or Heidenhain's iron hematoxylin).

Type locality
The top 10 cm of moderately coarse sand at 10 m depth near a reef known locally as Minnesota Reef (near Friday Harbor), San Juan Island, Washington (48°31′48″N, 122°58′55″).

Etymology
The generic designation recognizes the contributions to the study of both nemertines and mesopsammic animals by Prof. Nathan W. Riser. The specific designation refers to Puget Sound.

External features

Gliding specimens reach 15 mm in length and 0.15 to 0.2 mm in diameter. Those with numerous eggs are somewhat stouter. The epidermis is translucent white with transmitted light and opaque white with reflected light. The body is moderately flattened dorsoventrally when gliding but the cephalic region is cylindrical and has a cephalothricid appearance (Fig. 1). The cephalic region is relatively elongate; in a 10 mm specimen, the pair of conspicuous pits leading to the cerebral organs is about 500 μm from the cephalic tip. An epidermal furrow encircles the body 500 μm farther posteriad, but anterior to the mouth (Figs. 1–3). This furrow is directed posteriad dorsally and anteriad ventrally. The cephalic tip bears a small dimple, the rhynchodaeal pore, which is surrounded by a few elongate cilia, but these are shorter than typical sensory cilia. With gentle squeezing of living specimens under a coverslip one can occasionally observe the pair of dorsolateral nephric pores and also individual mucoid strings emanating from large gland cells. These appear to account for the sticky, multipoint haptic reaction exhibited by this species. When severely agitated, members of this species have a tendency to coil. There is no indication of a caudal cirrus.

Body wall

Epidermis in sections is 6–10 μm thick, with a ciliary border of 4–5 μm. Two cell types predominate and are about equally represented (Figs 4–11). One is a homoserous glandular cell (Norenburg, 1985), which forms slender to spherical goblets. The content of this cell is first an azanophilous or eosinophilous (non-metachromatic with toluidine blue) matrix containing many fine, more densely-staining granules. Upon contact with seawater the goblet content changes appearance, starting with the portion outside the goblet and proceeding inward. The exposed secretion is homogeneously metachromatic with toluidine blue, intensely azanophilous or olive-green to blackish-brown with hematoxylin. The goblet cells are separated by attenuate ciliated cells with an expanded distal surface. The goblet cells do not have a uniform distribution and there are numerous patches of ciliated cells without goblets.

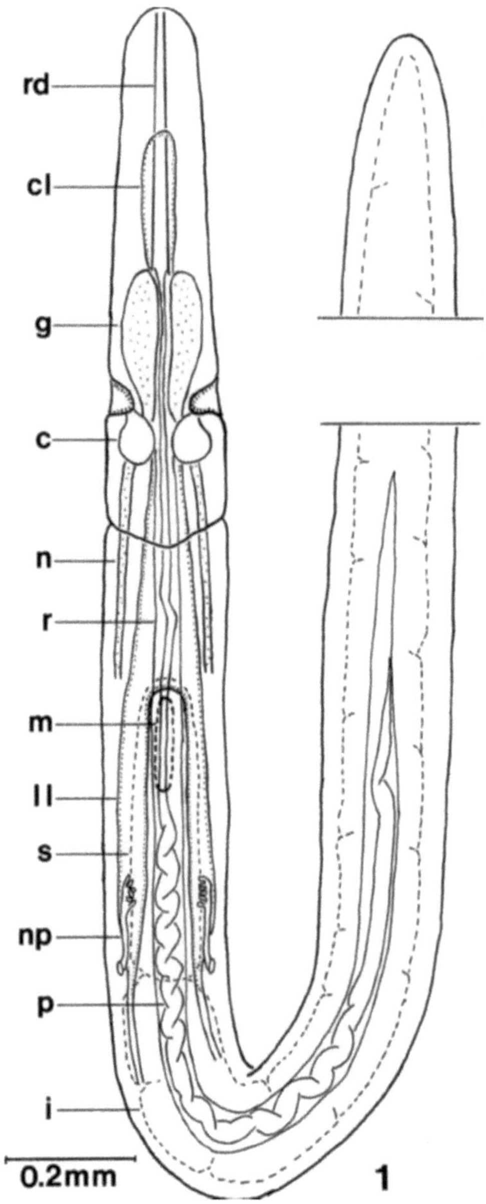

Fig. 1. Riserius pugetensis. Schematic figure of anterior one-quarter and posterior terminus of body showing organization ot major structures as seen from dorsal perspective. Abbreviations: c = cerebral organ; cl = cephalic blood lacuna; g = cerebral ganglion; i = intestine; ll = lateral blood lacuna; m = mouth; n = lateral nerve cord; np = nephridial complex; p = proboscis; r = rhynchocoel; rd = rhynchodaeum; s = stomach.

Goblet cells are lacking entirely in the first 15–20 μm of the cephalic tip. The ciliated cells of these patches are short (6 μm), columnar cells and can be seen to contain large spherical nuclei and granular cytoplasm. A third cell type, the bacillary cell, is irregularly scattered among the goblet cells but is present in much lower numbers. It is slender and contains small, discrete granules that are weakly azanophilous or eosinophilous. Beneath the epidermis is a thin zone (0.5 μm) of granules, followed by a thin dermis (1 μm), and then by a sparse, subepidermal, outer circular musculature (OCM) consisting of widely spaced single fibers (only poorly observable, even in tangential sections). The latter is followed by a relatively well developed outer longitudinal musculature (OLM) (5–10 μm); both are sparse precerebrally (Fig. 4). Precerebrally, the OLM gradually thins as it proceeds anteriorly by sending fibers into the cephalic medulla (Fig. 12); thus, there is not a discrete split of the OLM. Some of these fibers are closely associated with nerve tracts. Others bend in front of the cerebral ganglia, where some comprise a very diffuse proboscis insertion and others become confluent with either the inner longitudinal musculature (ILM) or the longitudinal muscles of the rhynchocoel. Posterior to the cerebral ganglia OLM is underlain by the lateral nerve cords and a nerve plexus (Figs 8–11). Below this is a single layer of middle circular musculature (MCM) followed by a well developed ILM. These layers surround the rhynchocoel, gut, blood vessels and gonads. The MCM extends anteriorly to about the rhynchodaeum. The ILM forms a prominent band along the dorsolateral surface of each of the two anterior lacunar portions of the lateral blood vessels (Figs 7–9, 11), but is lacking along a narrow, dorsomedian strip of the rhynchocoel wall between the two lacunae (Fig. 11). This strip extends for the entire length of the rhynchocoel and at this site the circular muscles of the rhynchocoel are confluent, but apparently not interwoven, with the MCM of the body wall. The rhynchocoel has separate outer circular and inner longitudinal musculature. The circular muscles form occasional slings which insert dorsolaterally in the OCM of the body wall. The stomach is invested by a sparse musculature of circular fibers. Some of these extend laterally to the OCM. In the vi-

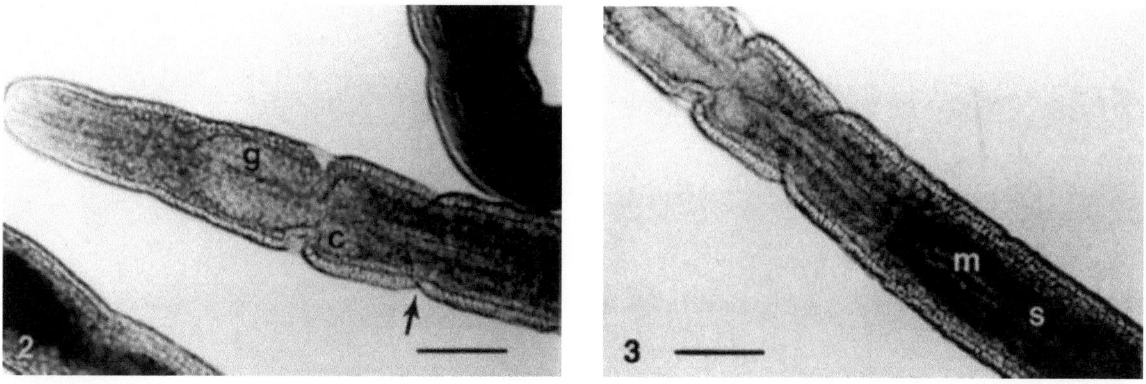

Figs 2–3. Riserius pugetensis. Photomicrographs of live specimen. 2. Anterior, cephalic region showing large lateral cerebral organ openings and dorsal aspect of annular cephalic furrow (arrow). 3. Ventral view of posterior cephalic region showing annular furrow and mouth. Abbreviations as in Fig. 1. Scales = 150 μm.

cinity of the stomach, some of the inner circular muscle fibers extend to the OCM and act as dorsoventral muscles. A longitudinal muscle plate between gut and rhynchocoel was not discerned.

In the region between the cerebral organs and the mouth, there are 8 to 10 large gland cells (10–12 μm) located in the parenchyma on each side of the body. These are reminiscent of cephalic gland cells of palaeonemertines, staining weakly with aniline blue and appearing pale yellow with hematoxylin. These may be the same as cells referred to as salivary cells in other species. A communication with the body surface could not be identified for these cells. In the vicinity of the two excretory ducts there are several clusters of various-sized droplets. These also are in the parenchyma and are reminiscent of neurosecretory material. The typical heteronemertine subepidermal glands are lacking entirely.

Nervous system and sensory structures

The central nervous system, consists of four cerebral ganglia, two lateral nerve cords and a dorsal nerve. Each of the two lateral lobes consists of a fused dorsal and ventral ganglion and is an elongate structure about 250–300 μm long (from anterior margin to cerebral organ). In transverse section, maximum combined width of the two lobes is about 60% of cephalic width (Fig. 5). Dorsal and ventral ganglia of a lobe are not demarcated from each other externally. Internally they are separated for the posterior two-fifths of their length by a horizontal wedge of neural cells. Their neuropils fuse anteriorly but are distinguishable throughout their length. A massive ventral commissure (40 μm minimum diameter in section) connects the two ventral neuropils (Fig. 5). The ventral neuropils are simple and continue posteriorly as the lateral nerve cords. The dorsal neuropils are complex, each presenting from 2 to 4 major fiber tracts in transverse section. The dorsal ganglia are united by a dorsal commissure (15 μm diameter) near the anterior margin of the ganglia, slightly anterior to the ventral commissure. The dorsal neuropil ends posteriorly with an abrupt lateral flexure which extends into the cerebral organ.

The lateral nerve cords have a pronounced ventromedial flexure below the cerebral organs, but then lie lateral for the rest of their posterior extent, ending in a supra-anal commissure. Each has a single neuropil. The dorsal commissure gives rise to a nucleated dorsal nerve (5–10 μm diameter) which extends posteriorly, in the neural plexus, along the dorsal midline of the rhynchocoel. It could not be detected beyond the rhynchocoel. A rhynchocoelic nerve was not observed.

The cerebral ganglia do not have well-defined anterior faces from which nerve tracts project.

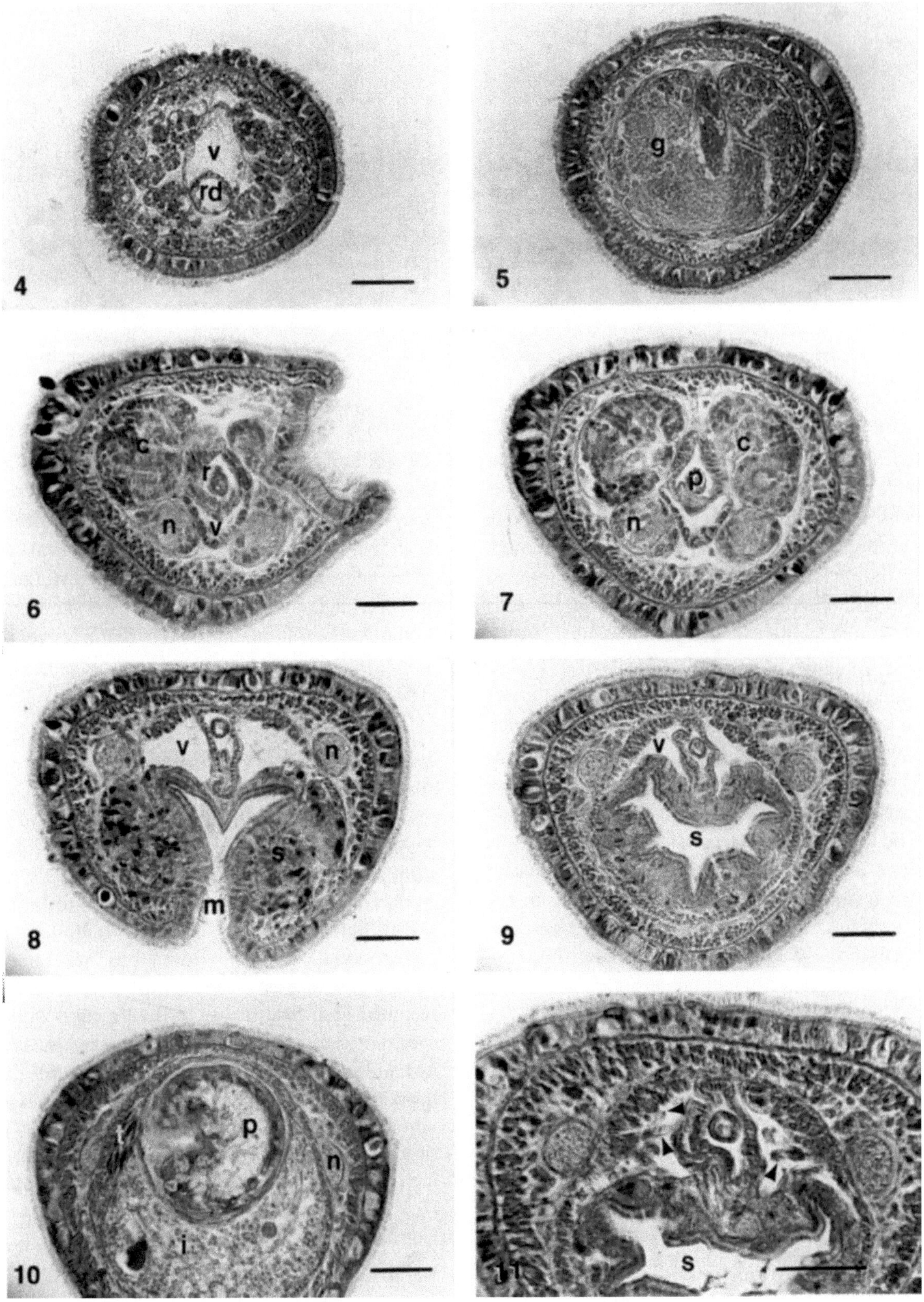
v
rd
4
g
5
c
r
n
v
6
c
p
n
7
v
n
s
m
8
v
s
9
p
n
i
10
s
11

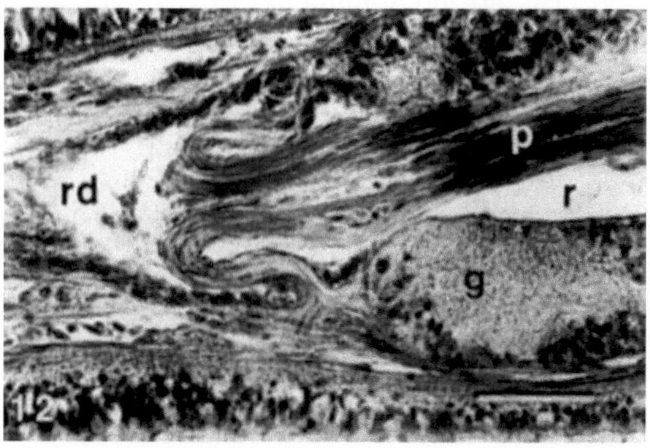

Fig. 12. Riserius pugetensis. Histological sagittal section through proboscis insertion Scale = 25 μm.

Rather, each lobe gradually dissolves to form 3 to 4 major tracts that extend anteriorly to the cephalic tip. Each tract is accompanied by muscle fibers and abundant neural cells that are continuous with the neural cell coat of the ganglia. Neurochord elements are lacking. A pair of nerves extends along the ventrolateral margins of the stomach and are usually somewhat embedded in its gland cell layer. There are at least two cross connections (commissures) between these. Their origin and distal extent were not determined. Presumably these are the homologues of oesophageal nerves.

Ocelli and statocysts are lacking. Besides the pair of cerebral sensory organs (CSO) and their associated lateral pits, the only other potential sensory structures are the few short cirri adjacent to the rhynchodaeal pore. A frontal organ (pore) was not observed. The CSO lie posterior to and are not fused with the cerebral ganglia and do not penetrate the ILM or the vascular system. Maximum length and diameter of the CSO is about 250 μm and 50 μm respectively. The entrance to each CSO is in a deep lateral pit in the body wall even with the posterior margin of the cerebral ganglia (Figs 1–3, 6). Each pit is shaped somewhat like a postero-medially directed funnel, with a surface opening of about 50 μm diameter and a gradual indentation of the body wall anteriorly but an abrupt margin or lip posteriorly. The pit is lined with specialized ciliated cells (12 μm tall) bearing few very elongate cilia (14 μm long) and narrows to give rise to the ciliated canal of the CSO. The CSO nerve lies immediately dorsal to the canal at this point. A little further posterior (5–7 μm) the canal twists abruptly back to the body wall and abuts the OLM. This portion of the canal is surrounded by neural (bipolar?) cells. The canal extends posteriorly along the outer longitudinal musculature for another 35–40 μm and is flanked medially by more neural cells that surround the cerebral organ nerve in this region. The ciliated canal ends blindly and the cerebral organ nerve ends against this blind terminus. There is a relatively large pool of non-stained secretory droplets accumulated at the terminus but there are very few glandular cells in evidence in the rather large cerebral organ (Fig. 7). These are scattered individually throughout the organ and their content did not stain with azan or hematoxylin techniques. Most of the bulk of this organ is made up of relatively small neural cells.

Proboscis complex

The proboscis and rhynchocoel extend for about 20 to 25% of body length. The proboscis is frequently somewhat coiled and is up to 40% of body diameter (Fig. 10), but about the first millimeter is extremely slender (Figs 6–9). The proboscis is described in its everted position. The lining epidermis is up to 10 μm thick and consists

Figs 4–11. Riserius pugetensis. Histological transverse sections. 4. Section through precerebral region, showing cephalic blood lacuna and rhynchodaeum flanked by longitudinal muscle bundles. 5. Section through cerebral ganglia and ventral commissure. 6. Oblique section showing opening to cerebral organ on the right and, more posteriorly, cerebral organ with ciliated canal on left. 7. Section showing glandular region of cerebral organ on right. 8. Section through mouth at bifurcation of medial blood lacuna into two lateral lacunae. 9. Section through stomach and lateral lacunae. 10. Section showing intestinal region and mature testis. 11. Section through lateral blood lacunae showing clusters of cyrtocytes (arrows). Abbreviations: t = testis; v = vascular system; others as in Fig. 1. Scales = 25 μm.

of: modified ciliary cells with few, stout bristles; large, goblet gland cells with characteristics similar to those of the epidermis; and cells containing pseudocnidae (terminology after Martin, 1914: 272), usually arranged in parallel in clusters of 4 or 5. The pseudocnide appears as a slender, clavate capsule (0.5 × 13 μm) enclosing an eversible, thread-like tubule (about 30 μm long). The epidermis is underlain by a sparse nerve plexus: in one specimen at least one longitudinal nerve could be consistently recognized. Underneath the neural plexus, in only a few longitudinal sections, occasional circular fibers seemed to be present. I am not convinced that this is musculature. The remaining neural plexus is underlain by a well-defined, single layer of longitudinal muscle fibers. Inner circular musculature and distinct endothelium were not detectable even in longitudinal sections, nor were muscle crosses. Rhynchocoel musculature is described earlier; a distinct endothelium was not recognized here either. The proboscis retractor muscle ends in the longitudinal muscles of the posterior extremity of the rhynchocoel. The rhynchodaeum is thin-walled (Fig. 4) and (about 50 to 100 μm posterior to its invagination) is invested posteriorly by longitudinal muscle fibers confluent with that of the rhynchocoel and some that extend into the body wall OLM.

Alimentary canal

The mouth is an elongate opening, 80 μm long, located about 300 μm posterior to the cerebral ganglia. Its interior margin (lip) is a zone of large columnar cells bearing a sparse border of relatively long cilia (Fig. 8). This zone is surrounded by glandular cells indistinguishable from those of the stomach proper. Immediately above this lip the buccal cavity is lined by a very dense 'brush border' (cilia or microvilli) and its epithelium rapidly grades into that of the stomach. The stomach's moderately convoluted wall is 10 to 20 μm thick and comprises large spherical gland cells whose necks project to the luminal surface through the proximal 'brush border' (Fig. 9). The gland cells are predominantly of one type, containing finely granular, yellowish-brown material that does not stain well with azan or hematoxylin techniques. A few cells are vacuolate and contain a variable quantity of small, unstained droplets. The stomach is 350–400 μm long. There is a gradual diminution in thickness of the wall and in number of gland cells posteriorly. The posterior limit of the stomach is demarcated by an abrupt change, without constriction, to an intestinal epithelium of tall, vacuolate, columnar cells with few cilia. The intestine (Fig. 10) is a simple tubular structure having no diverticula, although it may be slightly constricted at sites of mature gonads.

Blood vascular system

The vascular system is simple, consisting of a middorsal vessel and two lateral vessels; all three unite anteriorly and posteriorly. Its precerebral portion is a thin-walled, elongate lacuna that lies medially over the rhynchodaeum to near the cephalic tip (Fig. 4). This passes posteriorly through the cerebral ring and comes to lie ventral to the rhynchocoel; presumably it bifurcates around the rhynchodaeum and then re-anastomoses ventrally. The middorsal vessel arises between the cerebral ganglia and passes immediately through the musculature of the rhynchocoel wall (Figs 6–8), where it forms a relatively thick-walled construction of cells with dense cytoplasm. The latter may represent a rhynchocoelic villus, although a discrete villus could not be identified. The vessel lies completely within the rhynchocoel cavity until it reaches the posterior limit of the stomach. Here it passes back through the musculature of the rhynchocoel wall, becomes thin-walled and continues posteriorly, between the rhynchocoel and intestine, to a supra-anal anastomosis with the lateral vessels. A voluminous post-cerebral lacuna lies above the foregut, almost enveloping the rhynchocoel so that it appears to be suspended from the dorsum of the lacuna. The lacuna becomes bifurcated in the vicinity of the mouth by the descending rhynchocoel (Fig. 8). The resulting pair of lacunae continues posteriorly, bordered by inner longitudinal muscles, stomach, and rhynchocoel (Figs 9, 11). At the posterior limit of the stomach, which also is the posterior limit of the nephridia, the vessels narrow markedly and are difficult to detect for most of the rest of their course to the supra-anal anastomosis. The endothelial lining of the lacunae contains numerous vacuolate cells containing various quantities of minute droplets. Some of these form a more or less continuous longitudinal tract along the outer edge of each lacuna, adjacent to the lateral nerve cords. The inner longitudinal musculature, which lies along the outer-margin of the lacunae, is lacking along these two tracts. Valves were not found.

Nephridial system

The nephridial system is largely confined to the lumen of the two lateral blood lacunae. Its posterior limit is marked by a single exit pore on each side of the body just above the lateral nerve cord and slightly posterior to the stomach (Fig. 1). From each pore a ciliated duct (8–10 μm diameter) extends directly to and penetrates the blood lacuna. The duct extends anteriorly, lying along the dorsolateral surface of the lacuna. This entire portion of the duct (15–18 μm diameter) is relatively thick-walled (4–5 μm), densely ciliated and extends for about 140 μm betore abruptly narrowing (5–8 μm diameter) in cell thickness and lumen diameter. This thin duct extends anteriorly another 140 μm and at irregular intervals receives the terminal tubules of cyrtocytes. The cyrtocytes project into the vascular lumen singly and in clusters (Fig. 11). Each cyrtocyte is about 4 μm in diameter and 18 μm long, has a single lateral nucleus located proximally and a central cluster of long cilia.

Reproductive structure

The species is gonochoric with gonads arranged linearly along each side of the intestine, and bounded externally by the middle circular musculature. Testes open via individual pores located dorsolateral to the rhynchocoel and intestine. Maturation of the testes of a given individual is synchronous. In most testes there is a single cluster of mature sperm (Fig. 10) with all their heads parallel and their flagella extending dorsally into the gonoduct. Motile sperm have a head-piece about 16 μm long and 0.5 μm in diameter with a flagellum of about 30 μm.

Females obtained at the same time as mature males had ova in clusters of 2 to 6, each ovum 50 to 60 μm in diameter with a 20 μm germinal vesicle. A 10 mm female had about 50 irregularly spaced ovaries with a total of about 300 ova. No ovaries were observed in sectioned material.

Natural history

The species has been found intertidally at Richmond Beach on Puget Sound; at San Juan Island it was found subtidally. It occurs in the top 5 cm of relatively poorly sorted sand at both sites. The sand is moderately coarse (200–450 μm) but has a significant silt component. In August of 1979, six specimens, 10–15 mm long, were collected at Richmond Beach. Two were mature males, but the others lacked recognizable gonads. Of three specimens collected at Minnesota Reef in mid-July of 1981, one was a mature male and two were females with ova (50–60 μm).

Discussion

Currently, structure of the body wall is the principal diagnostic feature for the orders Palaeonemertina (*sensu* Iwata, 1960) and Heteronemertina. That is, the body wall of Heteronemertina includes between epidermis and nerve plexus an outer longitudinal musculature (OLM) and a well-developed cutis of connective tissue and glandular cells (Gibson, 1982); these three elements may have various structural relationships to each other (Riser, 1993). The primary feature defining, but not necessarily uniting, Palaeonemertina is lack of these elements. Position of the nervous system is often cited as a distinction; in fact, it is in homologous position in both groups – overlying the original body wall circular musculature (Stiasny-Wijnhoff, 1923). Transverse sections of the body wall of *Riserius pugetensis* are convincingly heteronemertinean in appearance until one realizes that there is no dermis and there are no subepidermal glandular cells. In the absence of other patently heteronemertinean features, this condition could represent loss of the latter two by a heteronemertine or gain of OLM by a palaeonemertine. The latter is not a novel suggestion. Cantell (1969), in showing that *Hubrechtella dubia* Bergendal, 1902, possessed a pilidium larva, suggested that this might demonstrate 'the pilidium larva as primitive among the Nemertini' or that *Hubrechtella* 'is a heteronemertine which has ... lost its outer longitudinal muscles.'

Few adult heteronemertines are as small as *Riserius pugetensis*; most are much larger. Possibly as a consequence of large size, heteronemertines tend to have relatively complex and diverse anatomies. Simplicity or absence of features will tend to characterize a taxon as relatively plesiomorphic unless one can infer reduction or loss. Smaller nemertines are frequently classified in groups that are generally considered to be 'primitive,' with the notable exception of the Tubulanidae, which includes some very large members. Meiobenthologists accept implicitly that reduction and/or loss of structure has a very appreciable probability for mesopsammon (see especially Swedmark, 1964). I conclude that *Riserius pugetensis* is mesopsammic (interstitial) based on its size, habitat, and its association with other typical mesopsammic species (these include *Cephalothrix cf. pacifica* Gerner, 1969, at Richmond Beach and one undescribed species each of *Ototyphlonemertes* Diesing, 1863, and *Annulonemertes* Berg, 1985, at Minnesota Reef). The visage of living *R. pugetensis* is strongly reminiscent of some smaller palaeonemertines, including *Hubrechtella*. Is this a function of convergence or recency of common ancestry? Evaluating this phylogeny problem is hampered by absence or ambiguity of information for many nemertinean characters. The problem is compounded by the prevalence of references to *ad hoc* assemblages of characters in constructing taxa. Phylogeny of *R. pugetensis* can be examined only with reference to phylogeny of palaeo- and heteronemertines. A full cladistic analysis of that relationship is not available and is outside the scope of this paper, but a start in that direction is made in the following discussion, wherein relevant characters are discussed, if somewhat broadly, in terms of Hennigian principles of homology analysis.

Most palaeonemertines have a short rhynchocoel lined only by circular muscles and most have a proboscis with outer circular musculature and inner longitudinal musculature (palaeotype). But the rhynchocoel of some, including *Hubrechtia* Bürger, 1892, *Hubrechtella* (Hylbom, 1957; Kirsteuer, 1967; Gibson, 1979a) and *Tetramys* Iwata, 1957, is invested by outer circular and inner longitudinal musculature as in *Riserius pugetensis* and heteronemertines. These also share the presence of a medial blood vessel. Therefore, it is not surprising that *Hubrechtia* and *Hubrechtella* lack the body-wall inner circular musculature (Kirsteuer, 1967; Gibson, 1979a) found in other palaeonemertines. I will postulate at the outset that 1) it is evident from a hierarchy of similarities that the cerebral sensory organ is a synapomorphy for all nemertinean groups where it is present or may be inferred to have been present ancestrally; 2) the medial (mid-dorsal) blood vessel is a synapomorphy for a smaller group of palaeo- and heteronemertines; 3) the pilidium larva is a synapomorphy uniting *Hubrechtella* and Heteronemertina; and 4) parallel losses are much more likely than parallel gains. These points are addressed to various extents in the remaining discussion.

Conspicuous external features of *Riserius pugetensis* are an annular cephalic furrow encircling the body anterior to the mouth and a pair of deep, lateral cephalic pits leading to the cerebral sensory organs. An annular cephalic furrow is characteristic of most hoplonemertines, at least one palaeonemertine, *Carinomella lactea* Coe, 1905 (personal observation), the heteronemertines *Cephalomastax* Iwata, 1957, *Diplopleura*

Stimpson, 1857 (Bürger, 1895), *Oxypolella* Bergendal, 1902, *Oxypolia* Punnett, 1901 (see also Cantell, 1972; Gibson, 1982), *Poliopsis* Joubin, 1890, and *Pontolineus* (Müller & Scripcariu, 1964). Coe (1895) reports that the head of *Zygeupolia rubens* (Coe, 1895) is 'marked off from the body by a slight constriction formed by a pair of shallow, lateral, transverse grooves', which are separate from the openings of the cerebral sensory organs. These grooves may correspond to the annular furrow, but neither Coe nor Thompson (1901) offers clear evidence for this. The furrow is anterior to the mouth in each of the heteronemertines (*cf.* Iwata, 1957; Cantell, 1972), including *R. pugetensis*, whereas it is intersected by the mouth in *Pontolineus*, and it is posterior to the mouth in *C. lactea*. This relationship in hoplonemertines is obscured by the enigmatic morphogenetic history of the mouth. The cerebral sensory organs of *Oxypolella* and *Oxypolia* open into the annular furrow (Cantell, 1972; Punnett, 1901), and open anterior to the furrow in hoplonemertines and the heteronemertines *R. pugetensis*, *Cephalomastax*, *Pontolineus*, and, I presume, the other heteronemertines cited above. Thus, this landmark bolsters the likelihood that this furrow is a synapomorphy for a group encompassing at least the hetero- and hoplonemertines; it may be a more plesiomorphic link for all three orders.

Nothing is obviously unique about the cerebral sensory organ (CSO) of *Riserius pugetensis*. They are not intimately joined to the cerebral ganglia, a feature shared with most other palaeonemertines, some baseodiscids and most hoplonemertines. Thus, this may be a plesiomorphic condition. Intimate apposition of CSO and cerebral ganglia is typical for lineids and Gibson (*in litt.*) notes that a similar degree of intimacy is found in some reptantic Polystilifera and some species of *Nipponnemertes*. Presumably this represents two or three instances of convergence, unless there is a definable shared structural expression of this intimacy. The CSO of *R. pugetensis* are not embedded in vascular lacunae, unlike those of *Hubrechtia* and some other palaeo- and heteronemertines. However, I am not aware of clear evidence that any CSO actually penetrates the vascular epithelium. This criterion seems only to describe a particular degree of contact with vasculature and could also be a function of vascular volume. The CSO of *R. pugetensis* opens into a large lateral pit, or depression of the body wall, which is similar to that of most palaeonemertines, baseodiscids (the pits tend to be more ventral in these) and other heteronemertines lacking horizontal fissures. The cerebral sensory openings of most hoplonemertines present relatively minor variation from this construction, which probably represents the plesiomorphic condition. The shape and relatively large size of these pits in *R. pugetensis* speaks for some degree of specialization. I do not believe this to be an artifact of perception caused by the small size of the animal; *Hubrechtella* tends to have similarly well-developed CSO pits. Müller & Scripcariu (1964) argue that the cephalic fissures of *Pontolineus* are reduced because of its endopsammic habit, but I believe this may as easily be plesiomorphic. This follows from my contention (Norenburg, 1985) that the primary ancestral radiation of heteronemertines was infaunal, and from the fact that the greatest development of cephalic fissures is seen in infaunal forms showing a high degree of apomorphy, such as species of *Cerebratulus* Renier, 1804. Further, there is no evidence of reduction in several macrofaunal lineiform species that live in the coarse sand of relatively high-energy beaches (personal observations), e.g, *Parvicirrus dubius* (Verrill, 1892) (see Riser, 1993). As there is no vestige of modified cephalic epidermis and there is no other obvious synapomorphy with lineiforms, I conclude that the CSO openings of *R. pugetensis* represent a unique form of a basically plesiomorphic condition.

The epidermis of *Riserius pugetensis* is simple, in terms of the elements present and structural complexity, when compared to hetero- and many palaeonemertines. There is some superficial similarity with species of *Hubrechtella*, but *R. pugetensis* lacks the diversity and localization of glandular cells reported from *Hubrechtella* (Hylbom, 1957; Gibson, 1979a). Lack of a definitive cutis, especially lack of subepidermal glandular cells, distinguishes the new species from all other heteronemertines except *Colemaniella albulus*, which appears otherwise to be a typical 'lineid' (Gibson, 1981). Norenburg (1985) concluded that presence of subepidermal mucous cells is the plesiomorphic condition for heteronemertines, leaving the epidermis dominated by homoserous cells (with isolated other kinds of glandular cells). This is supported by structure of the epidermis in *Hubrechtella*, in which it is also the mucous cells that tend to be submerged. It is the homoserous cell that dominates the epidermis of *R. pugetensis*; there appears to be a complete lack of integumentary mucous cells. In contrast, the epidermis of macrofaunal and mesopsammic hoplonemertines is dominated by mucous cells, as is the epidermis of *Arhynchonemertes* (Riser, 1988). Thus, I conclude that absence of subepidermal glandular cells (and perhaps a cutis) is a derived condition in *R. pugetensis*.

Pseudocnidae very similar to those of *Riserius pugetensis* are also found in the proboscis epidermis of interstitial cephalothricid archinemertines (Gerner, 1969; personal observations), in all but one of the five species of the palaeonemertine genus *Hubrechtella* (Hylbom, 1957; Gibson, 1979a; personal observations), and in the heteronemertines *Cerebratulus urticans*, *Micrura purpurea* (Bürger, 1895) and *Paralineus elisabethae* Schütz, 1911 (Schütz, 1912). It is not clear how or whether these structures relate to the much smaller heteronemertinean proboscidial 'spines' of other authors. This enigmatic character invites phylogenetic speculation, but its taxonomic distribution reduces its systematic value for our purposes, at least until there is a comparative study of pseudocnidae.

The nephridial system of *Riserius pugetensis* is simple, resembling that of several palaeonemertines (e.g., *Hubrechtia*) and heteronemertines (e.g., *Paralineus*). This is probably the plesiomorphic state for most, if not all nemertines (absence of nephridia in *Hubrechtella* must be considered apomorphic).

The presence of an outer circular musculature, albeit sparse, would by classical diagnoses set *Riserius pugetensis* apart from other heteronemertines except *Cephalomastax*, which does have well-developed subepidermal gland cells and cutis. However, outer circular muscles are found in the cutis or among the subepidermal glandular cells of many heteronemertines, as has been mentioned by many authors, including: *Baseodiscus delineatus* (Delle Chiaje, 1825) (*cf.* Gibson, 1979b), *Lineus coccinus* Bürger, 1892, and *Valencinia longirostris* Quatrefages, 1846 (*cf.* Bürger, 1895). It has been treated implicitly as 'different' from the so-called primary body wall musculature, in effect as a variable of the connective tissue. It seems to represent a plesiomorphic layer for at least the heteronemertines; its absence in some, if any, heteronemertines probably is secondary and more likely to be a homoplasious condition than is its presence. It is not apparent how this body wall musculature relates to that of palaeonemertines (with a subepidermal nervous system), especially a form such as *Tetramys* Iwata, 1957, which has the same sequence of body wall muscle strata as *R. pugetensis*.

The heteronemertines are currently divided into five 'convenient families' based primarily on the nature of cephalic furrows (Gibson, 1982, 1985). The Lineidae, the largest 'family' numerically, is characterized by lateral horizontal cephalic fissures (Friedrich, 1960) and generally seems to include the so-called 'most-advanced' heteronemertines. The lateral fissures are a fundamental synapomorphy for this large group of heteronemertines (it seems more appropriate, relative to the remaining 'families', to consider the otherwise morphologically-diverse lineids as a suborder, Lineiformes). The four remaining 'families' (Poliopsiidae, Pussylineidae, Valenciniidae and Baseodiscidae) are less-readily defined (*cf.* Gibson, 1982, 1985), largely because there are numerous information gaps about characters of the more obscure taxa, but some useful comparisons with *R. pugetensis* are possible.

Structure of the proboscis has been a traditional cornerstone for taxonomy of nemertinean, and especially heteronemertinean, genera (Wijnhoff, 1914; Friedrich, 1960; Gibson, 1985). Wijnhoff (1914), in analyzing probable phylogeny of the nemertinean proboscis, makes a strong argument that the plesiomorphic proboscis (termed 'palaeotype') has two muscle strata, outer circular and inner longitudinal, and a nerve zone between the epithelium and circular musculature. She argues that the addition of outer longitudinal musculature between the epithelium and nerve zone leads to a heterotype proboscis, thereby reflecting in part structure of the body wall. Gibson (1985) suggests that this heterotype is the logical ancestral condition for the heteronemertinean proboscis, whereas Riser (1993) states that the two-layer condition (palaeotype) is plesiomorphic for heteronemertines. I support the latter on the basis of outgroup comparison, but this may not be the case for lineiform nemertines. Wijnhoff (1914) argues that the plesiomorphic position of the nerve plexus and central nervous system also is subepidermal for the nemertinean body wall (see also Bürger, 1895; Bergendal, 1903); this still is widely accepted (Hylbom, 1957, 1993; Gibson, 1972). The nervous system of the Heteronemertina has not 'migrated' but is in its plesiomorphic position, displaced inward by the addition of outer musculature. However, Wijnhoff and others have since overlooked the importance of the nervous system as a positional reference in determining homologies among proboscidean muscle strata. If one assumes, as most authors imply, that there is a morphogenetic relationship between body wall and proboscis wall, proboscis outer longitudinal musculature (between nerve zone and epithelium) is the homologue of body wall outer longitudinal musculature, but proboscis inner circular musculature (between endothelium and inner longitudinal muscle) has no body wall homologue. Gibson's (1985) group A proboscis (outer longitudinal, middle circular, inner longitudinal) is not homologous to the body wall just because 'outer and inner longitudinal...layers sandwich a middle circular coat'; rather, it is because they have the correct anatomical relationship to other body wall components. For the remaining discussion proboscis muscle strata are defined as follows: outer longitudinal (OLM) is between epidermis and neural tissue, outer circular (OCM) is subneural and is followed by inner longitudinal (ILM), inner circular muscles (ICM) are subendothelial and have no body wall homologue. I am assuming, without rechecking, that all of the genera in Gibson's (1985) group B have a proboscis with OCM and ILM, and those of group A have OLM in addition. It should be noted that *Parapolia aurantiaca* Coe, 1895, is included in group A (parsimony favors this as the correct interpretation) but position of the proboscis nerves is unknown, nor is it clarified by Wheeler (1934, 1940) or Friedrich (1970) for a second species. Coe (1895) initially described the proboscis of *Zygeupolia rubens* (Coe, 1895) as having OLM, OCM and ILM in its main body, but lacking the OLM (inner by his definition at that time) anteriorly, whereas Thompson (1901), in her very detailed description, cites only OCM, ILM, and a thin 'endothelial' circular musculature (ICM?). Coe (1905) seems to accept Thompson's description by not offering further comment.

Gibson's (1985) group C is more problematical. It may be used to illustrate the organizing strength of phylogenetic analysis and the pitfalls of using parsimony uncritically; e.g., without homology analysis. Group C, although unintended, is founded on parsimony; that is, membership is based on having a proboscis with 'outer longitudinal, inner circular muscle'. *Uchidana parasita* Iwata, 1967, is described as having 'outer longitudinal and inner circular muscle' (Iwata, 1967), but it is clear from the rest of his description and figures that these are OCM and ILM, placing it in group B. *Cephalomastax*, according to Iwata's (1957) description, has OLM and ILM plus a discrete subendothelial circular musculature (ICM); that is, the apparent 'outer longitudinal' is split by the nervous system and OCM is actually absent. Likewise, *Valencinia longirostris*, according to Bürger's (1895) description and figure (Plate 23:9), has OLM, ILM and a subendothelial wickerwork of ILM and ICM comprising the main body of the proboscis. Bergendal (1902) suggests, from personal examination of sections, that Bürger's interpretation of this wickerwork was incorrect and that it is composed only of interlacing obliquely-orientated circular muscle fibers. This opinion is shared by Wijnhoff (1914); but, she considers this to be the 'result of the dissolving of muscle crosses', which could occur only if there were an intervening LM originally. In the 'middle section' of the proboscis of *Valencinura bahusiensis* Bergendal, 1902, the inner portion of the circular musculature consists of obliquely oriented fibers (Bergendal, 1902). The position of proboscis nerves is unknown for *Chilineus* Friedrich, 1970, but the proboscis is clearly described (using current convention) as containing 'inner circular and outer longitudinal' musculature. Our knowledge of the proboscis in *Poliopsis* consists of Friedrich's (1970) remark 'soweit er erhalten ist, äLm und Rm...'. In the absence of contrary data parsimony argues that these are OCM and OLM in the latter two genera (*i.e.*, the nervous system should be between OLM and OCM). Wijnhoff (1914) asserts that the same sequence in baseodiscids, 'outer longitudinal and inner circular' musculature, is derived from the heterotype proboscis by loss of the ILM. However, the 'outer' longitudinal muscles of *Baseodiscus* are subneural, therefore ILM, and the circular muscle is subendothelial (ICM). Unless the nervous system has relocated this proboscis may be derived by loss of OCM of a heterotype proboscis or by loss of OLM from a proboscis such as that of *Cephalomastax* or *Valencinia longirostris*. The proboscis of *Riserius pugetensis* has only the subneural longitudinal musculature, which on the basis of parsimony is likely to be ILM; it has lost at least the OCM. Baseodiscids have a tendency to 'lose' the ICM (*cf.* Gibson, 1979b); in its absence the two proboscides would be considered homologous.

Evolution of heteronemertinean proboscides is more complex than previously entertained. The morphotypes identified above are summarized graphically (Fig. 13) without any indication of relationship. Transformation series are dependent on presumed ancestral state, and many equally parsimonious or 'logical' orderings are possible if the root is unspecified. Recognition of ancestral states is confused significantly by the fact that the proboscides of *Hubrechtella queenslandica* Gibson, 1979, and *Tetramys* have three muscle strata (outer circular, middle longitudinal, and inner circular) (Iwata, 1957), whereas the four remaining species of *Hubrechtella* have proboscides with outer longitudinal (OLM) and inner circular musculature (ICM) (Hylbom, 1957; Kirsteuer, 1967), and that of *Hubrechtia* is palaeotype (OCM, ILM). Some 'logical' transformations present themselves on the basis of parsimony. I consider parallel or convergent losses as likely but would require significant corroboration from other characters to accept the same as gains. Thus, OLM may be plesiomorphic for heteronemertines of Gibson's groups A and B (except *Zygeupolia*). It seems plausible, and even likely, that some of the generic inconsistencies as shown in Gibson's (1985) groups A and B are due to convergent loss of OLM (bad taxonomy may account for other conflicts). Hylbom (1957) alluded to the possible relationship between muscle crosses in palaeo- and heteronemertines. The broad distribution of muscle crosses in heteronemertines favors a single origin over multiple ones. It seems evident that presence of muscle crosses infers existence of an ICM, as previously suggested by Thompson (1901) and Wijnhoff (1914), although it may be very weakly developed, as in *Zygeupolia*. It is tempting to view *Zygeupolia* as approximating an ancestral heteronemertinean condition, with its lateral sensory organ, and proboscis with OCM, ILM, rudimentary muscle crosses and ICM. Obviously, one would not expect muscle crosses in the absence of an 'inner' longitudinal musculature, but the oblique musculature in the proboscides of *Valencinia longirostris* and *Valencinura bahusiensis*, both of which are reported to have a weak 'endothelial' circular musculature (Bergendal, 1902), may betray its origin as a muscle cross. Although I do not necessarily subscribe to Wijnhoff's (1914) explanation for the developmental differentiation of rhynchocoel sheath and proboscis musculature, it is evident that a relatively minor morphogenetic shift during this differentiation could significantly alter the relative position of the nervous system, as is the case in hoplonemertines.

The proboscis of *Riserius pugetensis* can be derived from group B, *Baseodiscus*, *Zygeupolia*, or *Poliopsis* by one loss, and from group A or *Cepha-*

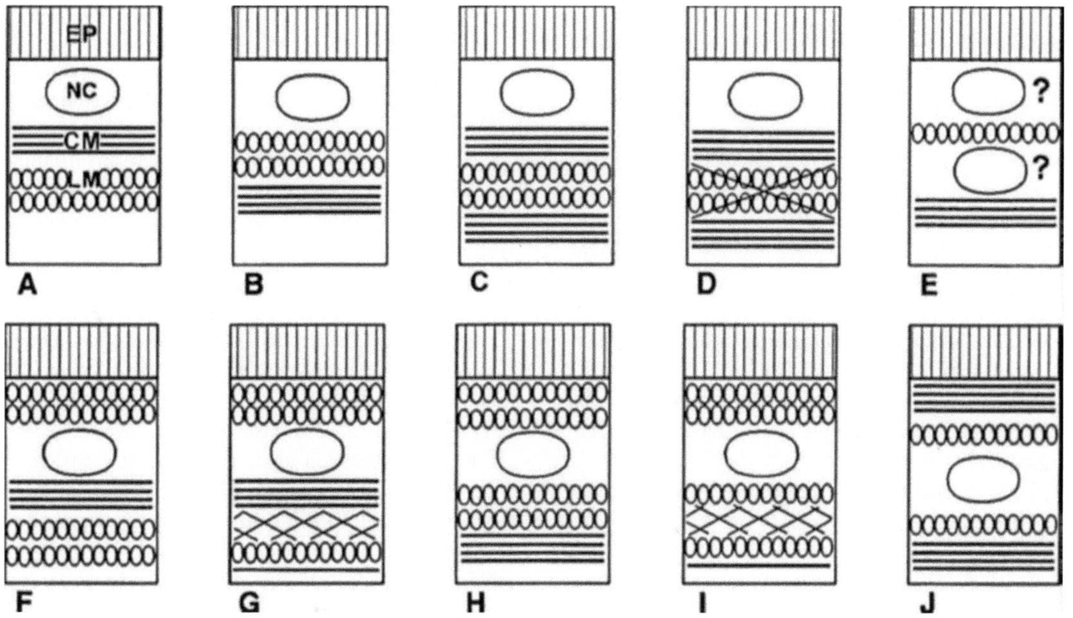

Fig. 13. Schematic representations of nemertinean proboscides in transverse section as found in: A. Palaeotype (e.g., Cephalothricidae, *Hubrechtia*, and Heteronemertina of Group B – see text). B. Baseodiscidae, *Hubrechtella* (in part). C. *Tetramys*, *Hubrechtella queenslandica*. D. *Zygeupolia*, mid-region showing the muscle cross between circular and so-called endothelial circular muscle. E. *Poliopsis* (position of nerve cord is unknown). F. Heterotype (e.g., *Valencinina*, Heteronemertina of Group A – see text). G. *Valencinura*, mid-region showing zone of oblique muscle fibers. H. *Cephalomastax*. I. *Valencinia longirostris*, mid-region showing 'obliquely-orientated circular muscle fibers.' J. Hoplonemertina. Abbreviations: CM = circular musculature; EP = outer epithelium; LM = longitudinal musculature; NC = nerve cord. No indication of relative thickness of musculature is intended in any of the figures; 'outer' and 'inner' are respectively to the top and bottom of figures.

lomastax by two losses. More importantly, to be derived from any of the advanced 'heterotype' proboscides, *R. pugetensis* must have lost subepidermal glandular cells (and cutis). Alternatively, it may have diverged from an ancestral heteronemertine that had not developed a cutis or subepidermal gland cells, and lost proboscis musculature in parallel with other heteronemertines.

Riserius pugetensis cannot be referred to any prior genus or higher taxon of heteronemertines. It shares plesiomorphic features with the palaeonemertines *Hubrechtia*, *Hubrechtella* and *Tetramys* that are also shared with heteronemertines. The putative loss of subepidermal mucous cells and the presence of OLM in the body wall suggest a more recent common ancestry with the heteronemertines, perhaps with the baseodiscids or *Paralineus*.

Acknowledgements

I acknowledge with gratitude the support of Sigma Xi for a Grant-in-Aid of Research and the American Museum of Natural History for an award from the Theodore Roosevelt Memorial Fund, both of which funded the field work. I also am grateful to Friday Harbor Laboratories of the University of Washington, the Marine Science Center of Northeastern University, and the Smithsonian Institution for use of facilities. My visits and work on the West Coast were facilitated and enriched by many people – notably, Stephen Stricker, Paul and Ruth Illg, M. Patricia Morse, Robert Fernald and Christopher Reed. This is Contribution No. 192 from the Marine Science Center, Northeastern University, Nahant, MA.

References

Bergendal, D., 1902. Studien über Nemertinen. II. *Valencinura bahusiensis* Bgdl., ein Beitrag zur Anatomie und Systematik der Heteronemertinen. Lunds Univ. Arskr. 38: 1–104.

Bergendal, D., 1903. Studien über Nemertinen. III. Beobachtungen über den Bau von *Carinoma* Oudemans nebst Beiträgen zur Systematik der Nemertinen. Lunds Univ. Arskr. 39: 1–86.

Bürger, O., 1895. Nemertinen. Fauna und Flora des Golfes von Neapel 22: 1–743.

Cantell, C.-E., 1969. Morphology, development, and biology of the pilidium larvae (Nemertini) from the Swedish West Coast. Zool. Bidr. Upps. 38: 61–111.

Cantell, C.-E., 1972. The anatomy of *Oxypolella bergendali* sp.n., a new nemertine from the West Coast of Sweden. Zool. Scr. 1: 79–84.

Coe, W. R., 1895. Descriptions of three new species of New England palaeonemerteans. Trans. Conn. Acad. Arts Sci. 9: 515–522.

Coe, W. R., 1905. Nemerteans of the West- and Northwest coasts of North America. Bull. Mus. comp. Zool. Harv. 47: 1–319.

Friedrich, H., 1936. Nemertini. Tierwelt N.-u. Ostsee 4: 1–69.

Friedrich, H., 1960. Bemerkungen über die Gattung *Micrura* Ehrenberg 1831 und zur Klassifikation der Heteronemertinen nebst vorläufigem Bestimmungsschlüssel. Veröff. Inst. Meeresforsch. Bremerh. 7: 48–62.

Friedrich, H., 1970. Nemertinen aus Chile. Sarsia 40: 1–80.

Friedrich, H., 1979. Mophogenese der Tiere, Lieferung 3:D_5-I, Nemertini. Gustav Fischer Verlag, Stuttgart, N.Y., 136 pp.

Gerner, L., 1969. Nemertinen der Gattungen *Cephalothrix* und *Ototyphlonemertes* aus dem marinen Mesopsammal. Helgoländer wiss. Meeresunters. 19: 68–110.

Gibson, R., 1972. Nemerteans. Hutchinson & Co. Ltd., London, 224 pp.

Gibson, R., 1979a. Nemerteans of the Great Barrier Reef 1. Anopla Palaeonemertea. Zool. Linn. Soc. 65: 305–337.

Gibson, R., 1979b. Nemerteans of the Great Barrier Reef 2. Anopla Heteronemertea (Baseodiscidae). Zool. J. Linn. Soc. 66: 137–160.

Gibson, R., 1981. Nemerteans of the Great Barrier Reef 3. Anopla Heteronemertea (Lineidae). Zool. J. Linn. Soc. 71: 171–235.

Gibson, R., 1982. Nemertea. In S. P. Parker (ed.), Synopsis and classification of living organisms. McGraw-Hill, N.Y.: 823–846.

Gibson, R., 1985. The need for a standard approach to taxonomic descriptions of nemerteans. Am. Zool. 25: 5–14.

Hylhom, R., 1957. Studies on palaeonemerteans of the Gullmar Fiord area (West coast ot Sweden). Ark. Zool. 10: 539–582.

Hylbom, R., 1993. A proposal for a check-list of characteristics to be used in the description of palaeonemertean species. Hydrobiologia 266: 169–174.

Iwata, F., 1957. Nemerteans from Sagami Bay. Publ. Akkeshi mar. biol. Stn 7: 1–31.

Iwata, F., 1960. Studies on the comparative embryology of nemerteans with special reference to their interrelationships. Publ. Akkeshi mar. biol. Stn 10: 1–51.

Iwata, F., 1967. *Uchidana parasita* nov. gen. et nov. sp., a new parasitic nemertean from Japan with peculiar morphological characters. Zool. Anz. 178: 122–136.

Kirsteuer, E., 1967. New marine nemerteans from Nossi Be, Madagascar. Zool. Anz. 178: 110–122.

Martin, C. H., 1914. A note on the occurrence of nematocysts and similar structures in the various groups of the animal kingdom. Biol. Zbl. 34: 248–273.

Müller, G. I. & D. Scripcariu, 1964. *Pontolineus arenarius* nov. gen. nov. sp. (Heteronemertini, Lineidae) und Diagnose der Gattung Antarctolineus nov. gen. Rev. Roum. Biol., Sér. Zool. 9: 313–319.

Norenburg, J. L., 1985. Structure of the nemertine integument with consideration of its ecological and phylogenetic significance. Am. Zool. 25: 37–51.

Norenburg, J. L., 1987. A new old, mesopsammic nemertine? Am. Zool. 27: 41A.

Norenburg, J. L., 1988a. Nemertina. In R. P. Higgins & H. Thiel (eds), Introduction to the study of meiofauna. Smiths. Inst. Press, Wash. D.C.: 287–292.

Norenburg, J. L., 1988b. Remarks on marine interstitial nemertines and key to the species. Hydrobiol. 156: 87–92.

Norenburg, J. L. & J. M. Barrett, 1987. Steedman's polyester wax embedment and de-embedment for combined light and scanning electron microscopy. J. Electr. Microsc. Tech. 6: 35–41.

Punnet, R. C., 1901. On two new British nemerteans. Q. Jl microsc. Sci. 44: 547–564.

Riser, N. W., 1988. *Arhynchonemertes axi* gen. n, sp. n. (Nemertinea) – an insight into basic acoelomate bilaterian organology. Fortschr. Zool. 36: 367–373.

Riser, N. W., 1993. Observations on the morphology of some North American nemertines with consequent taxonomic changes and a reassessment of the architectonics of the phylum. Hydrobiologia 266: 141–157.

Schütz, V., 1912. *Paralineus elisabethae* (nov. gen. et. sp.). Z. wiss. Zool. 102: 11–135.

Steedman, H., 1960. Section cutting in microscopy. Charles C. Thomas, Springfield, Illinois.

Stiasny-Wijnhoff, G., 1923. Die Entstehung des kopfes bei den Nemertinen. Acta Zool. 4: 223–240.

Swedmark, B., 1964. The interstitial fauna of marine sand. Biol. Rev. 39: 1–42.

Thompson, C. B., 1901. *Zygeupolia litoralis*, a new heteronemertean. Proc. Acad. nat. Sci. Philad. 53: 657–739.

Wheeler, J. F. G., 1934. Nemerteans from the South Atlantic and southern oceans. Discovery Rep. 9: 217–294.

Wheeler, J. F. G., 1940. Nemerteans of Kerguelen and the Southern Ocean. B.A.N.Z. Antarctic Res. Exped. 1929–31, Reports, Ser. B, 4: 233–256.

Wijnhoff, G., 1914. The proboscidean system in nemertines. Q. Jl microsc. Sci. 60: 273–312.

Hydrobiologia **266**: 219–238, 1993.
R. Gibson, J. Moore & P. Sundberg (eds), *Advances in Nemertean Biology*

Enzyme electrophoresis, genetic identity and description of a new genus and species of heteronemertean (Nemertea, Anopla) from northwestern Spain and North Wales

Alex D. Rogers [1], Juan Junoy [2], Ray Gibson [3] & John P. Thorpe [1]
[1] *Department of Evolutionary and Environmental Biology, University of Liverpool, Port Erin Marine Laboratory, Port Erin, Isle of Man, U.K.;* [2] *Departamento de Biologia Animal, Universidad de Alcala de Henares, 28871 Alcala de Henares, Madrid, Spain;* [3] *School of Natural Sciences, Liverpool Polytechnic, Byrom Street, Liverpool L3 3AF, U.K.*

Key words: Heteronemertea, genetics

Abstract

The anoplan order Heteronemertea, particularly the genera *Cerebratulus*, *Lineus* and *Micrura*, contains a very large number of nominate species, many of which are inadequately described. As a consequence, systematic difficulties are encountered with the identification of many taxa in this group, especially those originally established primarily on the basis of their external features. The present paper concerns heteronemerteans collected from two locations, the Foz Estuary (north-western Spain) and Llandudno (North Wales). The Spanish collection included specimens identified as *Lineus longissimus* (Gunnerus), whilst samples from Llandudno contained large numbers of *Lineus viridis* (Müller); samples of a third similar but apparently undescribed species were found at both locations. Starch gel electrophoresis showed that samples of the apparent third species were genetically almost identical from each of the two locations, but were clearly different from the two described *Lineus* species. Histological studies of the unknown specimens revealed anatomical characters, including the unique feature of a proboscis epithelium ciliated throughout its length, which exclude it from any known heteronemertean taxon; it is accordingly placed in a new genus and species, for which the name *Riseriellus occultus* is proposed.

Introduction

The study of nemertean systematics is plagued by a multitude of inadequately described taxa, at and below the family level (Gibson, 1982a, 1985). Many of the older species were established primarily on the basis of external features, many of which are now considered to be taxonomically unreliable (Gibson, 1985; Gibson & Crandall, 1989) and also of little value in determining generic affinities. Even where species have been erected after histological investigation, descriptions of their anatomy are frequently either incomplete or emphasise characteristics whose taxonomic significance is difficult to assess. This is especially true when problems of intraspecific variability are considered (Friedrich, 1960; Berg, 1972; Sundberg, 1979, 1980), and it is highly likely that many nemertean 'species' are either complexes of different taxa or are synonymous with other forms.

The genus *Lineus* Sowerby, 1806, is the oldest established nemertean taxon. It contains more than 90 nominate species (Gibson, 1982b) and typifies many of the fundamental problems outlined above. During September 1990 nemerteans of this genus were collected at Llandudno, North Wales, and in the Foz Estuary, north-western Spain. The collections included numerous examples of *Lineus longissimus* (Gunnerus, 1770) and *Lineus viridis* (Müller, 1774), as well as a third group of black to dark green heteronemerteans, in size and general appearance very similar to, but externally distinguishable from, both the *Lineus* species. Whether this third morphotype represented a separate species or was merely a variety of one or the other taxa was uncertain, since previous descriptions of the external features of both *Lineus longissimus* and *Lineus viridis* have been ambiguous. Histological differences between *Lineus* species are often small (Cantell, 1972, 1975) and may simply represent intraspecific variation; further, descriptions of the internal anatomy of species such as *Lineus longissimus* are frequently contradictory (Friedrich, 1935; Cantell, 1975, 1976). Preliminary studies on the systematic relationships between *Lineus longissimus*, *Lineus viridis* and the third morphotype found in Spain and Wales thus required a more objective approach, such as an investigation of enzyme variation by electrophoresis. This technique has been used to resolve taxonomic problems in many invertebrate groups (for reviews, see Ferguson, 1980; Thorpe, 1982, 1983; Ayala, 1983), including nemerteans (Cantell & Gidholm, 1977; Williams *et al.*, 1983; Sundberg & Janson, 1988). Enzyme electrophoresis was therefore employed to estimate the genetic distance between sympatric populations of *Lineus longissimus*, *Lineus viridis* and the third unknown heteronemertean morphotype; the results demonstrated conclusively that the unknown form was conspecific with neither of the lineids, and indicated that extensive histological investigations of specimens from both locations were needed to resolve the systematic status of this heteronemertean.

Materials and methods

Samples

Nine specimens of *Lineus longissimus* were collected from the Foz Estuary, Spain (approximately 43° 34′ N, 7° 14′ W) during September 1990, close to mean low water spring tide level. Ten examples of the third morphotype were also found at Foz, but these occurred much higher up the shore in mud banks, or amongst *Spartina* roots. Several additional specimens of the unknown taxon were obtained from the same location during May 1991. Collections made at Llandudno, also during September 1990, yielded 59 examples of *Lineus viridis* and 39 individuals apparently of the same unknown morphotype as those found at Foz. At Llandudno all samples were found on the upper shore beneath stones and rocks lying on damp, fine mud or silt. Further examples of the unknown species have subsequently been found (August 1991) at Rhosneigr and Trwyn du Point, Anglesey.

Representative living specimens of each of the three species were examined in detail under a binocular microscope, their maximum relaxed lengths and widths being recorded.

Electrophoresis

Horizontal starch gel electrophoresis was performed by standard methods (see Harris & Hopkinson, 1978; Ferguson, 1980; Richardson *et al.*, 1986) using 12.5% starch gels (Sigma Chemical Co. Ltd., Poole, Dorset, U.K.). Samples, were cut from the posterior ends of live specimens of each species and homogenized in a small volume of buffer (0.06M Tris/HCl pH 8.0). Two buffer systems were used:

Buffer system I (Ward & Beardmore, 1977): electrode buffer (30.29 g Tris, 11.98 g citric acid per litre, pH 8.0); gel buffer (192.5 ml electrode buffer diluted to 5 litres, pH 8.0).

Buffer system II (Williams *et al.*, 1983): electrode buffer (16.35 g Tris, 8.26 g citric acid, 0.46 g disodium salt of ethylene diamine tetra-acetic acid [Na_2EDTA] per litre, pH 7.1); gel buffer (16.35 g Tris, 9.035 g citric acid, 6.9 g Na_2EDTA per litre, diluted 1:15 with distilled water, pH 7.1).

Buffer system I was run at 140 V, 50 mA for 7 hours, buffer system II at 100 V, 50 mA for approximately 7 hours.

The following staining schedules, with the appropriate Enzyme Commission (E.C.) numbers indicated, were employed:

(i) Aminopeptidase A (APA) (E.C. 3.4.11.1) (Williams *et al.*, 1983): buffer system I; stain: 10 mg N-glycyl-L-leucine, 5 mg L-amino acid oxidase, 5 mg *o*-dianisidine, 10 mg horseradish peroxidase, 1 mg manganous chloride, 25 ml Tris HCl pH 8.0, 2% agar.

(ii) Glutamate-oxaloacetate transaminase (GOT) (E.C. 2.6.1.1) (Williams *et al.*, 1983): buffer system II; stain: 0.6 g Tris, 150 mg aspartic acid, 40 mg α-ketoglutaric acid, 20 mg polyvinylpyrolidone, 15 mg pyridoxal-5-phosphate, 40 ml distilled water. Incubate for 30 minutes, then add 100 mg Fast Blue RR salt in 10 ml distilled water.

(iii) Isocitrate dehydrogenase (ICD) (E.C. 1.1.1.42): buffer system II; stain: 80 mg isocitric acid, 285 mg magnesium chloride, 5 mg nicotinamide adenine dinucleotide phosphate (NADP), 5 mg MTT, 5 mg phenazine methosulphate (PMS), 25 ml 0.06M Tris HCl, pH 8.0.

(iv) Malate dehydrogenase (MDH) (E.C. 1.1.1.37): buffer system I; stain 100 mg L-malic acid, 970 mg Tris, 80 mg magnesium chloride, 15 mg nicotinamide adenine dinucleotide (NAD^+), 10 mg MTT, 1 mg PMS, 40 ml distilled water, 25 ml 2% agar.

(v) Mannose phosphate isomerase (MPI) (E.C. 5.3.1.9): buffer system II; stain: 40 mg mannose-6-phosphate, 10 mg NADP, 50 μl glucose-6-phosphate dehydrogenase, 40 u glucose phosphate isomerase, 5 mg MTT, 1 mg PMS, 10 mg magnesium chloride, 30 ml 0.06M Tris HCl, pH 8.0.

(vi) Octopine dehydrogenase (ODH) (E.C. 1.5.1.11) (Walsh & Somero, 1981): buffer system I; stain: 25 mg octopine, 15 mg NAD^+, 7 mg MTT, 2 mg PMS, 25 ml 0.06M Tris HCl, pH 8.0, 25 ml 2% agar.

(vii) 6-phosphogluconate dehydrogenase (6-PGDH) (E.C. 1.1.4.4): buffer system I; stain: 20 mg 6-phosphogluconate, 5 mg NADP, 5 mg MTT, 5 mg PMS, 10 mg magnesium chloride, 25 ml 0.06M Tris HCl, pH 8.0.

(viii) Phosphoglucose isomerase (PGI) (E.C. 5.3.1.9) (Harris & Hopkinson, 1978): buffer system I; stain: 20 mg fructose-6-phosphate, 5 mg NADP, 5 mg MTT, 5 mg PMS, 10 u glucose-6-phosphate dehydrogenase, 25 ml 0.06M Tris HCl, pH 8.0, 25 ml 2% agar.

(ix) Phosphoglucomutase (PGM) (E.C. 2.7.5.1): buffer system II; stain: 100 mg glucose-1-phosphate, 50 mg magnesium chloride, 10 mg NADP, 5 mg MTT, 5 mg PMS, 40 u glucose-6-phosphate dehydrogenase, 25 ml 0.06M Tris HCl, pH 7.1, 25 ml 2% agar.

Histological studies

Four specimens of the unknown morphotype, two from each location, were anaesthetized in 7.5% magnesium chloride, fixed in Bouin's fluid (made up in filtered seawater) and sectioned at 6 μm in 56 °C m.p. paraffin wax. Mallory's triple stain (Pantin, 1960) was used as standard.

Electrophoretic results and discussion

Allele frequencies for the ten enzyme loci which produced useful results for one or more of the species examined are presented in Table 1.

The results show that at nearly all loci there are large differences in allele frequency between sympatric samples of *Lineus longissimus* and the third morphotype from Spain, and between *Lineus viridis* and the unknown type from Llandudno, whereas between specimens of the form occurring both in Spain and North Wales differences in allele frequencies are very small. Over all the loci used, the genetic differentiation between the three species can be reduced to a single figure by using measures of genetic distance or similarity. Between *Lineus longissimus* and Spanish specimens of the new morphotype, and between *Lineus viridis* and Welsh examples of the new form, the value of Nei's (1972) genetic identity falls below 0.35 (Table 2). This indicates that not only are these likely to be separate species, but that they show genetic identity values lower than those usually found in congeneric comparisons (Ayala, 1975, 1983; Thorpe, 1982, 1983). However, similarly low values of genetic identity have been

Table 1. Allele frequencies at 11 loci for the species of heteronemerteans collected at Llandudno, North Wales, and in the Foz Estuary, northwestern Spain. Alleles are arranged in order of increasing mobility. Enzyme abbreviations are given in the materials and methods section.

Locus	Allele	*Lineus viridis*	*Lineus longissimus*	Unknown form	
				Llandudno	Foz
APA	1	0.000	0.940	0.000	0.000
	2	0.000	0.060	0.000	0.000
	3	1.000	0.000	1.000	1.000
GOT-2	1	0.000	0.000	1.000	1.000
	2	0.000	0.800	0.000	0.000
	3	0.000	0.200	0.000	0.000
	4	1.000	0.000	0.000	0.000
ICD-1	1	0.992	0.000	0.000	0.000
	2	0.000	1.000	0.000	0.000
	3	0.008	0.000	0.000	0.000
	4	0.000	0.000	1.000	1.000
ICD-2	1	0.052	0.000	0.000	0.000
	2	0.509	0.000	0.013	0.150
	3	0.431	0.833	0.154	0.300
	4	0.008	0.000	0.000	0.050
	5	0.000	0.167	0.820	0.450
	6	0.000	0.000	0.013	0.050
MDH-1	1	0.000	1.000	–	–
	2	0.008	0.000	–	–
	3	0.992	0.000	–	–
MDH-2	1	0.000	1.000	–	–
	2	0.297	0.000	–	–
	3	0.703	0.000	–	–
ODH	1	0.026	0.000	0.128	0.000
	2	0.000	0.000	0.115	0.250
	3	0.328	0.000	0.000	0.000
	4	0.000	0.944	0.757	0.750
	5	0.060	0.056	0.000	0.000
	6	0.224	0.000	0.000	0.000
	7	0.086	0.000	0.000	0.000
	8	0.233	0.000	0.000	0.000
	9	0.043	0.000	0.000	0.000
6-PGDH	1	0.000	–	0.743	0.700
	2	0.000	–	0.054	0.000
	3	0.000	–	0.176	0.300
	4	1.000	–	0.027	0.000
PGI	1	0.009	0.000	0.000	0.000
	2	0.966	0.000	0.000	0.000

Table 1. (Continued)

Locus	Allele	*Lineus viridis*	*Lineus longissimus*	Unknown form	
				Llandudno	Foz
	3	0.000	0.000	0.066	0.100
	4	0.025	0.000	0.000	0.000
	5	0.000	0.000	0.474	0.500
	6	0.000	0.100	0.000	0.000
	7	0.000	0.000	0.408	0.400
	8	0.000	0.900	0.000	0.000
	9	0.000	0.000	0.052	0.000
PGM	1	0.000	–	0.039	0.000
	2	0.000	–	0.132	0.150
	3	0.000	–	0.724	0.600
	4	0.000	–	0.079	0.250
	5	0.000	–	0.026	0.000
	6	0.042	–	0.000	0.000
	7	0.958	–	0.000	0.000

found between other species of nemerteans which, on morphological grounds, are considered congeneric (Williams *et al.*, 1983; Sundberg & Janson, 1988).

Distortion of gene frequencies due to inbreeding or the presence of clones is unlikely since there was substantial genetic variation within samples and two of the species showed good fits of allele frequencies to Hardy-Weinberg expectations (but note the statistical weakness of such tests on other than large sample sizes: Lewontin, 1958; Fairburn & Roth, 1980; Valenzuela, 1985). *Lineus longissimus* was not tested for fits of allele frequencies to Hardy-Weinberg expectations, since the sample size was too low.

Between specimens of the unknown morphotype from Spain and Wales the value of Nei's (1972) genetic identity is above 0.9. This value is within the range normal for conspecific populations and above the range expected for congeneric species, and indicates that the populations of this morphotype from the two geographically separate locations are very likely to be conspecific. The species represented by this morphotype thus apparently shows little genetic differentiation over a moderately large zoogeographical range. This suggests that it is likely to be sexually reproducing with a planktonic dispersive phase. Such a

Table 2. Pairwise comparisons based on isozyme data from *Lineus longissimus*, *Lineus viridis* and the unknown morphotype. Above diagonal, values for genetic distance; below diagonal, values for genetic identity (both calculated after Nei, 1972).

	L. longissimus	*L. viridis*	Unknown form	
			Llandudno	Foz
L. longissimus	+	2.364	1.386	1.280
L. viridis	0.094	+	3.912	2.847
Unknown (Llandudno)	0.250	0.020	+	0.018
Unknown (Foz)	0.278	0.058	0.982	+

suggestion is supported by the observations on one individual, which released tiny planktonic eggs into the surrounding seawater in a manner similar to that described for *Lineus lacteus* McIntosh, 1874, by Gontcharoff & Lechenault (1958).

Expected and observed values for mean heterozygosity per locus for these species were high (range 9.7–26.8%; Table 3), especially for *Lineus viridis* and the unknown form. Heterozygosity estimates for these two taxa are considerably higher than those typically found for most other eukaryotic species (generally about 5–15%; see Selander, 1977; Nevo, 1978; Nevo *et al.*, 1984) High values of heterozygosity have also been found previously in *Lineus torquatus* Coe, 1901 (Balakirev & Manchenko, 1984) and the hoplonemertean *Oerstedia striata* Sundberg, 1988 (Sundberg & Janson, 1988). These values are at variance with those of Williams *et al.* (1983), who described a complete absence of allelic diversity from *Lineus viridis*. This can partially be explained by the fact that these authors did not stain for ODH or ICD, the two enzymes which showed the largest allelic diversity in *Lineus viridis* during the present studies (observed heterozygosity for ODH = 0.775, for ICD = 0.553).

It would seem that at least some species of nemerteans show heterozygosity values higher than most other groups of organisms (see Nevo, 1978; Nevo *et al.*. 1984). Such high values are apparently common in sea anemones and sponges (Solé-Cava & Thorpe, 1990, 1991). The reasons for such high levels of genetic variation are uncertain, but it has been suggested that genetic diversity may be related to environmental heterogeneity or predictability (Hedrick *et al.*, 1976). It may be of significance that, of the three species studied, *Lineus longissimus,* a predominantly subtidal taxon, had the lowest figures for heterozygosity. *Lineus viridis* and the new form are both intertidal taxa, and show exceptionally high values of heterozygosity. It could be argued that this difference occurs because the intertidal environment is far more heterogeneous and unpredictable than the subtidal, but it must be noted that in two similar species of *Oerstedia*, studied by Sundberg & Janson (1988), the subtidal species showed the greatest genetic diversity.

Table 3. Estimates of genetic variation in populations of *Lineus longissimus*, Lineus viridis and the unknown morphotype. H_e and H_o: mean expected and observed heterozygosities per locus respectively; $P_{(0.95)}$ and $P_{(0.99)}$: proportions of polymorphic loci with frequency of most common allele < 0.95 or < 0.99 respectively. n_e: mean effective number of alleles per locus.

	Lineus longissimus	*Lineus viridis*	Unknown form	
			Llandudno	Foz
H_e	0.126	0.178	0.217	0.268
H_o	0.097	0.145	0.163	0.254
$P_{(0.95)}$	0.375	0.273	0.636	0.545
$P_{(0.99)}$	0.375	0.545	0.636	0.545
n_e	1.625	2.454	2.500	2.128

Comparison in external features between *Lineus longissimus*, *Lineus viridis* and the third morphotype

The external features of the unknown morphotype resemble both *Lineus longissimus* and *Lineus viridis*, but close examination of living specimens reveals that all three taxa can be consistently distinguished from each other. Accounts of the external appearance of each of the three taxa are given below.

Lineus longissimus *(Gunnerus, 1770)*

Living specimens vary in length from 150–590 mm, in maximum width from 0.8–1.8 mm. Body width tends to remain uniform for up to about 70 mm from the tip of the head, but then gradually tapers posteriorly to end in a blunt tail. The body is typically somewhat dorsoventrally compressed.

The rather flattened head (Fig. 1A) is spatulate but appears short (length = twice width); it is characteristically slightly wider than the succeeding body region. The anterior tip of the head usually appears slightly bilobed. The dorsal cephalic surface is dark chocolate brown to black in colour, the anterior margins are white. Five pale and imprecisely defined longitudinal stripes occur on the dorsal cephalic surface, one median flanked on either side by two dorsolateral; the median stripe is typically white whereas the remaining stripes are pale brown. The cerebral ganglia may colour the posterior half of the head reddish. The ventral cephalic surface is usually paler than the dorsal and is marked by a single pale median stripe. Each lateral margin of the head bears a distinct horizontal cephalic furrow. There are 10–40 black or silvery-black eyes on each side of the head, forming an irregular patch near the anterior margin. In one specimen the ocelli were situated along the lower margin of the cephalic furrows. The slit-like mid-ventral mouth is long (5–6 mm) and commences just before or immediately after the posterior end of the cephalic slits.

The general body coloration is variable. The upper surface may be brown, dark chocolate brown or black; under artificial light it may appear maroon or exhibit a blue or purple iridescence. Five pale longitudinal stripes, continuous with those on the head, extend the full length of the body on the dorsal surface. The ventral surface may be the same colour as or slightly paler than the dorsal.

Lineus viridis *(Müller, 1774)*

Living specimens vary in length from 35–109 mm, in width from 0.4–1.5 mm, and exhibit a very variable body shape. Behind the cephalic region the body may be more or less uniform in width for up to about 60% of its length, then gradually narrow posteriorly to end in a finely pointed tail. Alternatively, the body width may increase behind the head for 7–26% of the body length, and

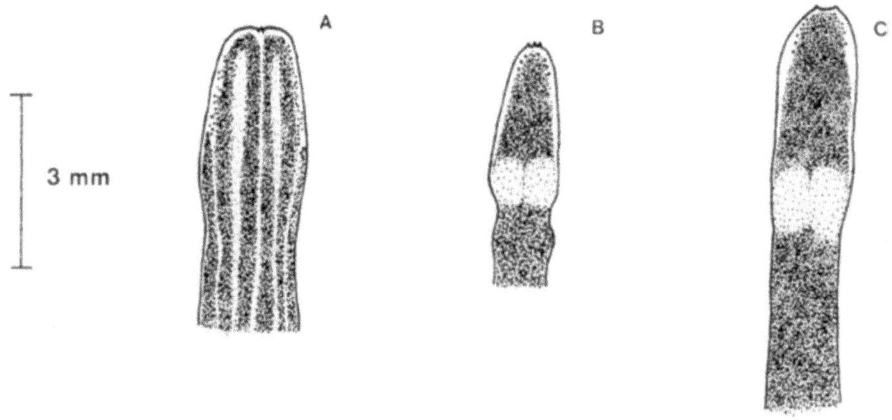

Fig. 1. The appearance of the heads, viewed in dorsal aspect, of A, *Lineus longissimus*, B, *Lineus viridis*, and C, the unknown morphotype, described in the present paper as *Riseriellus occultus* gen. et sp. nov.

then remain uniformly wide up to about 60% of its length before gradually narrowing to the tail. The body is compressed dorsoventrally, becoming progressively more flattened from the anterior to the posterior end.

The dorsoventrally flattened head (Fig. 1B) is generally spatulate, although its posterior portion is often markedly wider than the anterior to give it a spade-like shape. There is normally a shallow but distinct constriction between the head and the remainder of the body. The cephalic colour is variable, ranging from pale green through dark olive green to a deep bronze green or black. The cerebral ganglia are externally visible in the posterior part of the head as a pair of more or less distinct red patches. The margins of the head, which bear a single pair of lateral horizontal cephalic furrows, are pale. Two to eight black ocelli form an irregular and often asymmetrical dorsolateral row along either side of the anterior head region. The ocelli may be obscured by pigmentation, especially in darker coloured individuals. The small, mid-ventral and slit-like mouth commences just in front of or close behind the posterior end of the cephalic furrows. The oral margins are generally pale and often appear somewhat swollen.

The general body colour is also very variable. The dorsal surface may be black, deep bronze green, dark to light olive green or pale green. The colour intensity often shades from darker anteriorly to paler posteriorly. Pale annulations are normally distinguishable along the full length of the body and a pale mid-dorsal line may be present. In sexually mature specimens the gonopores show as an irregular dorsolateral row of white spots on each side of the body, commencing some distance behind the rear of the head. The ventral body surface is usually paler than the dorsal. When touched, the worms characteristically contract without coiling.

The new morphotype

Living specimens are 150–700 mm long and 1.0–1.3 mm wide. Their body width remains uniform for up to half the length, but then decreases posteriorly to terminate in a pointed tail.

The head (Fig. 1C) is spatulate and continuous with the succeeding body region, with no intervening constriction. In colour the head is very dark brown or green to black, with pale narrow lateral margins bearing a pair of horizontal cephalic slits. Three to eleven black ocelli form a dorsolateral row on each side of the anterior portion of the head. The posterior third of the head is often reddish due to the colour of the cerebral ganglia. The mid-ventral mouth begins far behind the brain lobes, approximately 1 mm beyond the rear end of the cephalic furrows; it is long, slit-like and possesses pale margins.

The body colour is variable. Anteriorly it may be black, greenish-black or dark olive, but the colour is nearly always lighter towards the posterior, shading to a dark or pale olive green or greenish-brown. The ventral surface may be similarly coloured or slightly paler. Under artificial light the body surface often shows a purple iridescence. Towards the posterior end of the body an irregular dorsolateral row of white spots on each side marks the position of the gonopores in sexually mature individuals. When touched, the worms tend to coil up into a tight spiral. The appearance of a complete specimen is shown in Fig. 2.

Description of the new taxon

Genus Riseriellus gen. nov.

Type species
Riseriellus occultus sp. nov.

Etymology
The genus is named in honour of Professor Nathan W. Riser as a tribute to his work on the morphology of nemerteans, in particular the Heteronemertea. The specific epithet, the Latin *occultus* (very secret), was chosen to indicate how the new taxon has previously been 'hidden' amongst *Lineus* species which externally it resembles.

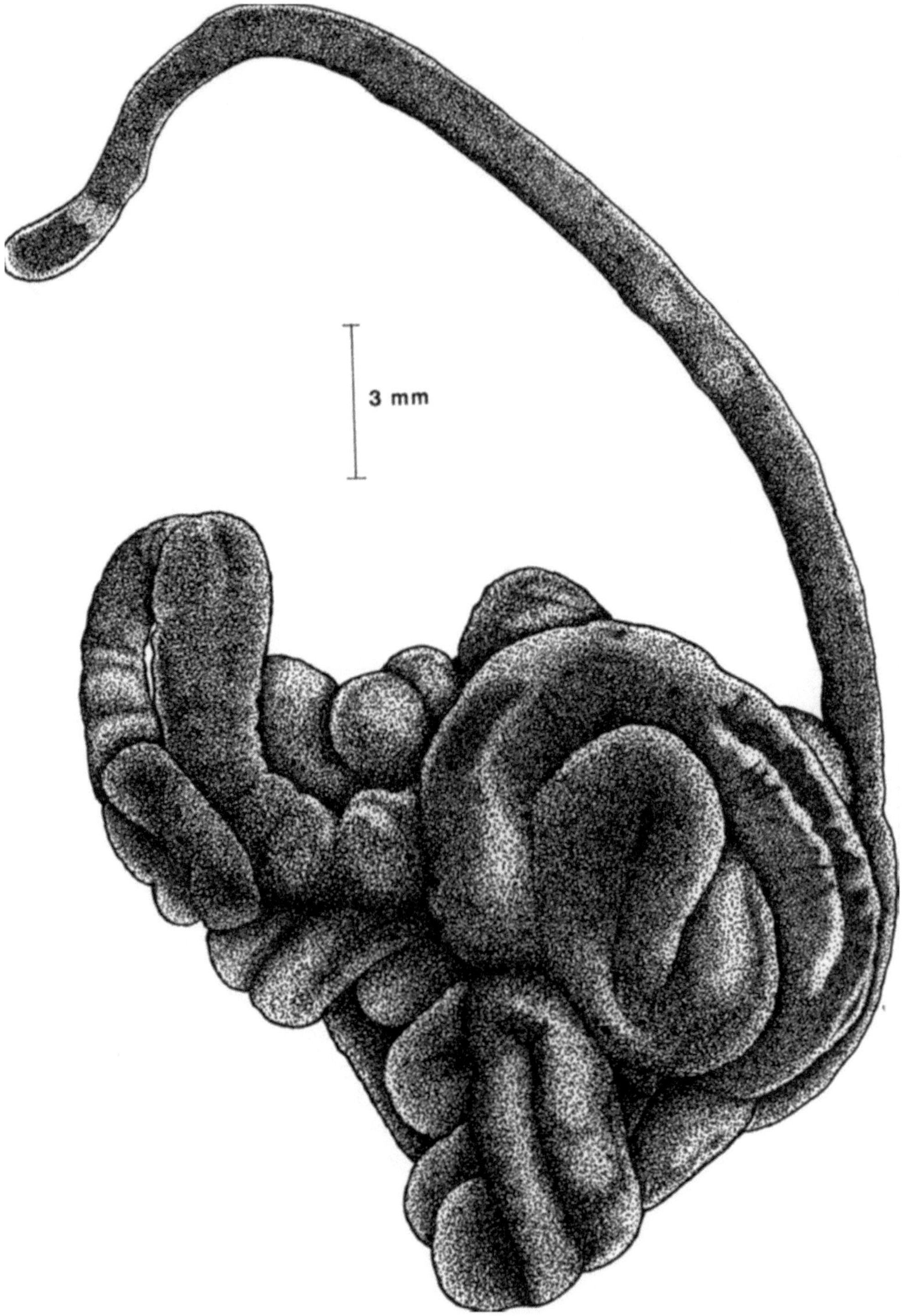

Fig. 2. The general appearance of a complete and somewhat contracted specimen of *Riseriellus occultus* gen. et sp. nov.

Diagnosis

Heteronemertea with a single pair of horizontal lateral cephalic slits; proboscis unbranched, containing two (outer circular, inner longitudinal) muscle layers and two muscle crosses; rhynchocoel circular muscles not interwoven with adjacent body wall inner longitudinal muscle fibres; dorsal fibrous core of cerebral ganglia forked only at rear into upper and lower branches; nervous system without neurochordal elements; foregut with neither somatic musculature nor subepithelial gland cell layer; dermis composed of well developed gland cell zone abutting directly against body wall outer longitudinal muscle layer, without distinct connective tissue stratum; body wall musculature without diagonal layer; caudal cirrus absent; intestinal diverticula indistinct; dorsoventral muscles missing from intestinal region; apical organ consisting of three separate sensory pits situated on tip of head; eyes present, arranged in row on either side of head; blood system comprising single spacious cephalic lacuna, foregut vascular plexus and three longitudinal vessels in intestinal region which are linked by pseudometameric transverse connectives; cephalic glands well developed but confined to dorsal half of head and not posteriorly reaching brain; excretory system situated between brain and mouth, collecting tubules penetrating post-cerebral blood lacunae; sexes separate.

Riseriellus occultus sp. nov.

Type specimens

The holotype, a female consisting of 168 slides of transverse sections, is deposited in the Liverpool Museum, William Brown Street, Liverpool L3 8EN, Registration Number 1992-7(01-168). (LIV); two unsectioned specimens are registered under number 1992. (LIV).

Type locality

Ría de Foz, north-western Spain (43° 34′ N, 7° 14′ W), first discovered by J. J. 20 September 1984 in seagrass beds (*Zostera noltii* Hornemann) in soft mud, upper shore.

Additional locations

Ría de Foz, in consolidated mud among roots of *Spartina* spp, upper shore, or in muddy sand with *Zostera noltii*; Llandudno, North Wales (53° 19′ N, 3° 49′ W), upper shore in damp fine mud or silt under stones and rocks; Rhosneigr, Anglesey (53° 15′ N, 4° 30′ W), upper shore in sandy-mud beneath boulders; Trwyn du Point, Anglesey (53° 18′ N, 4° 02′ W), upper shore in silty-sand, beneath boulders. In the Foz region recorded densities range from 6–131 worms m^{-2}; dominant associated fauna includes the polychaetes *Capitella capitata* (Fabricius), *Streblospio benedicti* Webster, *Pygospio elegans* (Claparède), *Heteromastus filiformis* (Claparède), *Alkmaria romijni* Horst, *Nereis* (= *Hediste*) *diversicolor* O. F. Müller and *Malacoceros* (= *Scolelepis*) *fuliginosus* (Claparède), the molluscs *Hydrobia ulvae* (Pennant) and *Scrobicularia plana* (da Costa), and the crustaceans *Melita palmata* (Montagu) *Idotea chelipes* (Pallas), *Chaetogammarus marinus* Leach, *Hyale nilssoni* (Rathke) and *Carcinus maenas* (Linnaeus), as well as unidentified species of Oligochaeta and Chironomidae. Mean grain sizes of sediments inhabited by the nemerteans are in the range 0.047–0.17 mm, whilst the silt-clay fraction varies from 16.8–73.15% and the organic content from 3.72–9.02%.

Body wall, musculature and parenchyma

The richly glandular epidermis, 25–30 μm in maximum thickness, closely agrees with the generalised heteronemertean form described by Norenburg (1985). Throughout most of the body it is dominated by large numbers of acidophilic rhabditoid glands (Fig. 3) which appear to correspond to the serous cells identified by Norenburg (1985). Rhabditoid density decreases markedly in the anterior ventral cephalic region, and the cells are completely missing from the epidermis on either side of the median furrow leading to the proboscis pore (Fig. 4); some reduction in their number is also evident in the posterior body regions. Throughout the body the epidermal basement lamina is thin but distinct.

Below the epidermis the subepidermal muscle layers (Figs 3, 5, 6) are well developed. The outer

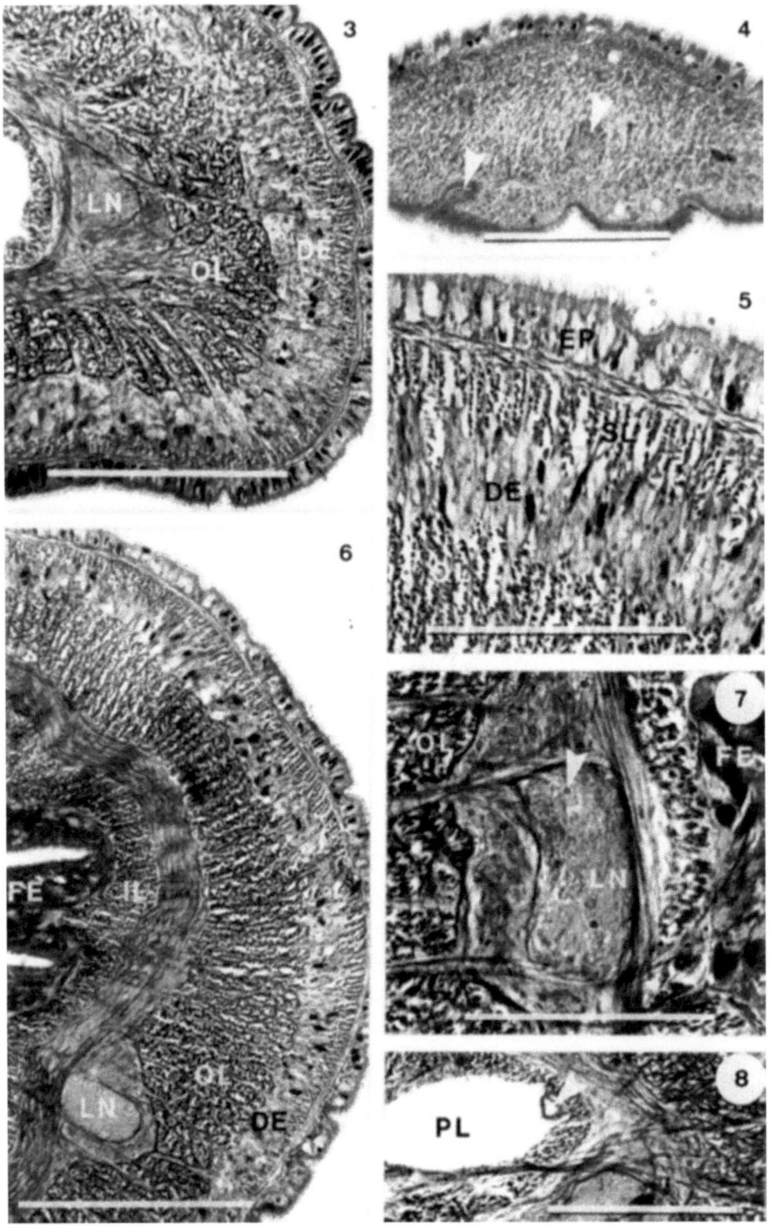

Fig. 3. Riseriellus occultus gen. et sp. nov. *3.* Transverse section between the brain and mouth to show the epidermal rhabditoid glands and body wall layers. Scale bar = 250 μm. *4.* Oblique transverse section near the tip of the head, showing two of the apical organ chambers (arrowheads) and absence of epidermal rhabditoid glands from the ventral surface. The ventral median furrow leads back to the proboscis pore. Scale bar = 200 μm. *5.* Part of the body wall in transverse section to show the structure of the epidermis, subepidermal muscle layers, dermis and outer portion of the main body wall outer longitudinal muscle layer. Scale bar = 100 μm. *6.* Transverse section through the foregut region to show the organisation of the body wall musculature. Scale bar = 250 μm. *7.* Transverse section through a lateral nerve cord in the foregut region, showing some of the radial muscle and connective tissue fibres which extend between the foregut wall and proximal portion of the epidermis. A myofibrillar (neuromuscular) strand in the fibrous tissue of the lateral nerve is indicated by an arrowhead. Scale bar = 100 μm. *8.* Transverse section through one of the post-cerebral lateral lacunae to show some of the horizontal transverse muscle bands extending peripherally above a lateral nerve cord. The arrowhead points to an excretory tubule running in the lumen of the blood lacuna. Scale bar = 200 μm. DE = dermis; EP = epidermis; FE = foregut epithelium; IL = body wall inner longitudinal muscle layer; LN = lateral nerve cord; OL = body wall outer longitudinal muscle layer; PL = post-cerebral blood lacuna; SL = subepidermal longitudinal muscle layer. All photomicrographs of sections stained with Mallory.

circular layer, 3–5 μm in maximum thickness, extends to the tip of the head. In contrast, the inner longitudinal fibres, particularly in the anterior ventral cephalic region, are barely distinguishable in front of the proboscis pore, whereas for most of the body length they are 15–30 μm deep (Fig. 5) and proximally abut directly against the well developed dermal gland cell layer.

The dermis is mostly 45–60 μm thick and composed entirely of gland cells (Fig. 5); a laminated connective tissue layer, typical of many heteronemertean taxa, is completely missing. Internally the dermal glands are bordered by the fibres of the main outer longitudinal body wall muscle layer (Fig. 5). Both bacillary and mucous dermal glands (Norenburg, 1985) are abundant, their distal portions in many places extending peripherally to discharge on the body surface. In the cerebral region and further forward the dermal gland cell layer becomes progressively reduced and, in front of the brain, it loses its integrity such that the glands cannot with certainty be distinguished from the cephalic and other subepidermal glands of the head. Behind the mid-foregut region the dermis is only about 20 μm thick and gradually decreases posteriorly.

The main body wall muscles comprise the typical heteronemertean arrangement of outer longitudinal, middle circular and inner longitudinal layers (Fig. 6), respectively 60–135 μm, 20–40 μm and 15–45 μm thick in the foregut region. There is no diagonal muscle layer. All three muscle layers are considerably reduced in thickness in the posterior portion of the body. The bundles of outer longitudinal muscle fibres are distinguishable from those of the subepidermal longitudinal layer by being enclosed in well developed connective tissue membranes which stain an intense blue colour with Mallory; the subepidermal muscles lack these membranes. Throughout the foregut region large numbers of radially oriented muscle and connective tissue fibrils (Fig. 7) extend peripherally from the wall of the gut, pass through the body wall layers and penetrate the extreme proximal portion of the epidermis. On either side of the mouth these radial muscle strands are reinforced by oblique muscles which have their origin at the outermost surface of the circular muscle coat.

The musculature between the mouth and brain is complex. Close in front of the mouth, where the anterior margins of the buccal chamber bulge forwards, several radial muscle fibrils merge to form thicker transverse bands of horizontal muscles which cross the body below the rhynchocoel. Laterally some of the fibres of these bands extend towards the body margins, passing either above (Fig. 8) or below the lateral nerve cords to terminate among the dermal gland cells. Further forwards, in front of the buccal region, the transverse muscle bundles split up to provide a meshwork of fibres criss-crossin the central part of the body between the lateral nerves and below the rhynchocoel. At this level the main circular muscle layer, which ventrally separates behind the mouth, remains incomplete but laterally gives off isolated fibres which merge with those of the meshwork. Close behind the brain, however, the circular layer for a short distance again forms a complete band of muscle fibres, but then begins to break up as it nears the rear of the cerebral sensory organs and in the cerebral ring is evident only as isolated fibres with a predominantly circular orientation.

The cephalic musculature consists of a loose meshwork of diagonal and oblique fibres, between which the various glands of the head are distributed (Fig. 9). These muscle fibres originate either in the body wall outer longitudinal layer, some of whose fibres turn obliquely inwards throughout the length of the head, or in the circular muscle layer which encloses the rhynchodaeum and cephalic blood lacuna. A layer of longitudinal muscle fibres, about 15–20 μm thick, is situated between the blood lacuna and circular muscles, the two muscle coats together forming a muscle cylinder similar to that described for many other heteronemerteans (Figs 9, 10).

Parenchymatous connective tissues are nowhere extensive; they show their greatest development in the intestinal region adjacent to the blood vessels and between these and the gut wall.

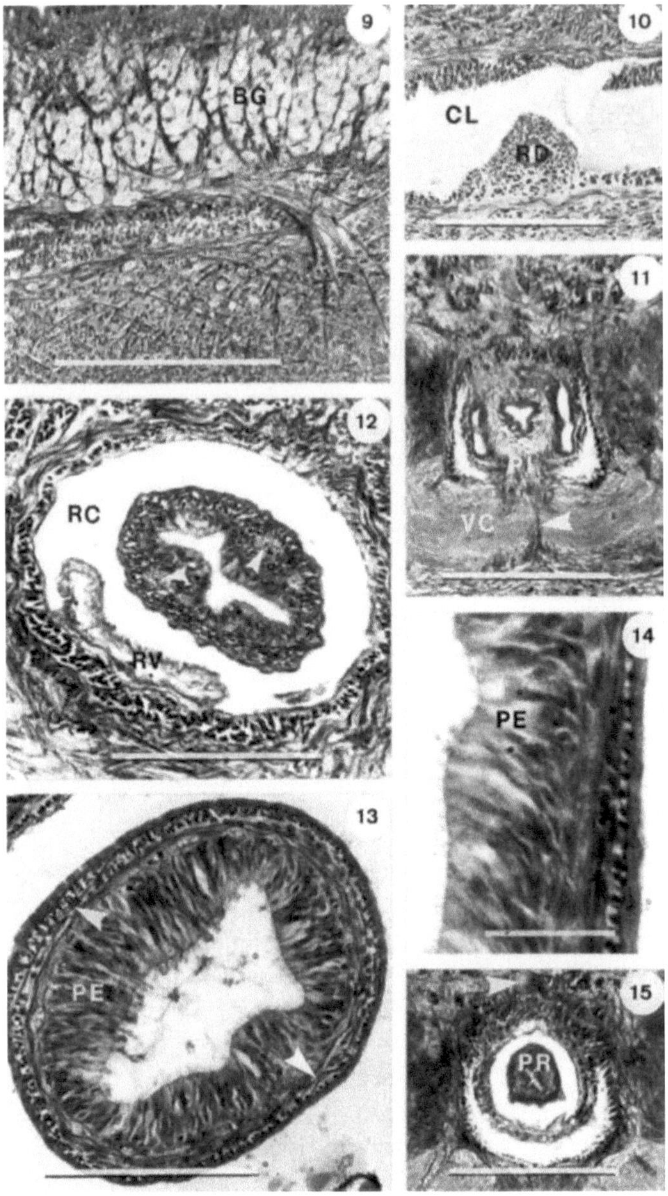

Fig. 4. Riseriellus occultus gen. et sp. nov. *9*. Transverse section through the cephalic region to show the basiphilic cephalic gland lobules and fibres of the central muscle cylinder. Scale bar = 250 μm. *10*. Transverse section through the middle cephalic region to show the single spacious blood lacuna and the rhynchodaeum. Scale bar = 200 μm. *11*. Transverse section through the brain region above the ventral cerebral commissure to show fibres leading to the proboscis insertion, indicated by an arrowhead, passing through the commissure. Scale bar = 200 μm. *12*. Transverse section to show the construction of the anterior portion of the proboscis, the two distinct muscle layers in the rhynchocoel wall and the rhynchocoelic villus. Arrowheads indicate the two large proboscis nerves. Scale bar = 100 μm. *13*. Transverse section to show the construction of the main portion of the proboscis. The position of the two muscle crosses are indicated by arrowheads. Scale bar = 100 μm. *14*. An oil immersion photomicrograph through part of the proboscis in longitudinal section to show the ciliation of its epithelium. Scale bar = 25 μm. *15*. Transverse section through the cerebral region to show the U-shaped blood channel arching below the rhynchocoel. The median dorsal nerve extending back from the dorsal cerebral commissure is indicated by an arrowhead. Scale bar = 200 μm. BG = basiphilic cephalic gland lobules; CL = cephalic lacuna; PE = proboscis epithelium; PI = proboscis insertion; PR = proboscis; RC = rhynchocoel; RD = rhynchodaeum; RV = rhynchocoelic villus; VC = ventral cerebral commissure. All photomicrographs of sections stained with Mallory.

Proboscis apparatus

The ventral, subterminal proboscis pore commences as a longitudinal invagination of the body wall, which internally expands to form an open tubular canal extending back for a short distance before closing off ventrally. For most of the cephalic length the rhynchodaeum appears as a simple tubular duct running below the blood lacuna, lined by a ciliated but non-glandular epithelium 10–12 μm thick, enclosed by longitudinal fibres of the cephalic muscle cylinder (Fig. 10). In the posterior part of the head, however, it begins to expand dorsoventrally, for a while partially protruding into the lumen of the blood channel, until the latter divides posteriorly to form a pair of spacious lacunae located on either side of the rhynchodaeum. The posterior portion of the rhynchodaeum is surrounded by a sphincter-like layer of circular muscles immediately in front of the proboscis insertion; the insertion is located in the brain ring above the ventral cerebral commissure (Fig. 11), muscle fibres leading to it from the body wall outer longitudinal muscle layer passing through a fenestration in the commissure and effectively splitting it into anterior and posterior portions.

The rhynchocoel reaches almost to the posterior tip of the body. Its wall contains separate longitudinal and circular muscle layers (Fig. 12), respectively some 7–8 μm and 15–18 μm in maximum thickness. The circular muscle layer is not interwoven with the fibres of the adjacent body wall inner longitudinal musculature, as in some heteronemertean genera.

The proboscis is unbranched and small; it is less than one-third the length of the body and has a maximum retracted diameter of about 180–200 μm. For most of its length it consists of a glandular epithelium 15–40 μm thick, a delicate nerve layer in which up to 10–12 distinct neural swellings can be distinguished, an outer circular muscle layer 4–6 μm in maximum thickness, an inner longitudinal muscle coat 6–8 μm across and an endothelium 4–5 μm deep (Fig. 13). There are two weakly developed muscle crosses extending from the circular musculature to the endothelial lining. The organisation of the proboscis musculature identifies the nemerteans as members of the family Lineidae, as defined by Gibson (1985). A short anterior region, extending from the proboscis insertion, presents a simpler construction. This region (Fig. 12) comprises an epithelium, 6–10 μm tall, in which there are fewer gland cells than in the main portion of the organ, a longitudinal muscle layer 6–20 μm thick in which two large nerves, 12–15 μm in diameter, can be distinguished, and an endothelium. An unusual feature of the proboscis, not reported for any other nemertean species, is that throughout its length the epithelium is ciliated (Figs 13, 14); basal bodies of the cilia can just be made out as minute dark spots under oil immersion. Riser (pers. comm.) reports that in all the lineid species he has examined, the proboscis epithelium contains proboscidial spines (Riser, 1990) but none have been found in the present species.

Blood system

The cephalic blood supply comprises a single spacious thin-walled median lacuna (Fig. 10), extending from near the tip of the head back to just in front of the proboscis insertion. As the rhynchodaeum posteriorly begins to expand dorsoventrally, the cephalic lacuna becomes increasingly constricted mid-dorsally and eventually divides into a pair of spacious lateral channels which flank the posterior portion of the rhynchodaeum. These lacunae become progressively smaller as they approach the brain and enter the cerebral ring. Near the front of the ventral cerebral commissure the lacunae meet ventrally to form a U-shaped duct (Fig. 15) which continues back beyond the commissure, at one point becoming temporarily subdivided by the ventral muscle fibres leading to the proboscis insertion (Fig. 11). Near the rear of the brain the duct separates medially to form two compressed channels, each of which arches dorsolaterally outwards over, but not bathing, the cerebral sensory organs. The inner ventral portions of these channels again meet medially, so that between the brain and mouth there are essentially three longitudinal blood spaces, the lateral pair on either side of the rhynchocoel being the larger (Fig. 8). Both these and the median ventral channel are irregularly traversed by oblique and radial fibres derived from the median muscle meshwork; all three lacunae sometimes rejoin each other for a short distance, but their subdivision is such that in some sections there may appear to be four or five separate channels. Near the front of the mouth the various blood spaces fuse to form two spacious dorsolateral lacunae which, above the buccal region, begin to split up to form the origin of the well developed foregut vascular plexus (Fig. 16). Throughout the length of the foregut the dorsolateral lacunae flanking the rhynchocoel remain larger than any of the plexal branches.

The mid-dorsal blood vessel arises as a median dorsal branch of the U-shaped duct close to the rear of the ventral cerebral commissure. It immediately penetrates the floor of the rhynchocoel to form the rhynchocoelic villus (Figs 12, 17), which reaches back for most of the foregut length before emerging to continue to the end of the body.

In the intestinal region there are three spacious longitudinal vessels, irregularly linked by transverse channels arching around the intestinal wall. These vessels possess distinct walls in which occasional muscle fibres can be distinguished.

Nervous system
The brain is well developed; dorsal and ventral lobes are similar in volume but the ventral are much more elongate and extend some distance behind the dorsal, ending near the posterior limits of the cerebral sensory organs. Fibrous tissues of the dorsal ganglia are forked only at their rear into upper and lower branches. The dorsal cerebral commissure, about 15 μm in diameter, is located slightly anterior to the much thicker (90 μm) ventral commissure; an unusual feature of the latter is that it is effectively divided into anterior and posterior portions by muscle fibres leading to the proboscis insertion (Fig. 11). Most of the brain contains no trace of either an inner or outer neurilemma, but delicate connective tissue membranes can be distinguished in the posterior cerebral regions near the sensory organs and origins of the lateral nerve cords; there are no neurochord cells in the ganglionic tissues. The outer surface of the brain lobes is irregularly defined as a consequence of the intermeshed radial and oblique muscle fibres which freely penetrate and pass through the neuroganglionic cerebral components.

The lateral nerve cords lead directly from the rear of the ventral ganglionic lobes; they do not contain neurochords but isolated muscle fibrils (myofibrillae) can be discerned in their fibrous cores (Fig. 7). Throughout their length the lateral nerves possess thin but distinct inner and outer connective tissue neurilemmae.

The peripheral nervous system is well developed and generally similar to that of many other heteronemertean taxa. Large numbers of cephalic nerves lead forwards into the head and can be followed for some distance between the fibres of the cephalic muscle meshwork (Fig. 9). The neural layer extending around the outer surface of the body wall circular muscle coat (Fig. 6) contains a large mid-dorsal nerve, originating from the rear of the dorsal cerebral commissure. From the inner margin of each ventral ganglionic lobe a distinct nerve separates off, runs back some distance among the neuroganglionic tissues and then emerges to continue between the fibres of the median muscle meshwork. Behind the brain the two nerves, which lead to the foregut, are linked by several slender transverse connectives before they move apart to pass on either side of the mouth just below the 'salivary glands' which surround the oral aperture (Fig. 18). Behind the mouth the two nerves begin to branch profusely, forming a nerve network adjacent to the foregut epithelium; there is no post-oral commissure connecting the foregut nerves and the ultimate fate of the network could not be traced.

Alimentary canal
The long, slit-like ventral mouth commences about 1 mm beyond the posterior limit of the cephalic furrows, and is situated some distance behind the brain. It opens into a buccal region whose epithelium is surrounded by large numbers of subepithelial 'salivary glands' of several types (Figs 18, 19); below the ventrolateral buccal bor-

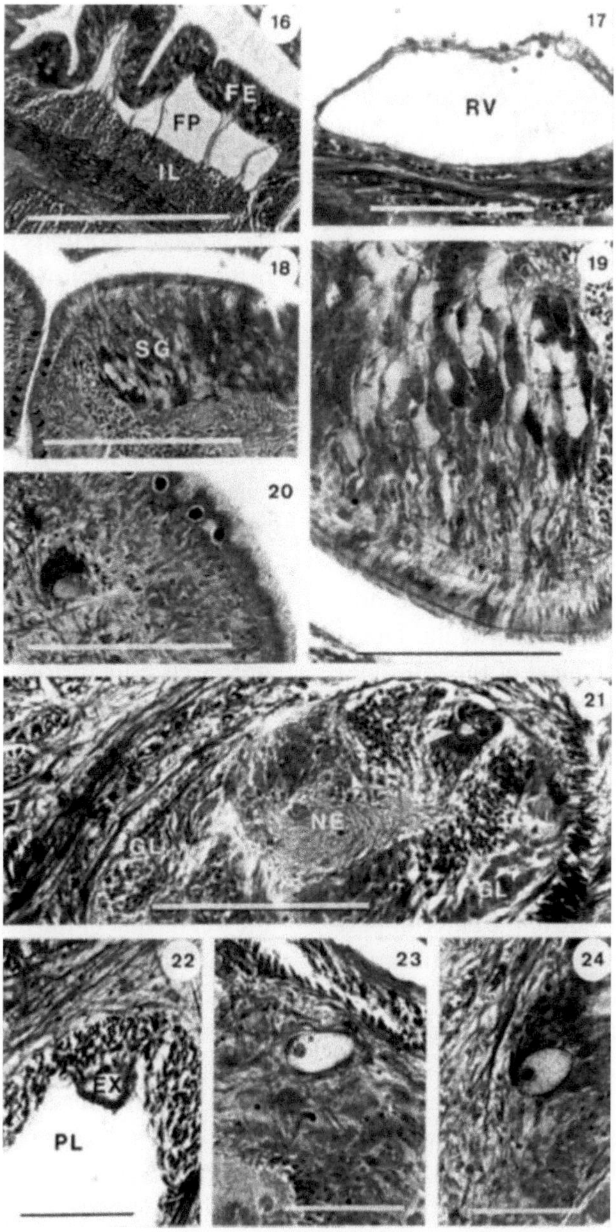

Fig. 5. Riseriellus occultus gen. et sp. nov. *16.* Part of the foregut region in transverse section to show branches of the vascular plexus. Scale bar = 250 μm. *17.* Transverse section through the posterior region of the rhynchocoelic villus. Scale bar = 50 μm. *18.* Transverse section through the mouth region to show the 'salivary glands' situated below the buccal epithelium. Scale bar = 250 μm. *19.* Enlargement of the 'salivary glands' to show the different types of constituent cells. Scale bar = 100 μm. *20.* Part of the head in transverse section to show one of the pigment cup ocelli. Scale bar = 100 μm. *21.* Transverse section through the middle portion of a cerebral sensory organ. The ciliated cerebral canal, in which major and minor canals can just be made out, is indicated by an arrowhead. Scale bar = 100 μm. *22.* Transverse section through one of the post-cerebral blood lacunae to show an excretory tubule running in its lumen. Scale bar = 50 μm. *23, 24.* Transverse sections through parts of the cerebral ganglia to show two of the parasites, possibly sporozoan, described in the text. Scale bars = 50 μm. EX = excretory tubule; FE = foregut epithelium; FP = branch of foregut vascular plexus; GL = glandular components of cerebral sensory organ; IL = body wall inner longitudinal muscle layer; NE = neural component of cerebral sensory organ; PL = post-cerebral blood lacuna; RV = rhynchocoelic villus; SG = 'salivary glands'. All photomicrographs of sections stained with Mallory.

ders these glands may extend 90 μm or more below the epithelium. The anterior portion of the buccal region, possibly as a consequence of contraction, bulges forwards, in sections appearing as a pair of large pouches reaching below the post-cerebral blood lacunae. A similar arrangement has been described for several other heteronemertean taxa.

The main foregut (Figs 6, 16) is lined by a richly glandular, ciliated and folded epithelium, up to 100 μm or more in maximum height. There is no subepithelial gland cell layer, nor somatic musculature, associated with the main foregut region.

The intestine is typically heteronemertean in appearance, its gastrodermal lining being up to 75–80 μm or more thick and containing enormous numbers of acidophilic glands. The lateral margins of the intestine are pouched but do not, as in many taxa, form distinct diverticula. Dorsoventral muscle bundles, typical of the intestinal region in many nemerteans, are completely missing from the present species. The anus opens at the posterior tip of the body.

Sensory organs

Three to eleven black ocelli are arranged in a dorsolateral row on either side of the anterior part of the head. Individual eyes are 30–35 μm in diameter and possess a simple pigment cup construction (Fig. 20). In sections the pigment shows as dense accumulations of minute dark brown granules.

The shallow lateral cephalic furrows extend from the tip of the head back to the rear cerebral region. As they pass the brain they begin to expand internally to form longitudinal chambers whose epithelium lacks gland cells but is densely ciliated. At about the level where upper and lower branches of the dorsal brain lobes separate, the inner wall of the chambers develops longer cilia (20–25 μm) and are associated with a zone of neuroganglionic tissue 25 μm or more deep. The ciliated cerebral canals lead obliquely inwards from this point, extending towards the upper branches of the dorsal cerebral lobes. As the canals reach the brain they meet further neuroganglionic masses and ventral accumulations of glandular tissue. Each canal, about 20 μm in diameter, then turns posteriorly to run along the outer dorsolateral border of the appropriate cerebral sensory organ (Fig. 21). The organs are intimately attached to the brain tissues and have a typically heteronemertean construction. Major and minor canals can clearly be distinguished in the posterior portions of each cerebral canal, just before they turn inwards to end amongst an accumulation of vacuolate glands. Each cerebral organ has a maximum diameter of about 120–150 μm.

Apical organ and cephalic glands

The apical sensory organ consists of three separate ciliated chambers, each about 45 μm in diameter, opening ventrally near the tip of the head (Fig. 4). The median chamber is in front of the ventrolateral pair and opens at the front of the proboscis pore furrow, the other two open on the ventrolateral cephalic surface slightly further back. The epithelium lining the chambers is 5–6 μm thick, densely ciliated and completely lacks gland cells.

Typical basophilic cephalic gland lobules are well developed and extensive, but are confined to the dorsal half of the head (Fig. 9) and commence behind the level of the proboscis pore. They form a more or less well defined mass of glands interspersed between the fibres of the cephalic muscle meshwork. Posteriorly the glands do not quite reach the anterior borders of the brain.

Excretory system

The thick-walled excretory tubules are located between the brain and mouth regions. They are 20–25 μm in diameter and mostly run immediately adjacent to the dorsal walls of the post-cerebral lateral blood lacunae. Up to two or three tubules can be distinguished on either side of the body. The posterior portions of the collecting tubules usually penetrate and may extend freely into the lumen of the blood lacunae (Figs 8, 22). A single efferent canal leads from the anterior region of the excretory system on either side of the body to open via a dorsolateral nephridiopore.

Reproductive system

The sexes are separate. Gonads are distributed in dorsolateral rows throughout most of the intestinal region of the body, but are missing from the extreme anterior portion. Testes of specimens collected during late September possess open gonoducts on their dorsolateral body surface. Each testis appears as a somewhat bilaterally compressed pouch lined by an undifferentiated epithelium up to 15 μm thick; in the lumen of many testes small numbers of spermatozoa can be distinguished and it appears that the animals had discharged their gonadal contents shortly before they were found.

Ovaries appear as large pouches containing masses of eggs, most of which are in a similar stage of development. More than 20 such eggs can be counted in any one ovary, and smaller numbers of immature ova can also be found attached to the ovarian wall. Larger eggs, 100 μm or more in diameter, contain a nucleus about 40 μm across in which a single nucleolus 10–12 μm wide is distinguishable. Females collected with mature males also possess open gonoducts, and in the laboratory one individual released eggs. The discharge of the eggs is very similar to that described for *Lineus lacteus* (Rathke) by Gontcharoff & Lechenault (1958); strings of eggs are released from the gonopores but quickly break up so that individual eggs are free. This differs strikingly from the egg-laying shown by *Lineus ruber* (O. F. Müller) and *Lineus viridis* (O. F. Müller), both of which lay their eggs in mucous sheaths attached to rock or algal surfaces.

Parasites

Several of the specimens examined contain parasites of various types. In the intestine gregariniform protozoans 45 μm or more long, containing a small nucleus in which two nucleoli are distinguishable, are often partially embedded in the gastrodermal wall. Other, possibly sporozoan, parasites are irregularly distributed in a variety of body tissues, including the brain, muscle layers and cephalic glands; these parasites are oval in shape, 30 μm long by 20 μm wide, and filled with a finely particulate cytoplasm in which a distinctly excentric acidophilic nucleus 5 μm wide is visible (Figs 23, 24). The parasites appear to be encapsulated in a thin but distinct membrane. A third type of parasite was found in smaller numbers in the epidermis and dermis; this appears as a large (60 μm or more diameter) more or less spherical cavity containing a membrane-bound amorphous mass of tissue in which no particular structure can be discerned. The appearance of these parasites bears some resemblance to the morula-like masses produced by the plasmodial generation of orthonectids, which are known to occur in a number of different invertebrates including nemerteans (Stunkard, 1982; Vernet, 1990).

Ecology

In the Foz Estuary the nemerteans are found in fairly compacted but wet muddy sediments in the upper shore zone, especially among the roots of *Spartina* or among *Zostera* beds, whereas at the same locality *Lineus longissimus* occurs at the mouth of the estuary under stones at low tide level. At Llandudno and on Anglesey *Lineus viridis* and the new species do not exhibit a similar spatial separation, both being found on the upper shore in silty mud beneath boulders.

Systematic discussion

Gibson (1985), in emphasising the need for complete descriptions of nemerteans, discussed the anatomy of the heteronemertean proboscis and proposed that all those genera with outer circular and inner longitudinal muscle layers should be grouped in the family Lineidae. Like many other authors (e.g., Wijnhoff, 1914), however, Gibson overlooked the significance of the nervous system in the determination of proboscis muscle layer homologies. Norenburg (1993) elegantly demonstrates that the evolution of the proboscis among heteronemerteans is more complex than previously appreciated and both discusses and illustrates several different proboscis morphotypes. His 'palaeotype', in which the two muscle layers (outer circular, inner longitudinal) are both subneural, is characteristic of the Lineidae as defined by Gibson, although Norenburg states that it seems more appropriate to consider the morphologically diverse lineids as a suborder, the Lineiformes. The proboscis of the new species possesses a 'palaeotype' organisation, and on this basis the species is placed in the Lineidae *sensu* Gibson (Lineiformes *sensu* Norenburg).

Some 29 heteronemertean genera are known or

Table 4. Some of the morphological features commonly used for distinguishing between heteronemertean genera, summarised for those forms that are known to, or probably, possess a 'palaeotype' proboscis; data from Gibson (1990b), Gibson & Qi Sang (1991), Iwata (this volume) and Riser (this volume).

Genus	1	2	3	4	5	6	7	8
Amniclineus	0	0	0	c	+	0	0	+ [a]
Antarctolineus	2	0	0	1	+	?	+	+
Apatronemertes	2	0	0	1	0	0	0	+ [a]
Australineus	2	+	0	1	0	0	0	+
Corsoua	2	+	0	c	+	0	+	0
Craticulineus	0	+	+	c	0	0	0	0
Eousia	2	0	0	1	+	+	+	+
Euborlasia	1–2	+	0	c	0	0	+ ?	+
Flaminga	1	0	0	c?	+	+	0	0
Hinumanemertes	2	0	0	c + 1?	0	+	0	+
Kirsteueria	1	0	0	1	0	0	0	+
Lineopsella	0–2	0	0	c	0?	0	0?	+
Lineopselloides	0	0	0	c	0	0	+	+
Lineus	0–2	0– +	0	c/1	0	0	+	0– +
Micrella	2	0	0	?	0	+	0	+
Micrellides	2	0	0	c	0	?	0	+
Micrura	0–2	0– +	0	0/c/1	0	0– +	0	+
Micrurimorpha	2?	0	0	?	0	0	0?	+
Micrurinella	2	0	+	?	0?	?	0?	?
Neolineusz	0	0	0	?	0	0	+	+
Paralineopsis	0	0	0	0	0	0	+	0
Paralineus	2	0	0	?	0	0	0	0
Paramicrurinella	0	0	+	c	0	0	0	0
Planolineus	1	0	0	?	0?	?	?	?
Pontolineus	2	0	0	?	+	0?	0	0
Pussylineus	2	0	0	?	+	0	0?	?
Siolineus	2	0	0	1	0	0	0?	+
Tenuilineus	?	0	0	0?	0	0	+	+
Utolineus	0	0	0	c	+	0	+	+
Riseriellus gen. nov.	2	0	0	0	0	0	0	+

1 = Number of muscle crosses in proboscis.
2 = Dermis with (+) or without (0) distinct connective tissue layer between gland cells and body wall outer longitudinal musculature.
3 = Circular muscles of rhynchocoel wall interwoven between fibres of adjacent body wall inner longitudinal musculature (+) or quite separate (0).
4 = Foregut somatic musculature absent (0), circular (c) or longitudinal (1).
5 = Nervous system with (+) or without (0) neurochord cells.
6 = Caudal cirrus present (+) or absent (0).
7 = Foregut with (+) or without (0) subepithelial gland cell layer.
8 = Blood system in foregut region developed into plexus (+) or consiting of only t<wo vesels running in parenchyma (0).
[a] The foregut vascular plexus in *Amniclineus* (Gibson & Qi Sang, 1991) and *Apatronemertes* (Wilfert & Gibson, 1974) is intimately associated with components of the excretory system, the unusual arrangement in these taxa being interpreted as a consequence of their freshwater habits.

suspected to possess a 'palaeotype' proboscis (Gibson, 1985, 1990a, b; Gibson & Qi Sang, 1991; Iwata, this volume; Riser, this volume), although in a few (e.g., *Tenuilineus*) the proboscis is regarded as transitional to the 'heterotype' because it possesses a short additional outer longitudinal muscle layer just behind the proboscis insertion. Commonly several morphological features have been used as a means of distinguishing between the numerous heteronemertean genera and these are summarised for the 'palaeotype' group in Table 4.

The present specimens have a foregut vascular plexus and proboscis containing two muscle crosses, their rhynchocoel circular muscles are quite separate from the adjacent body wall inner longitudinal muscle fibres, they have no dermal connective tissue layer, neurochords nor caudal cirrus, and their foregut possesses neither somatic muscles nor subepithelial glands. Two further features of the present taxon have not been described for any other nemertean species; these are the ciliated nature of the proboscis epithelium, and the manner whereby the ventral cerebral commissure is divided by muscle fibres leading to the proboscis insertion. It is concluded that the combination of morphological characters shown by the nemerteans is such that they cannot be placed in any of the known heteronemertean genera with a 'palaeotype' proboscis and the new genus *Riseriellus* is accordingly established for them.

The morphological evidence used to establish the new genus and species is fully vindicated by the electrophoretic data discussed earlier; indeed, it was these biochemical investigations which first showed that the taxon, though in its external features very similar to both *Lineus longissimus* and *Lineus viridis*, was in fact significantly different from them. This is the first time that the two approaches have been used in conjunction with each other to establish a new nemertean taxon; other 'lineid' species are being similarly investigated and results will be reported in future publications.

Acknowledgements

Most of the work was supported by Natural Environment Research Council grant number GR3/7168, which is gratefully acknowledged. We would also like to thank Dr J. D. Bishop for reviewing the manuscript, and Tim Holden and Eric McEvoy for assistance with the collection of specimens.

References

Ayala, F. J., 1975. Genetic differentiation during the speciation process. In T. Dobzhansky, M. K. Hecht & W. C. Steere (eds), Evolutionary biology. Plenum Press, New York 8: 1–78.

Ayala, F. J., 1983. Enzymes as taxonomic characters. In G. S. Oxford & D. Rollinson (eds), Protein polymorphism: adaptive and taxonomic significance. Academic Press, London: 3–26.

Balakirev, E. S. & G. P. Manchenko, 1984. Vnutrividovaya geneticheskaya izmenchivost belkov u nemertinui *Lineus torquatus*. Genetika 20: 2070–2071.

Berg, G., 1972. Taxonomy of *Amphiporus lactifloreus* (Johnston, 1828) and *Amphiporus dissimulans* Riches, 1893 (Nemertini, Hoplonemertini). Astarte 5: 19–26.

Cantell, C.-E., 1972. Studies on the morphology, taxonomy, and larval development of heteronemertines (Nemertina). Acta Univ. Uppsala 218: 1–7.

Cantell, C.-E., 1975. Anatomy, taxonomy, and biology of some Scandinavian heteronemertines of the genera *Lineus*, *Micrura* and *Cerebratulus*. Sarsia 58: 89–122.

Cantell, C.-E., 1976. Complementary description of the morphology of *Lineus longissimus* (Gunnerus, 1770) with some remarks on the cutis layer in heteronemertines. Zool. Scr. 5: 117–120.

Cantell, C.-E. & L. Gidholm, 1977. Isozyme studies of two related nemertines, *Micrura fasciolata* Ehrenberg 1831, and *Micrura baltica* Cantell 1975. Zoon 5: 87–89.

Fairburn, D. J. & D. A. Roth, 1980. Testing genetic models of isozyme variability without breeding data: can we depend on the X^2? Can. J. Fish. aquat. Sci. 37: 1149–1159.

Ferguson, A., 1980. Biochemical systematics and evolution. Blackie & Son, Glasgow.

Friedrich, H., 1935. Studien zur Morphologie, Systematik und Ökologie der Nemertinen der Kieler Bucht. Arch. Naturgesch. 4: 293–375.

Friedrich, H., 1960. Bemerkungen Uber die Gattung *Micrura* Ehrenberg 1831 und zur Klassifikation der Heteronemertinen nebst vorläufigem Bestimmungsschlüssel. Veröff. Inst. Meeresforsch. Bremerh. 7: 48–62.

Gibson, R., 1982a. British nemerteans. University Press, Cambridge.

Gibson, R., 1982b. Nemertea. In S. P. Parker (ed.), Synopsis and classification of living organisms. McGraw-Hill, New York 1: 823–846.

Gibson, R., 1985. The need for a standard approach to taxonomic descriptions of nemerteans. Am. Zool. 25: 5–14.

Gibson, R., 1990a. The macrobenthic nemertean fauna of Hong Kong. In B. Morton (ed.), Proceedings of the second international marine biological workshop: the marine flora and fauna of Hong Kong and southern China. University Press, Hong Kong 1: 33–212.

Gibson, R., 1990b. The macrobenthic nemertean fauna of the Albany region, Western Australia. In F. E. Wells, D. I. Walker, H. Kirkman & R. Lethbridge (eds), Proceedings of the third international marine biological workshop: the marine flora and fauna of Albany, Western Australia. Western Australian Museum, Perth 1: 89–194.

Gibson, R. & F. B. Crandall, 1989. The genus *Amphiporus* Ehrenberg (Nemertea, Enopla, Monostiliferoidea). Zool. Scr. 18: 453–470.

Gibson, R. & Qi Sang, 1991. A new freshwater heteronemertean from the Zhujiang (Pearl River), People's Republic of China. Hydrobiologia 220: 167–178.

Gontcharoff, M. & H. Lechenault, 1958. Sur le déterminisme de la ponte chez *Lineus lacteus*. C. r. hebd. Séanc. Acad. Sci., Paris 246: 1630–1632.

Harris, H. & D. A. Hopkinson, 1978. Handbook of enzyme electrophoresis in human genetics. North-Holland, Amsterdam.

Hedrick, P. W., M. E. Ginevan & E. P. Ewing, 1976. Genetic polymorphism in heterogeneous environments. Annu. Rev. Ecol. Syst. 7: 1–32.

Lewontin, R. C., 1958. A general method for investigating the equilibrium of gene frequencies in a population. Genetics 43: 419–434.

Nei, M., 1972. Genetic distance between populations. Am. Nat. 106: 283–292.

Nevo, E., 1978. Genetic variation in natural populations: patterns and theory. Theoret. Pop. Biol. 13: 121–177.

Nevo, E., A. Beiles & R. Ben-Schlomo, 1984. The evolutionary significance of genetic diversity: ecological, demographic and life history correlates. In G. S. Mani (ed.), Evolutionary dynamics of genetic diversity. Springer-Verlag, Heidelberg: 13–212.

Norenburg, J. L., 1985. Structure of the nemertine integument with consideration of its ecological and phylogenetic significance. Am. Zool. 25: 37–51.

Norenburg, J. L., 1993. *Riserins pugetensis* gen. n. sp. n. (Nemertina: Anopla), a new mesopsammic species, and comments on phylogenetics of some anoplan characters. Hydrobiologia 266: 203–218.

Pantin, C. F. A., 1960. Notes on microscopical technique for zoologists. University Press, Cambridge.

Richardson, B. J., P. M. Baverstock & M. Adams, 1986. Allozyme electrophoresis Academic Press, Sydney.

Riser, N. W., 1990. New Zealand nemertines from kelp holdfasts: Heteronemertinea 1. *Adenorhagas aurantiafrons* gen. n., sp. n. N.Z. Jl Zool. 17: 597–606.

Selander, R. K., 1977. Genetic variation in natural populations. In F. J. Ayala (ed.), Molecular evolution. Sinauer Associates, Sunderland, Mass.: 21–45.

Solé-Cava, A. M. & J. P. Thorpe, 1990. High levels of genetic variation in marine sponges. In K. Rützler (ed.), New perspectives in sponge biology. Smithsonian Institution Press, Washington, D.C.: 332–337.

Solé-Cava, A. M. & J. P. Thorpe, 1991. High levels of genetic variation in natural populations of marine lower invertebrates. Biol. J. linn. Soc. 44: 65–80.

Stunkard, H. W., 1982. Mesozoa. In S. P. Parker (ed.), Synopsis and classification of living organisms. McGraw-Hill, New York 1: 853–855.

Sundberg, P., 1979. Statistical analysis of variation in characters in *Tetrastemma laminariae* (Nemertini), with a redescription of the species. J. Zool., Lond. 189: 39–56.

Sundberg, P., 1980. Intraspecific variation in *Tetrastemma laminariae* (Nemertini): an examination of one of its possible causes. J. Zool., Lond. 192: 137–141.

Sundberg, P. & K. Janson, 1988. Polymorphism in *Oerstedia dorsalis* (Abildgaard, 1806) revisited. Electrophoretic evidence for a species complex. Hydrobiologia 156: 93–98.

Thorpe, J. P., 1982. The molecular clock hypothesis: biochemical evolution, genetic differentiation and systematics. Annu. Rev. Ecol. Syst. 13: 139–168.

Thorpe, J. P., 1983. Enzyme variation, genetic distance and evolutionary divergence in relation to levels of taxonomic separation. In G. S. Oxford & D. Rollinson (eds), Protein polymorphism: adaptive and taxonomic significance. Systematics Association, London: 131–152.

Valenzuela, C. Y., 1985. Algebraic and epistemological restrictions in studies on Hardy-Weinburg equilibrium. Am. Nat. 125: 744–746.

Vernet, G., 1990. *Rhopalura linei*, Orthonectide parasite de l'Hétéronémerte *Lineus ruber*. Cah. Biol. mar. 31: 251–255.

Walsh, P. J. & G. N. Somero, 1981. Temperature adaptation in sea anemones. Physiological and biochemical variability in geographically separate populations of *Metridium senile*. Mar. Biol. 62: 25–34.

Ward, R. D. & J. A. Beardmore, 1977. Protein variation in the plaice (*Pleuronectes platessa*). Genetic Res. 30: 45–62.

Wijnhoff, G., 1914. The proboscidian system in nemertines. Q. Jl microsc. Sci. 60: 273–312.

Williams, K. A., D. O. F. Skibinski & R. Gibson, 1983. Isoenzyme differences between three closely related species of *Lineus* (Heteronemertea). J. exp. mar. Biol. Ecol. 66: 207–211.

Hydrobiologia **266**: 239–246, 1993.
R. Gibson, J. Moore & P. Sundberg (eds), Advances in Nemertean Biology

Morphological variation in the palaeonemertean *Tubulanus annulatus* (Montagu, 1804)

Mikael Svensson
University of Gothenburg, Department of Zoology, Medicinaregatan 18, S-41390 Göteborg, Sweden

Key words: Nemertea, Anopla, intraspecific variation

Abstract

Previous studies on the palaeonemertean *Tubulanus annulatus* have hinted on possible geographic variation, or even a species complex. This study describes the variation in morphology, focusing on the characters that have been supposed to vary geographically. Specimens from the Swedish west coast (Tjärnö and Kristineberg), Scottish west coast (Millport), and the Mediterranean (Naples and Split) were included in the study. It is concluded that the pattern of variation in the studied characters do not correspond to geography and most of the character states previously thought to vary geographically are in fact found within one and the same population.

Introduction

Tubulanus annulatus is a nemertean with a wide geographic distribution, reported from the Mediterranean, from the coasts of Great Britain, Norway, Sweden and further north to Greenland and Alaska (Hylbom, 1957; Gibson, 1982). Specimens from these localities are externally similar, but they vary geographically in internal morphology according to Bergendal (1900, 1903), Stephenson (1911), Stiasny Wijnhoff (1912) and Hylbom (1957) who all reported deviations from Bürger's (1895) original description of the species from Naples, Italy.

In Bürger's description the cerebral ganglia were situated internal of the body wall musculature. The 'normal' position (as described by all others, including Bergendal's (1903) study on the Naples form) is between the outer circular muscle layer and the basement membrane. Furthermore, Bürger's description showed distinct and relatively well developed cerebral sense organs innervated by a thick nerve from the dorsal ganglia, which contrasts with the less well developed organs innervated by several thin nerves reported in other studies. Bergendal (1903) also mentioned a difference between his and Bürger's specimens in the structure of the basement membrane. Bürger illustrated the basement membrane as homogeneous (though relatively thick), while Bergendal's illustrations show a two-layered structure with an outer homogeneous and inner mesh-like layer. Neither did Bürger mention the projections from the basement membrane into the epidermis seen by Bergendal in the Swedish specimens. Bürger also reported a thin inner circular muscle layer and the presence of a dorsal muscle cross, which do not agree with the north European specimens.

This pattern of morphological variation, seemingly associated with geographic distribution of the populations (especially the differences between specimens from Sweden and Great Britain versus specimens from the Mediterranean), have raised the question whether the name *Tubulanus annulatus* contains more than one species or if it is just geographic variation (Hylbom, 1957). In this paper I will focus on (1) the morphological variation in *Tubulanus annulatus*, and (2) the question if it is geographic variation or not. The study of this variation will also shed light on the question whether *T. annulatus* encompasses more than one species. For these purposes specimens from Sweden (Tjärnö and Kristineberg), Great Britain (Millport), and the Mediterranean (Naples and Split) are included in the study.

Material and methods

The worms (Table 1) were sampled from most seasons of the year (March, July, and October) to avoid a bias of seasonal influence. Included are both small/young (5 cm) and large/old (> 20 cm) specimens. The specimens from Kristineberg Marine Biological Station (KMBS) and Millport were collected by Hylbom during the 1950's and the specimens from Naples and Split are from the collections of the Museum of Natural History in Stockholm. The material from Tjärnö Marine Biological Laboratory (TMBL) were collected by dredging and the worms were extracted with the oxygen depletion method described by Sundberg (1980). External features were observed on live specimens.

The worms were anaesthetized in MS 222 (Sandoz) and fixed in Bouin's fluid before sectioning. The Bouin fluid was washed out with

Table 1. Collection sites for *Tubulanus annulatus*, and the number of specimens collected. TMBL = Tjärnö Marine Biological Laboratory, KMBS = Kristineberg Marine Biological Station.

Site	No. of specimens
TMBL, west coast of Sweden	4
KMBS, west coast of Sweden	12
Millport, west coast of Scotland	2
Naples, Mediterranean	1
Split, Mediterranean	1

70% alcohol and the worms were dehydrated in THF (Tetrahydrofuran) before embedded in paraffin (m.p. 58 °C) for sectioning (7 μm). The staining methods used were Erlich's hematoxylin-eosin, Fe-hematoxylin and Mallory's trichrome method.

The characters

In this section I will mainly discuss the characters reported to vary geographically but also include information on other variable characters.

External features

Tubulanus annulatus is reported to reach lengths of > 75 cm, with a diameter of 3–4 mm (Gibson, 1982). The specimens I have studied averaged a length of 20–30 cm, when more or less fully extended, with a diameter of 1.5–3 mm. Bürger (1895) reports lengths between 8 and 10 cm, with a diameter of 1.5–2 mm from the Mediterranean. The specimen from Naples that I had the opportunity to study was within the range given by Bürger and so was the specimen from Split. However, the sample size in this study is far too small for any conclusions to be made about differences in length between sites.

The basic colour is bright red (but sometimes darker, Bürger (1895) mentions a chocolate brown form), with one dorsal and two lateral (one on each side) white longitudinal stripes. The tip of the head is white with a red dot in the middle. Transverse white rings are spread out all along the body, except in the head region. In small/young specimens the region between the first ring and the head is much darker than the rest of the body, and the head is completely white. A large specimen possessed white dots (rings) on the posterior part of the body just above the lateral white stripes.

An additional ventral white stripe and the somewhat darker (brownish) colour separates *T. superbus* from *T. annulatus* in external characters. Even though Bürger (1895) reports on a brown form of *T. annulatus* and Beaumont (1900) has described small specimens of *T. superbus* lacking a ventral white stripe, the red colour with the white rings and only a dorsal white stripe is a possible synapomorphy for the taxon *T. annulatus*.

Transverse cephalic furrows are present between the head and the first white ring.

Basement membrane

The structure of the basement membrane was one character that differed between Bürger's and Bergendal's descriptions. Bergendal (1903) stated that *T. annulatus* (= *Carinella annulata* in Bergendal (1903)) always possesses a two-layered basement membrane. However, both structure and thickness varies substantially even within a single individual, and according to Norenburg (1986) the basement membrane is sensitive to fixation procedures. Normally the basement membrane is most developed dorsally to the cerebral ganglia. There is a more or less continuous variation from a thin, homogeneous, one-layered structure to a more complex two-layered structure. The two layers are made up of a thin outer, homogeneous layer and a thick, inner mesh-like layer (Fig. 1). Both types are present in Swedish and British populations and the variation cannot be an effect of geographical separation. Bergendal (1903) found the two-layered structure in one Naples specimen; I have not seen it in the Naples specimen included in this study, although in the specimen from Split. This two-layered structure is not specific for *T. annulatus*, I have also seen it in *T. superbus*, *T. punctatus*, and *T. polymorphus*.

The projections (Fig. 2) into the epidermis are more or less pronounced in different specimens. In some specimens they are restricted to the brain region while in others they are more widely distributed; some specimens even lack them completely. There is no correspondence between basement membrane projections and geographic site and these projections are not restricted to *T. annulatus*, they are also present in other *Tubulanus* species. In *T. punctatus*, for instance, the projections are extremely big (present only in the head region).

Cephalic gland

The cephalic gland (Fig. 4) consists of mucous cells, mostly foamy and homogeneous even though more flocculent or granular cells are present in some specimens. Most of the space inside the basement membrane, from the tip of the head to the cerebral ganglia, is filled by the cephalic gland. The glands discharge through several, mainly lateral, ducts. Cells of cephalic gland type are also present in the rhynchodeum.

In other *Tubulanus* species (e.g. *T. superbus* and *T. punctatus*), which are reported to lack a cephalic gland, cells of cephalic gland type are present outside the basement membrane in the head region, in a position between the projections from the basement membrane. Whether these cells are homologous to the cephalic gland in *T. annulatus* or not needs further study.

Nervous system

The nervous system is situated between the basement membrane and the outer circular muscle layer in all examined specimens (Figs 1–3, 5–6) in accordance with the descriptions by Bergendal (1900, 1903) and Hylbom (1957), but never inside all muscle layers as indicated by Bürger (1895). The fact that the cerebral ganglia was situated internal of the body wall musculature in Bürger's illustration is enigmatic. However, this illustration is probably incorrect, since Bürger (1895: 99) in his general text on the genus *Tubulanus*, stated that: 'Das Centralnervensystem der Carinellen [= *Tubulanus*] liegt unter der Grundschicht, zwischen dieser und dem Hautmuskelschlauch'. In connection with this statement he also refers to the figure (Bürger, 1895: Fig. 5, Plate 12) illustrating *T. annulatus*.

According to Hylbom (1957) there should be both an upper and a lower dorsal nerve in *T. annulatus*, the latter situated between the longitudi-

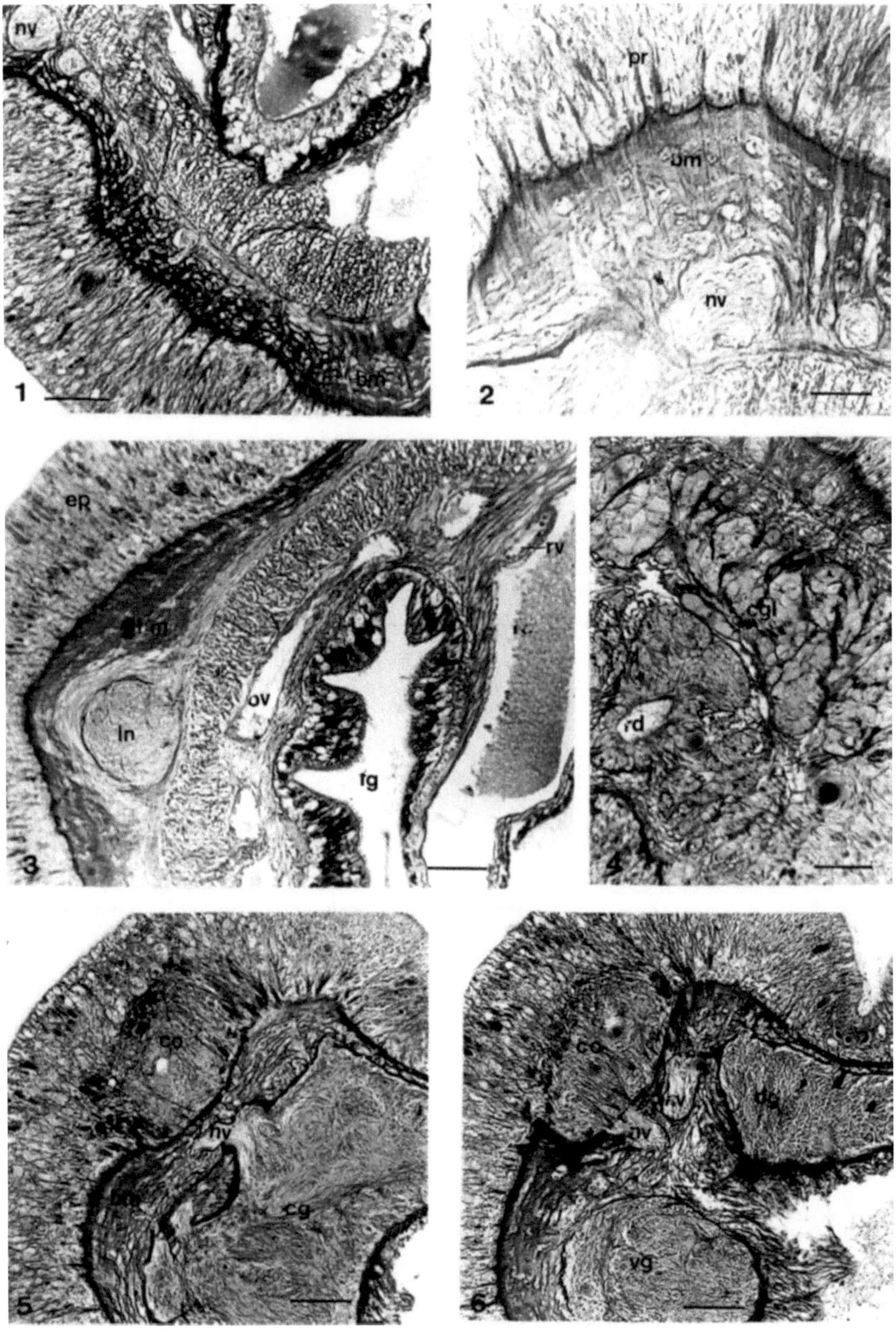
nv
bm
1
pr
bm
nv
2
ep
ln
ov
fg
rv
3
cgl
rd
4
co
nv
bm
cg
5
co
nv
vg
6

nal and inner circular muscle layer. This is also confirmed by Bürger (1895), who says that all *Tubulanus* species possess both an upper and lower dorsal nerve. I have found this in only one specimen of *T. annulatus*, and then it was very short and small and of doubtful identity. It seems to be more common with only an upper dorsal nerve.

Cerebral sense organs

The cerebral sense organs in *T. annulatus* are simple ciliated canals from the exterior to the basement membrane where they end with more or less well developed rounded organs, sometimes surrounded by gland cells. From the exterior they run backwards to the brain from where they are innervated. Hylbom (1957) states that they are less developed in Swedish specimens, than in Bürgers specimens from Naples. But, also Swedish specimens possess well developed cerebral sense organs, and the description by Bürger (1895) fits well on many Swedish and British specimens.

The cerebral sense organs show a considerable variation in their innervation and their relative position to the brain. The relative position to the brain is similar to the one in *Tetrastemma laminariae* (Hoplonemertea) described by Sundberg (1979).

The cerebral sense organs are innervated by either several thin nerves or by one or two thick nerves (Figs 5–6), generally more than ten times thicker than the thin ones. In some specimens, where the nerves are extra thick, the cerebral sense organs are almost incorporated into the cerebral ganglia. Furthermore, these organs can be innervated from either the dorsal or ventral lobes, or from smaller nerves posterior of the dorsal lobe, and both from the anterior and posterior part of the brain. This variation is found within one and the same population; for instance the two British specimens show different states. There is little information on the innervation in Bürger's description but in the general text on the genus *Tubulanus* he specifically discussed *T. annulatus* and stated that the cerebral sense organs are innervated by several small nerves from the dorsal ganglia. On the other hand, his illustration of the species indicates an innervation by one thick nerve from the same ganglia.

Body wall musculature

The body wall is fully developed posterior of the brain where it consists of three layers of variable thickness: one outer circular (OCM), one intermediate longitudinal (ILM) and one inner circular (ICM). The ILM is the thickest. In other species such as *T. superbus* and *T. punctatus* the muscle layers are fully developed in the precephalic region. These two species give a more compact impression than most (but not all) *T. annulatus*.

Stephenson (1911) and Hylbom (1957) pointed out that the ICM differ in thickness between Swedish and British and Naples specimens; the Swedish and British specimens have a thicker ICM than the Naples form. The muscle layers are highly variable when it comes to thickness, even though the ICM mostly is thin. This character contracts during fixation procedures, and there are no clear-cut difference between Swedish and British specimens versus Mediterranean specimens.

According to Bürger (1895) the Naples form posses a dorsal muscle cross while the Swedish and British specimens lacks such a muscle cross

Fig. 1. Transverse sections of *Tubulanus annulatus*. *1.* The two-layered basement membrane. *2.* The projections from the basement membrane. *3.* A rhynchocoel vessel and the position of the nervous system. *4.* The cephalic gland and the rhynchodeum. *5.* The innervation of a cerebral sense organ. *6.* Same as Fig. 5 but more posterior. Scale bars: 79 μm (Figs 1, 3–6) and 15 μm (Fig. 2). bm = basement membrane; cg = cerebral ganglia; cgl = cephalic gland; co = cerebral sense organ; dg = dorsal ganglion; ep = epidermis; fg = foregut; ln = lateral nerve; nv = nerve; ov = oesophagus vessels; pr = projections; rc = rhynchocoel; rd = rhynchodeum; rv = rhynchocoel vessel; vg = ventral ganglion.

according to Stephenson (1911) and Hylbom (1957). However, I have not seen any dorsal muscle cross in neither British or Swedish specimens nor in the specimens from Naples and Split. Muscle crosses are found in the related species *T. superbus* and *T. polymorphus*. A dorsal muscle cross seems to be more prevalent in species with both an upper and a lower dorsal nerve. The absence of a dorsal muscle cross is a possible synapomorphy for *T. annulatus*.

Blood vascular system

There is a large blood lacuna in the tip of the head, surrounded by the cephalic gland. The blood lacuna is more or less homogeneous but in some specimens interwoven by collagen and gland cells which gives one the impression of a branched lacuna. The lacuna forms an arch dorsal of the rhynchodeum. Hylbom (1957) states that there is a ventral arch as well, but this has not been observed in any of the specimens included in this study.

In the oesophagus and foregut region the lateral vessels form the oesophagus vessels which surrounds the oesophagus and foregut latero-ventrally. These vessels vary in number and the variation is found within a single population.

There is one vessel on each side of the rhynchocoel in the posterior part of the foregut in some specimens (both British and Swedish, and the specimen from Split). They are situated in a ventro-lateral position in the rhynchocoel (Fig. 4). This position coincides with the one given by Hylbom (1957) for the rhynchocoel vessels in *Carinina coei*, but both Bürger (1895) and Hylbom (1957) states that they are missing in *T. annulatus*, even though Hylbom (1957) mentions that he have seen something similar in a large specimen. However, also small (5 cm) specimens can possess such vessels. Oudemans (1885), on the other hand, mentions a similar condition in some, but not all, *T. annulatus*. Rhynchocoel vessels are also present in other *Tubulanus* species.

The blood vascular system is described by Oudemans (1885), but the specimens Oudemans described as *Carinella annulata* (Mont.) McIntosh include *T. superbus* and *T. polymorphus* (Bergendal 1903) which means that Oudemans description covers interspecific variation.

Discussion

With the re-evaluation of the distribution of the character states discussed above, and in previous studies, I think it is safe to say (even though the sample size in this study is relatively small) that they do not vary geographically, as previously thought. Rather, all variation can be found within one and the same population. *Tubulanus annulatus* is a highly variable species, even in qualitative characters at an intra-populational level. Finding a certain character state is often closely related to the number of specimens examined as evident from the two Swedish populations, which represents the largest samples in this study, where almost all character states discussed by previous authors are found.

Bürger's description of the species is within the range of variation mentioned in this paper with a few exceptions. These are (1) the position of the cerebral ganglia and (2) the possession of a dorsal muscle cross. The first exception is, as mentioned before, probably due to an incorrect illustration. The second exception, though, is more difficult to explain. One explanation could be that Bürger did not clearly distinguish between *T. annulatus* and *T. superbus* since he mentions a brown form of *T. annulatus*. The doubtful identity of the name around 1900 is strengthened by Beaumont's (1900) description of a small *T. superbus* lacking a ventral white stripe. But, on the other hand, *T. superbus* posses both a dorsal and ventral muscle cross. The fact remains though, that the present study, which includes specimens of *T. annulatus* from both the Mediterranean and north Europe, does not support the presence of a dorsal muscle cross.

In a cladistic analysis (Svensson, pers. obs.) with geographic populations of *T. annulatus*, *T. superbus*, *T. polymorphus* and *T. punctatus*, as terminal units, *T. annulatus* turned out as monophyletic even though the group was rather unresolved due to polymorphic populations. Monophyletic species are essential if we aim to anchor our species descriptions in the phylogenetic theory of systematics, *i.e.*, a phylogenetic species concept *sensu* Mishler & Donoghue (1982). The synapomorphies identifying the taxon in the analysis are (1) the red colour in combination with the white pattern and (2) the lack of a dorsal muscle cross. So, for the time being there is no evidence to be found in qualitative characters for splitting *T. annulatus* into two or more species, nor is there any evidence for geographic varieties in the sense of difference in morphology.

Several papers during the last twenty years have reported on the morphological variability in nemerteans (e.g., Berg, 1972; Sundberg, 1979, 1984) and these studies have shown that the variation is considerable. Although part of this variation is biological variability, a major part is probably due to artefacts. Nemerteans are very contractile and sensitive to fixation procedures and the possibility of artefacts should always be taken into account (Sundberg, 1979). Artefacts can, and should be minimized by careful preparations before and during fixation. Still, variation due to artefacts is unavoidable and it is therefore essential to consider all variation when describing species due to the difficulty of separating natural variation from variation caused by artefacts. If natural variation is neglected the risk of false assumptions of geographic variation (as in this case) or new unsupported taxa are apparent. However, even if all information on variation is included in a description of a species, *i.e.*, the species is defined as a cluster of continuous variation, delimited from others by gaps in variation, the risk of species based on overall similarity rather than genealogical relationships among populations is evident (Mishler & Donoghue, 1982; Donoghue, 1985).

Many nemertean taxa were established by the end of last century (Gibson, 1985), during a time when information on variation was not considered as important as today in taxonomy. This rather typological era has left behind taxa that might obscure the phylogenetic pattern we now try to systematize (*sensu* De Queiroz, 1988). As a result, nemertean taxonomy is at present undergoing a process of redescription, while at the same time attempts to elucidate phylogenetic relationships are being made. A combination that, hopefully, will turn out fruitful and vastly improve nemertean systematics.

Acknowledgements

I would like to thank Dr Per Sundberg for all valuable comments on this manuscript and his encouraging support during this research, and Mr Richard Hylbom for introducing me to the palaeonemerteans and for giving me access to his collection of nemerteans. I also thank Barbro Löfnertz for help with the histological work, and Mr Lennart Sandberg and Dr Barbara Kulikova for providing specimens.

References

Beaumont, W. I., 1900. Report on the Nemertea. Fauna and flora of Valencia harbour XI. Proc. r. Irish Acad. 3: 815.

Berg, G., 1972. Studies on *Nipponnemertes* Friedrich, 1968 (Nemertini, Hoplonemertini). I. Redescription of *Nipponnemertes pulcher* (Johnston, 1837) with special reference to intraspecific variation of the taxonomical characters. Zool. Scr. 1: 211–225.

Bergendal, D., 1900. Bör ordningen Palaeonemertini Hubrecht uppdelas i tvänne ordningar Protonemertini och Mesonemertini? Kungl. Vet. Akad. Förhandl. 6: 728–733.

Bergendal, D., 1903. Studien über Nemertinen. III. Beobachtungen über *Carinoma* Oudemans. Fysiogr. Sällsk. Lund Handl. 14: 66–72.

Bürger, O., 1895. Die Nemertinen des Golfes von Neapel. Fauna und Flora des Golfes von Neapel 22: 91–103, 523–525.

De Queiroz, K., 1988. Systematics and the Darwinian revolution. Phil. Sci. 55: 238–259.

Donoghue, M. J., 1985. A critique of the biological species concept and recommendations for a phylogenetic alternative. Bryologist 88: 172–181.

Gibson, R., 1982. British Nemerteans. University press, Cambridge, 212 pp.

Gibson, R., 1985. The need for a standard approach to taxonomic descriptions of nemerteans. Am. Zool. 25: 5–14.

Hylbom, R., 1957. Studies on palaeonemerteans of the Gullmar Fiord area (west coast of Sweden). Ark. Zool. 10: 539–582.

Mishler, B. D. & M. J. Donoghue, 1982. Species concepts: a case of pluralism. Syst. Zool. 31: 491–503.

Norenburg, J., 1986 Redescription of a brooding nemertine, *Cyanophthalma obscura* (Schultze) gen. et comb. n., with observations on its biology and discussion of the species of *Prostomatella* and related taxa. Zool. Scr. 15: 275–293.

Oudemans, A. C., 1885. The circulatory and nephridial apparatus of the Nemertea. Q. J. Microsc. Sci. (Suppl.) 25: 1–80.

Stephenson, J., 1911. The nemertines of Millport and its vicinity. Trans. r. Soc. Edinb. 48: 1.

Stiasny Wijnhoff, G., 1912. List of nemerteans collected in the neighbourhood of Plymouth from May to September 1910. J. mar. biol. Ass. 9: 407.

Sundberg, P., 1979. Statistical analysis of variation in characters in *Tetrastemma laminariae* (Nemertini), with a redescription of the species. J. Zool., Lond. 189: 39–56.

Sundberg, P., 1980. Intraspecific variation in *Tetrastemma laminariae* (Nemertini): an examination of one of its possible causes. J. Zool., Lond. 192: 137–142.

Sundberg, P., 1984. Multivariate analysis of polymorphism in the hoplonemertean *Oerstedia dorsalis* (Abildgaard, 1806). J. exp. mar. Biol. Ecol. 78: 1–22.

Hydrobiologia **266**: 247–254, 1993.
R. Gibson, J. Moore & P. Sundberg (eds), Advances in Nemertean Biology

Evidence for a FMRFamide-like peptide in the heteronemertine *Cerebratulus lacteus* Leidy

W. R. Kem[1] & C. Östman[2]
[1]*Department of Pharmacology and Therapeutics, University of Florida College of Medicine, Gainesville, FL 32610, USA;* [2]*Department of Zoology, University of Uppsala, Uppsala, Box 561, S-572 21 Sweden*

Key words: FMRFa, immunolocalization, neuropeptides, nemertine, muscle

Abstract

Circular body wall muscles of *Cerebratulus lacteus* respond to micromolar concentrations of the neuropeptide Phe-Met-Arg-Pheamide (FMRFa), first isolated from molluscan nervous tissue. Comparison of the relative body wall contractural potencies of various FMRFa analogs indicates that the Arg-Pheamide group is necessary for this activity, but the remaining N-terminal region can be altered considerably without loss of activity. The circular muscle failed to respond to met-enkephalin and many other vertebrate neuropeptides. Acetone-soluble extracts of *Cerebratulus* contained two FMRFa antibody immunoreactive components separable by reversed phase liquid chromatography. Neither component had the same retention time as FMRFa. Bouin's fixed and paraffin embedded *Cerebratulus* nervous tissues displayed specific immunofluorescence when incubated with FMRFa polyclonal antibody but not monoclonal antibody specific for the molluscan neuropeptide SCP-B. Some giant neuron somas in the lateral nerve cords and in the ventral cerebral ganglia were immunochemically reactive as were axons in the lateral nerve cord and in the circular and transverse body wall muscles. Pre-exposure of the antibody with FMRFa prevented the reaction. Thus several types of evidence suggest the presence of FMRFa-like neuropeptides in *Cerebratulus lacteus.*

Introduction

The neuropeptide Phe-Met-Arg-Pheamide (FMRFa) was initially found in nervous tissue of a clam (Price & Greenberg, 1977). Subsequently it was found in other molluscs (Price, 1982) and then other phyla (Grimmelikhuijzen, 1983; Dockray, 1985). In the past five years, the FMRFamide family of neuropeptides has been found in practically every metazoan phylum analyzed, including Cnidaria, Nematoda, Annelida and Chordata. Before the widespread distribution of this peptide family was recognized, we observed the presence of two FMRFamide immunoreactive constituents in the heteronemertine *Cerebratulus lacteus*. The body wall muscles, particularly the circular ones, were responsive to FMRFa as well as several of its analogs. More recently, we have extended the previous observations of Jennings *et al.*, (1987) demonstrating the occurrence of FMRFamide immunoreactive neurons in the nervous system of nemertines. Our study provides three different types of evidence for the existence of as yet unidentified neuropeptides related to FMRFamide in the phylum Nemertinea.

Materials and methods

Animals

Mature, living *Cerebratulus* (5–20 g fresh weight) were collected by the Department of Marine Resources, Marine Biological Laboratory (Woods Hole, MA 02543) and shipped to Florida, where they were kept in Instant Ocean artificial sea water at 10 °C before use.

Pharmacological experiments

Circular muscle strips were dissected from 1–2 mm thick pharyngeal cross-sections of *Cerebratulus* which had been relaxed for several minutes in cold 7.5% $MgCl_2$. The delicacy and limited dissectability of the preparation made it impractical to remove all of the outer and inner longitudinal muscle layers without damaging the thin circular layer. Muscle preparations were suspended in 10 ml temperature (10 °C) controlled tissue baths containing artificial sea water continuously circulated by a constant flow of bubbled air. One end of the muscle was connected to an isometric force transducer and Grass Polygraph, whereas the other end was tied to a metal wire holder. A resting tension of about 0.5 g was maintained with frequent adjustment during the experiment. The artificial sea water composition was 500 mM NaCl, 10 mM KCl, 10 mM $CaCl_2$, 20 mM $MgSO_4$, 8 mM $NaHCO_3$, and 10 mM dextrose (pH 7.6).

At the beginning and end of the experiment, the muscle was exposed to a high K saline (250 mM KCl and 250 mM NaCl; other ions were present in the same composition as above) in order to measure its maximal contractile ability. The isometric contractile responses to test substances were recorded beginning with low concentrations and then challenging the tissue with higher concentrations. After the contracture had fully developed, the preparation was washed twice with artificial sea water and then allowed to equilibrate for at least 20 min before the next challenge. Muscles routinely generated tensions between 2–5 g.

Neuropeptide isolation procedure

612 g of frozen *Cerebratulus* were suspended in 1 liter of 1% acetic acid containing 10–6 M diisopropylfluoridophosphate (an irreversible serine protease inhibitor) for 5 min at 5 °C. After removal of the toxic slime by filtering through cheesecloth, the worms were freeze-dried, pulverized with mortar and pestle, and then stirred in 90% acetone overnight. The precipitate was removed by suction filtration through Whatman # 1 paper. The acetone-soluble supernatant was rotary evaporated to a 50 ml aqueous phase and subsequently extracted with two volumes of petroleum ether. The resulting aqueous fraction was further evaporated to dryness and suspended in 20 ml of 20 mM ammonium acetate (AmAc), pH 6.0. This solution was diluted to 500 ml (in order to lower its conductivity to that of 20 mM AmAc) and adsorbed to 100 ml of CM 52 (Whatman) carboxylmethylcellulose in a beaker. Subsequently this was eluted with 50, 100, 200 and 500 mM AmAc solution (200 ml each) by suction filtration. Samples of the various phases were kept for muscle bioassays.

A high pressure liquid chromatographic separation of *Cerebratulus* acetone-soluble materials utilized a Novapak C18 reversed phase column, equilibrated with 0.1 M trifluoroacetic acid in methanol. The eluate fraction were evaporated to dryness and then subjected to radioimmunoassay as described by Price (1982).

Immunolocalization experiments

The second antibody (indirect) fluorescence method was used to identify FMRFa-immunoreactive substances. Transverse sections of *Cerebratulus lacteus* were fixed in Bouin's, then dehydrated and embedded in paraffin. Sections (10 microns thickness) were mounted on gelatin coated slides. After rehydration through xylene and a series of decreasing ethanol concentrations to 50% followed by incubation in phosphate buffered saline (20 mM Na phosphate, pH 7.4, 140 mM NaCl) containing 10% normal goat serum, the slides were initially exposed to 1:50 dilution of the primary antibody (anti FMRFa-serum, prepared by Dr E. Weber, University of Oregon) dissolved in phosphate buffer saline containing 10% normal goat serum. In some slides, the primary antibody was omitted; in others, primary antibody which had been preincubated with excess FMRFa was used. After 1 h, the primary antiserum was removed by exposing sections 3 times (5 min each time) with phosphate buffered saline containing 2% goat serum, and then a 1:50 dilution of Sigma fluorescein isothiocyanate conjugated goat anti-rabbit immunoglobin G containing 2% normal goat serum was added. After at least 1 h incubation, this antibody solution was removed as before, and the section was washed with phosphate buffered saline 4 times, 5 min incubation each time. Then a drop of 50% glycerin (plus a trace of phenylenediamine to stabilize fluorescence) in phosphate buffered saline was added, and finally a coverslip. Some sections were stained with hematoxylin-eosin.

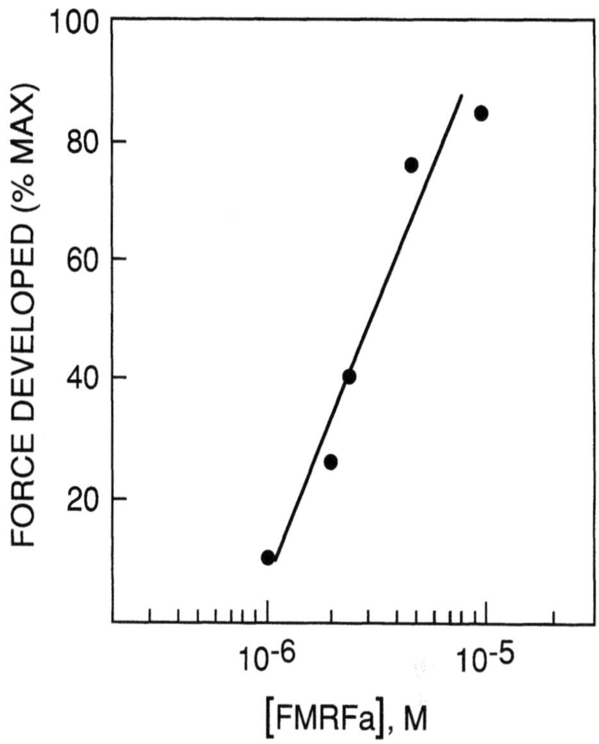

Fig. 1. Concentration dependence of the *Cerebratulus* circular muscle contractile response to FMRFa. Contractions were measured isometrically.

Results

Contracting activity of FMRFa and other peptides

The most common response to FMRFa was a slow contraction, which rarely exceeded 50% of the maximal tension produced in response to the high potassium saline. Occasionally a muscle would display a relaxation response or even a mixed contraction-relaxation response to FMRFa. *Cerebratulus* body wall circular muscles were not as sensitive to FMRFa as were molluscan heart (Price & Greenberg, 1980; Painter *et al.*, 1982; Kobayashi & Muneoka, 1986) and skeletal (Muneoka & Saito, 1986) muscles. The threshold concentration for response was slightly less than 10^{-6} M and the median effective concentration was 2.8×10^{-6} M (Fig. 1). Although the nerve plexus on the outer surface of the circular muscle was often present, other muscles completely lacking nerve plexus were just as sensitive to FMRFa. Other neuropeptides at concentrations of 10^{-5} M, including eledoisin, metenkephalin, bradykinin, angiotensin II, and proctolin, failed to contract or relax the circular muscle. It was also insensitive to 2×10^{-6} M *C. lacteus* polypeptide toxins B-IV (Kem, 1976) and A-III (Kem & Blumenthal, 1978). Unlike the outer longitudinal muscle, the circular muscle did not contract when exposed to the stable cholinergic agonist carbamylcholine at 1.2×10^{-3} M.

Comparison of the relative potencies (peptide median effective concentration divided by FMRFa EC_{50}) of various FMRFa analogs revealed a structure-activity relationship (Table 1) similar to that previously reported for molluscan heart and retractor muscles. The carboxy terminal end (Arg-Pheamide) could not be altered without drastic loss of activity. Removal of amide group reduced potency at least 50-fold. Replacing the phenylalanyl residue at position 4 with the similar tyrosyl group or with a leucyl residue also abolished activity at the concentrations we tested. Replacement of the arginyl group with either lysyl or ornithyl residues caused essentially complete

Table 1. Relative contracting activities of FMRFa-like peptides on *Cerebratulus* body wall circular muscle.

Peptide	Relative potency
pGlu-Asp-Pro-Phe-Leu-Arg-Phe-Nh_2	0.6
Tyr-Gly-Gly-Phe-Met-Arg-Phe-NH_2	0.8
Try-Gly-Gly-Phe-Met-Arg-Phe-CO_2H	<0.02*
Phe-Met-Arg-Phe-NH_2	1.0
Phe-Met-Arg-Phe-CO_2H	<0.01
Phe-Met-Arg-Tyr-NH_2	<0.03
Phe-Met-Arg-Leu-NH_2	<0.06
Phe-Met-Lys-Phe-NH_2	<0.02
Phe-Met-Orn-Phe-NH_2	<0.02
Phe-Ile-Arg-Phe-NH_2	3.0
Phe-Pro-Arg-Phe-NH_2	0.3
Arg-Phe-NH_2	0.002

* Where a maximum activity limit is given, the peptide has not yet been found active.

loss of activity. In contrast, the amino-terminal phenylalanyl and methionyl groups could be replaced with other amino acids without a great decrease in potency. Replacing the methionyl group with an isoleucyl residue significantly enhanced activity relative to FMRFa. Elongation of the amino terminal portion by addition of several residues was tolerated by the *Cerbratulus* circular muscle FMRFa- receptive system.

Isolation of Cerebratulus *peptide*

A whole body extract of freeze-dried *Cerebratulus* in artificial sea water was shown by circular muscle bioassay to possess about 300 nanomol of FMRFa equivalents per g fresh weight of *Cerebratulus*. This was much higher than the levels of immunoreactive FMRFa found in the whole snail (*Helix aspersa*) tissue, which was of the order of one nanomol per g fresh weight (Price, 1982). All of the *Cerebratulus* activity survived boiling, but it is uncertain at this time how much of the activity was due to FMRFa-like peptides.

In an attempt to more selectively extract FMRFa-like peptides, we also tried two organic solvent extraction procedures, utilizing 90% acetone or 75% ethanol. The results of these two experiments are summarized in Table 2. Both

Table 2. Extraction of circular muscle contracting activity from *Cerebratulus lacteus*.

Fraction	Dry Wt (g)	FMRFa equivalents	
		μmol g Dry Wt	Total μmol
90% Acetone method			
Whole body	114	1.6	182*
Acetone precipitate	101	2.0	202
Petroleum ether	0.6	0.5	0.3
Aqueous phase	1.0	25	25
CM cellulose eluates			
20 mM AmAc	14	0.28	3.9
50 mM AmAc	5.0	2.6	13
100 mM AmAc	3.0	0.17	0.51
200 mM AmAc	2.0	0.55	1.1
500 mM AmAc	0.8	2.0	1.6
75% Ethanol method			
Whole body	113	1.6	181*
Ethanol precipitate	80	0.19	15
Petroleum ether	10	2.6	27
Aqueous phase	8.5	3.3	28

* Standard error limit was ± 58 μmol.

procedures extracted amounts of muscle contracting activity similar to that observed with aqueous extract. Although FMRFa is soluble in acetone, most of the initial *Cerebratulus* contracting activity was insoluble in acetone. During the defatting step with petroleum ether, only a small portion of the contracting activity was removed, as would be expected for FMRFa or most other peptides. The petroleum ether insoluble contracting activity resulting from acetone extraction was adsorbed to the weak cation exchanger CM-cellulose, as expected for FMRFa, but was eluted with only dilute (0.05 M) ammonium acetate buffer.

In contrast with the acetone extraction, much of the whole body *Cerebratulus* contracting activity was soluble in 75% ethanol. About as much of this 75% ethanol soluble contracting activity partitioned into the petroleum ether phase as into the aqueous phase.

In another experiment, the 1% acetic acid soluble component of the acetone soluble fraction was placed on a G15 Sephadex column and the zone possessing circular muscle contracting activity at 2.1 × the column void volume was freeze-dried. An aliquot was injected into a C18 reversed-phase high pressure column equilibrated with 0.1 M trifluoroacetic acid and then developed with a gradient of increasing methanol. Two peaks of FMRFa immunoreactivity were detected in the eluted fractions (Fig. 2). Neither peak had a retention time similar to that of FMRFa. Due to the small amounts of peptide present, we were not able to characterize these fractions any further.

Immunolocalization of FMRFa-like reactivity

Our most extensive observations were upon the lateral nerve cords and the nerve plexus lying at the interface between the outer longitudinal and circular muscle layers. Observations on the brain of *Cerebratulus*, although incomplete, indicate that the ventral ganglia of the brain contain a significant number of relatively large neuronal soma visualized with FMRFa antiserum (Figs 3, 4). On

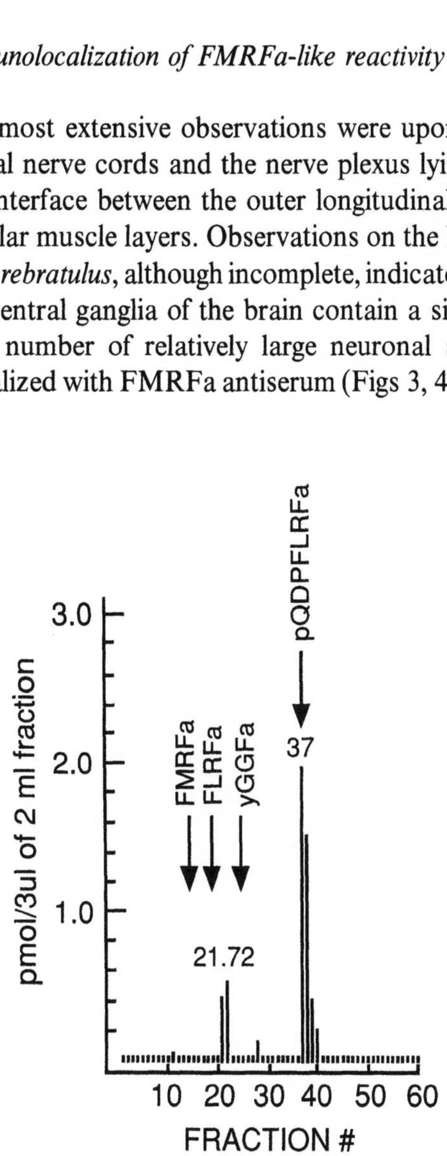

Fig. 2. Separation of two FMRFa antibody immunoreactive peaks from the acetone soluble whole body fraction of *Cerebratulus lacteus*.

the other hand, the dorsal ganglia displayed very little staining. Each lateral nerve cord (Figs 5, 6) contained two sites of FMRFa staining, the most obvious one being numerous axons in the neuropil (perhaps more than 20% of the axons there fluoresced). The second site was a small population of moderate sized neuronal soma lying outside the neuropil, in the region of the nerve cord ventral to the neuropil. Occasionally we observed large immunoreactive neural soma in the lateral cord outside the neuropil (Fig. 7).

Since the body wall circular muscle layer was stimulated by FMRFa, we were particularly interested in finding immunoreactive neurons in this muscle. However, except for occasional neurons in the circular muscle layer, immunoreactive neurons seemed to be very concentrated in the nerve plexus lying between the outer longitudinal and circular muscle layers (Fig. 8). Occasionally immunoreactive neurons were observed entering the circular muscle layer (Fig. 8). Only rarely did we observe any axon processes running parallel to the muscle fibers within either layer. However, we did observe small numbers of immunoreactive axons traversing the numerous muscle trabeculae which cause flattening along the dorso-ventral axis of *Cerebratulus*.

If the FMRFa primary antibody was omitted or preincubated with a saturating concentration of FMRFa, no fluorescent staining was observed in the *C. lacteus* tissue section. We also failed to observe any staining when *Cerebratulus* brain and nerve cord sections were treated with antibodies to two other related neuropeptides: molluscan neuropeptide SCP-B and met-enkephalin.

Discussion

Immunohistological evidence for two types of neuropeptides has been reported for nemertines and turbellarians: (1) met-enkephalin (Reuter *et al.*, 1984; Remy & Brossard, 1985; Wikgren *et al.*, 1986), and (2) FMRFa peptides (Wikgren *et al.*, 1986; Jennings *et al.*, 1987). However, no neuropeptide has yet been isolated from either phylum. There are difficulties obtaining enough

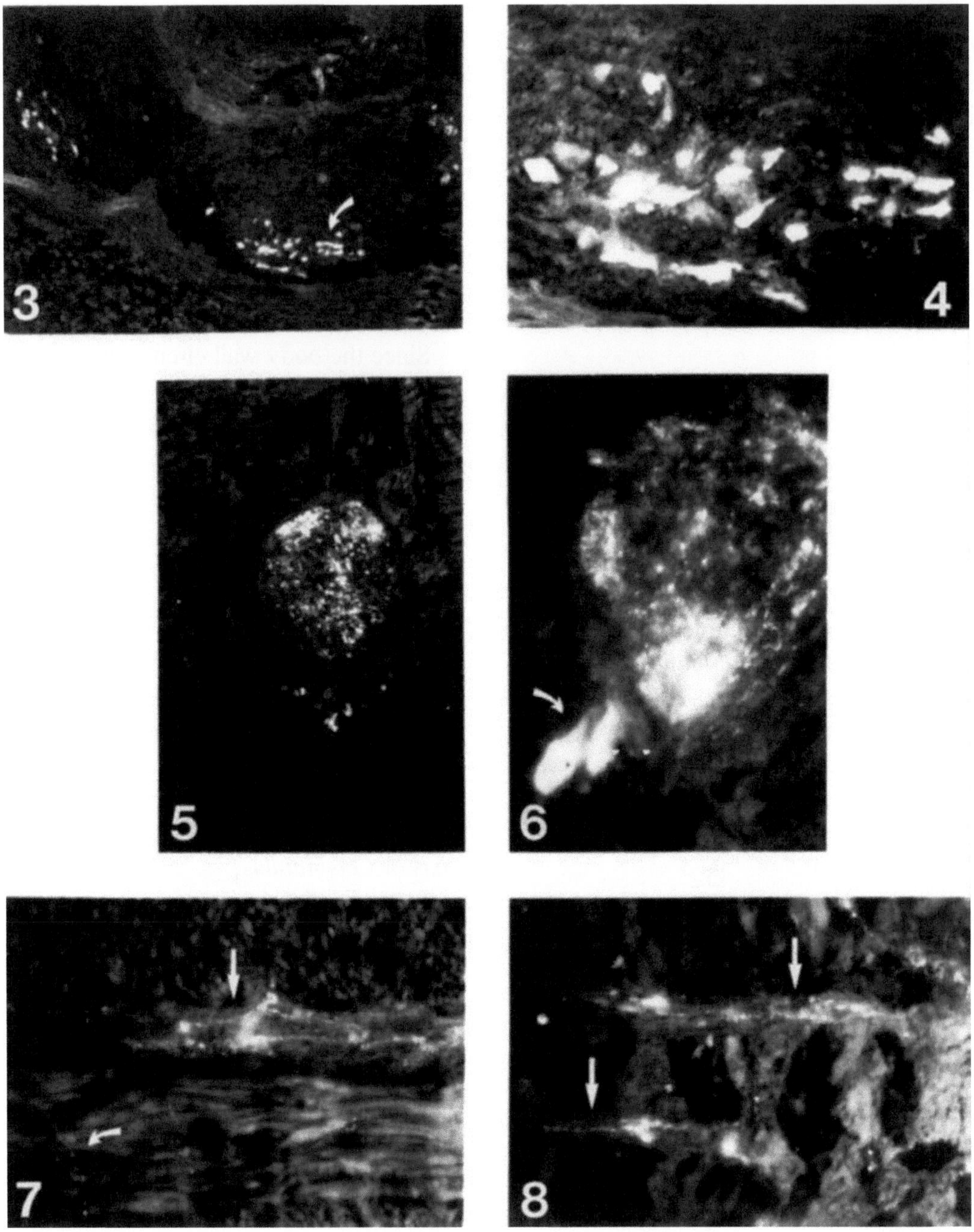

Figs 3–8. Cerebratulus lacteus. 3. FMRFa antibody immunoreactive staining of the ventral ganglia of the brain, 20 ×. Arrow indicates the region evident in Fig. 4. 4. FMRFa antibody immunoreactive neuronal soma in brain ventral ganglia, 40 ×. 5. FMRFa antibody immunoreactive cells in lateral nerve cord 20 ×. Note that only a few of the neuronal soma in the ventral region of the nerve cord, outside the neuropi, are fluorescent. 6. FMFRa immunoreactive soma in lateral nerve cord, 40 ×. Arrow identifies two large immunoreactive ganglionic neuronal cell bodies. 7. FMRFa immunoreactive neurons in nerve plexus between outer longitudinal (upper) and circular body (lower) wall muscle layers, 40 ×. The upper arrow points out the nerve plexus, whereas the lower arrow identifies an immunoreactive neuron which has entered the circular muscle layer. 8. FMRFa immunoreactive neurons in muscle trabeculae responsible for dorsal-ventral axis flattening of pharyngeal region, 40 ×. The body vertical plane is shown in the horizontal plane of photograph.

tissue for peptide isolation from these animal groups except for certain large nemertines such as *Cerebratulus*. Several FMRFa-like peptides have been isolated and characterized from relatively large amounts of nematode tissue, and these may become of potential therapeutic and veterinary use.

In this study, we have obtained several types of evidence for existence of FMRFa-like neuropeptides in a nemertine. The structure-activity relation for FMRFa-like peptide activity upon the *Cerebratulus* body wall circular muscle is quite similar to that previously reported for molluscan muscles, where the evidence for innervation by FMRFa-ergic neurons is excellent.

Evidence for existence of a peptide based upon immunoreactivity assumes that the antiserum is reacting specifically with the peptide or similar peptides. The existence of similar antigenic determinants on unrelated peptides or proteins in a tissue section is certainly possible. If this were the case in this study, one would have to be impressed with the very narrow localization of the antigenic determinants to a subpopulation of neurons in the ventral lobes of the brain, lateral nerve cords and the plexus between outer longitudinal and circular muscles. Since the FMRFa antibody preparation we used recognized the carboxy-terminal end of this peptide, our results further suggest that whatever peptides are present contain the Arg-Pheamide moiety.

Our peptide isolation results are not an unequivocal demonstration of a FMRFa-like peptide in *Cerebratulus*. The muscle bioassay we used cannot be considered specific for FMRFa since we did not have an antagonist available to assess receptor specificity. Furthermore, the immunoreactive fractions eluted from the reversed-phase column were not sufficient for bioassay, so we have not shown that the contractile activity we bioassayed corresponds to FMRFa-immunoreactive peptides. In future experiments, it might be possible to establish whether FMRFa-like peptides are contributors of contractile activity by carrying out the contraction bioassays in the absence and then in the presence of desensitizing FMRFa concentrations.

The concentration range in which FMRFa acted upon *Cerebratulus* circular muscle was perhaps 10–50 × higher than is effective upon molluscan muscles where FMRFa has been shown to be neurotransmitter. It seems likely that the endogenous FMRFa-like peptides of *Cerebratulus* would be more potent than FMRFa upon the muscles of this nemertine.

Further attempts to isolate endogenous contractile factors from *Cerebratulus* are planned. However, the small amounts of peptide present may make this approach difficult. Alternatively one could screen a *Cerebratulus* cDNA library with RNA probes for the Arg-Phe-Gly sequence expected to be present, probably as several copies per mRNA. In addition to isolation and characterization of the endogenous *Cerebratulus* peptides, we plan to investigate the ionic channel and second messenger systems of *Cerebratulus* muscles affected by this class of neuropeptide.

Acknowledgements

We thank Drs Paul Lindser, David Price, and Michael Greenberg at the University of Florida Whitney Laboratory, Marineland, for making available their facilities, peptides, and antisera. Dr Kemal Payza carried out the *Cerebratulus* peptide extractions and related bioassays in Dr Kem's laboratory. Dr Price carried out the experiment shown in Fig. 2. Dr A. O. Dennis Willows, Friday Harbor Laboratories, graciously provided the mouse monoclonal antibody for molluscan SCP-B peptide. We thank Ms. Judy Adams for word-processing the manuscript. This project was partially supported by the Swedish Institute and the Faculty of Science at Uppsala University.

References

Dockray, G. J., 1985. Characterization of FMRFamide-like immunoreactivity in rat spinal cord by region-specific antibodies in radio-immunoassay and HPLC. J. Neurochem. 45: 152–158.

Grimmelikhuijzen, C. J. P., 1983. FMRFamide immunoreactivity is generally occurring in the nervous system of coelenterates. Histochem. 78: 361–382.

Jennings, J. B., T. R. B. Davenport & I. M. Varndell, 1987. FMRFamide-like immunoreactivity and arylamidase activity in turbellarians and nemerteans-evidence for a novel neurovascular coordinating system in nemerteans. Comp. Biochem. Physiol. 86C: 425–430.

Kem, W. R., 1976. Purification and characterization of a new family of polypeptide neurotoxins from the heteronemertine *Cerebratulus lacteus* (Leidy). J. Biol. Chem. 251: 4184–4192.

Kem, W. R. & K. M. Blumenthal, 1978. Purification and characterization of the cytolytic *Cerebratulus* A toxins. J. Biol. Chem. 253: 5752–5757.

Kobayashi, M. & Y. Muneoka, 1986. Structural requirements for FMRFamide-like activity on the heart of the prosobranch *Rapana thomasiana*. Comp. Biochem. Physiol. 84C: 349–352.

Muneoka, Y. & H. Saito, 1986. Pharmacology of FMRFamide on molluscan catch muscle. Comp. Biochem. Physiol. 85C: 207–214.

Painter, S. D., J. S. Morley & D. A. Price, 1982. Structure-activity relations of the molluscan neuropeptide FMRFamide on some molluscan muscles. Life Sci. 31: 2471–2478.

Price, D. A., 1982. The FMRFamide-like peptide of *Helix aspersa*. Comp. Biochem. Physiol. 72C: 325–328.

Price, D. A. & M. J. Greenberg, 1977. Structure of a molluscan cardioexcitatory neuropeptide. Science 197: 670–671.

Price, D. A. & M. J. Greenberg, 1980. Pharmacology of the molluscan cardioexcitatory neuropeptide FMRFamide. Gen. Pharmac. 11: 237–241.

Rémy, C. & D. Brossard, 1985. Immunohistological detection of met-enkephalin-like neuropeptides in the brain of nemerteans. In G. B. Stefano (ed.), CRC Handbook of Comparative Opioid and Related Neuropeptide Mechanisms, Vol. 1. Publisher, place: CRC Press, Boca Raton, Florida. 155–163.

Reuter, M., T. Karhi & L. P. C. Schot, 1984. Immunocytochemical demonstration of peptidergic neurons in the central and peripheral nervous system of the flatworm *Microstomum lineare* with antiserum to FMRF-amide. Cell. Tiss. Res. 238: 431–436.

Wikgren, M., M. Reuter & M. Gustafsson, 1986. Neuropeptides in free-living and parasitic flatworms (Platyhelminthes). An immunochemical study. Hydrobiol. 132: 93–99.

Hydrobiologia **266**: 255–265, 1993.
R. Gibson, J. Moore & P. Sundberg (eds), *Advances in Nemertean Biology*

Molecular approaches to the study of evolution and phylogeny of the Nemertina

Joan D. Ferraris
Laboratory of Kidney and Electrolyte Metabolism, National Heart, Lung, and Blood Institute, National Institutes of Health, Bethesda, Maryland, 20892, USA

Key words: gene expression, nucleotide sequence, invertebrate

Abstract

Molecular biological tools currently available to us are revolutionizing the way in which we can address questions in evolutionary biology. The purpose of this article is to provide an overview of molecular techniques and applications available to biologists who are interested in evolutionary studies but who have little acquaintance with molecular biology. In evolutionary biology, techniques designed to determine degree of nucleic acid similarity are in common use and will be dealt with first. Another approach, namely gene expression studies, has strong implications for evolutionary biology but generally requires substantial familiarity with molecular biological tools. Expression studies provide powerful tools for discerning processes of speciation, as in the selection of genetic variants, as well as discerning lineages, e.g., expression of specific homeobox genes during segment formation. For investigations where either nucleic acid identity or gene expression are the ultimate goal, detailed information, protocols and appropriate controls are beyond the scope of this work but, where possible, recent review articles are cited.

Molecular biology and the Nemertina

Nemertine biologists should be able to take particular advantage of molecular biology. These worms comprise a difficult taxon to study since there are few external characters with which to identify species. Nemertines are often highly abundant and are found primarily in marine, but also in estuarine, freshwater and terrestrial habitats. The phylum contains approximately 1000 described species. In soft bottom communities and in the rocky intertidal, where these worms can be particularly numerous, they prey on polychaetes and crustacea. Parasitic forms are successful predators of decapod embryo masses and exact a substantial toll on crab and lobster fisheries.

Phylogenetically, nemertines may be the first of the acoelomate phyla in which the alimentary tract is complete and the circulatory system is closed; their position in the phylogeny of invertebrate lineages has long been debated. To identify species, it is almost essential to see specimens alive since fixation can have undesirable affects. Body wall and eye pigmentation is usually lost and internal characters of these highly contractile worms become distorted. The latter is particularly true if preservation is not performed by a nemertine specialist using appropriate relaxation and fixation techniques. Phylogeny reconstruction of the

Nemertina is at present based largely upon internal characters. Thus, the use of molecular biological techniques to generate additional phylogenetically useful characters is of obvious advantage.

DNA-DNA hybridization

Several approaches are available to determine the degree of identity between the DNAs or RNAs of individuals. In DNA-DNA hybridization studies, determination of nucleotide similarity is based on the thermal stability of double-stranded DNA molecules. The greater the number of nucleotide matches between the two DNA strands, the higher will be the thermal stability of the duplex molecule. The method consists of the following principal steps. The single-copy genomic DNA sequences from the two individuals to be compared are heat-denatured to separate the strands. A portion of denatured DNA from each individual is then combined and cooled slowly to allow strands to reanneal, forming heteroduplex DNA. The heteroduplex DNA is then heated gradually to dissociate the strands. A melting curve is generated by measuring the percent of the DNA that is single-stranded at each temperature. The temperature at which 50% of the DNA in solution is single-stranded (T_m) is the standard measure of thermal stability. The thermal stability of the heteroduplex DNA is compared with the average of the thermal stabilities of the homoduplex DNAs of the two individuals being compared. The difference in the T_m values is linearly related to the proportion of nucleotide mismatches in the DNA strands (Britten *et al.*, 1974). While DNA-DNA hybridization studies have played an important role, the technique has limitations. Since the experimental error associated with determining T_m is high, substantial amounts of DNA are required for replicate measurements. More importantly, resolution is low and the technique does not generate character data. It provides a quantitative measure of sequence divergence but no information on the nature of the differences. For a detailed discussion of DNA-DNA hybridization, see Springer & Krajewski (1989) and Werman *et al.* (1990).

Restriction site analysis and restriction fragment patterns

DNA-DNA hybridization, restriction site analysis and restriction fragment patterns each provide an indirect assessment of nucleotide substitutions. However, unlike DNA-DNA hybridization, the latter analyses provide both a quantitative and a qualitative assessment of sequence divergence. Base substitutions, sequence rearrangements, and insertions or deletions occurring since the time of divergence will result in differences in the size and number of restriction fragments (Upholt, 1977). Restriction enzyme methods are based on the cleavage of double-stranded DNA by restriction endonucleases or restriction enzymes. Restriction enzymes recognize specific double-stranded DNA sequences, usually 4 to 6 base pairs in length. The position where the enzyme cleaves the DNA is called the restriction site; the pieces of DNA generated are the restriction fragments. Following digestion with restriction enzymes, the DNA fragments are separated according to size by gel electrophoresis and visualized. Complex maps of overlapping fragments can be generated by sequential and reciprocal digestion by different enzymes. Either the 5′ or the 3′ end of a given fragment can be radioactively labelled and thereby tracked through subsequent digestions. Alternatively, a specific DNA sequence such as a cloned gene can be radiolabelled and hybridized to the digest fragments. Another approach would be to isolate a specific DNA sequence and generate a restriction fragment map. The approach, including the template, will depend on the question addressed, the taxonomic level and the organismal group. In general, a restriction site map is more difficult to generate but more accurate than a restriction fragment map (Li & Graur, 1991). Limitations of restriction analysis include the following: 1) alignment of maps is error prone, 2) sampling of nucleotides is biased, 3) application has a narrow taxonomic range and limited resolution, and 4) milligram quantities of DNA are required (Wilson *et al.*, 1989). Additional information on this technique may be found in Dowling *et al.* (1990).

RNA sequencing

The aforementioned techniques are limited in their resolution. For the determination of degree of nucleotide identity, the highest resolution will, of course, be obtained in RNA or DNA sequencing. RNA can be sequenced directly by using a modification (Qu *et al.*, 1983; Lane *et al.*, 1985) of the dideoxynucleotide chain termination method (Sanger *et al.*, 1977). In this reaction RNA, which is single-stranded, serves as the template. The enzyme, reverse transcriptase, in the presence of deoxyribonucleoside triphosphates (dNTPs) and a deoxynucleotide primer (a 15–30 base strand of DNA that is complementary to a region of the RNA), will extend the primer to create a strand of DNA that is complementary (cDNA) to the RNA strand. This technique has been most widely applied to small subunit ribosomal RNA (16S, 16S-like, 18S rRNA). Small subunit rRNA has several regions of highly conserved sequences, the complements of which can be used as 'universal' oligodeoxynucleotide primers (Lane *et al.*, 1985; Field *et al.*, 1988). The procedure is basically as follows. First, total cellular RNA is isolated; ribosomal RNA comprises the bulk of total cellular RNA thereby ensuring abundant template. Four separate reactions are run simultaneously. In each, the RNA is combined with the same 'universal' primer, four deoxynucleotides, one of four dideoxynucleotides (ddNTP = ddGTP, ddATP, ddTTP, or ddCTP), and reverse transcriptase. Each reaction proceeds such that the primer is elongated until a ddNTP is incorporated. Incorporation of a ddNTP terminates strand elongation. Thus, in a given reaction, the cDNA strands are of various lengths and end in a ddGTP, ddATP, ddTTP, or ddCTP. The resulting cDNA strands are radiolabelled during the reaction either by the incorporation of a radiolabelled dNTP or the primer is radiolabelled prior to performing the chain-termination reaction. The reactions are run side-by-side in a polyacrylamide sequencing gel to separate molecules differing in length by a single nucleotide; sequences are visualized by autoradiography. This method offers high resolution but also has its limitations. For one, substantial quantities of RNA are required. Using RNA as a template for direct sequencing, as opposed to the gene or DNA that codes for the RNA, introduces technical problems since RNA is easily degraded by RNases; RNA also has complex secondary structures through which it is difficult to sequence. Accuracy will also be limited since only a single complementary strand can be sequenced. However, as pointed out by Field *et al.* (1988), when RNA is the template, the derived sequences will represent commonly transcribed RNA genes as opposed to minor or inactive genes.

The polymerase chain reaction

The development of the polymerase chain reaction (PCR) (Saiki *et al.*, 1985, 1988; Mullis *et al.*, 1986) has largely eliminated the problem of limited quantities of material. PCR is a process whereby one can take small amounts of DNA (theoretically a single molecule) and amplify a specific segment, of defined size, to yield milligram quantities. Double-stranded DNA is the usual template but it does not necessarily have to be isolated from the cells first. When DNA isolation is not practicable, cells can be lysed with detergents to make the DNA available to the reaction. A PCR mixture comprises the DNA template, two oligonucleotide primers (one upstream sense primer and one downstream antisense primer) that flank the DNA region of interest, a heat-stable DNA polymerase such as Taq polymerase, and dNTPs. In the reaction (Fig. 1), the DNA is heated to denature or separate the strands, the temperature is lowered allowing the primers to anneal to complementary DNA sequences on opposite strands, the primers are extended by the action of the polymerase, and the cycle is repeated multiple times. This results in

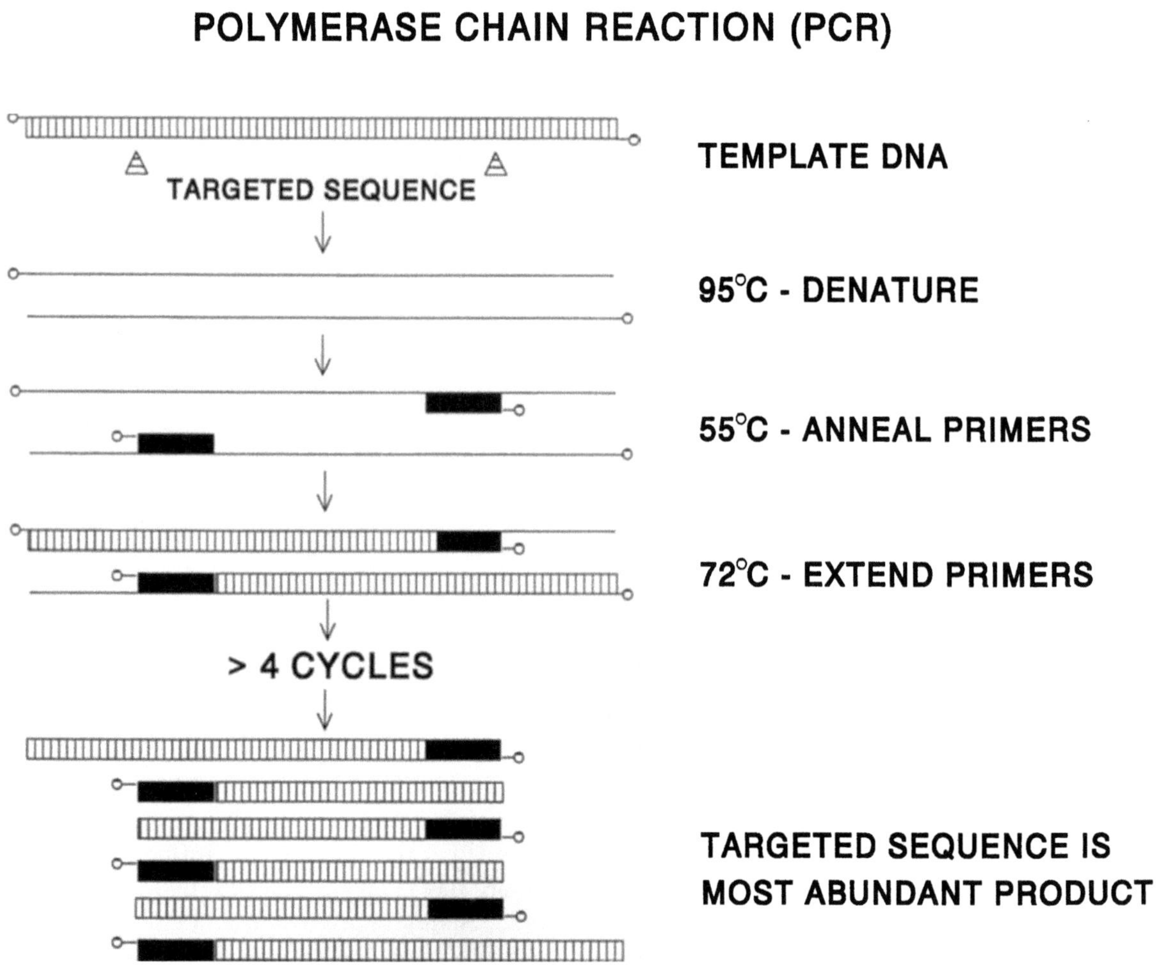

Fig. 1. Schematic representation of a typical polymerase chain reaction. The template is double-stranded DNA, the 5′ ends are indicated by open circles, the vertical lines represent bonds between complementary bases on the two strands, the sequence to be amplified lies between the two open triangles. The reaction mixture is heated to denature the DNA; this separates the DNA strands. The reaction is cooled to allow antisense and sense primers to anneal to complementary sequences. The primers are extended, each forming a complementary strand of DNA. At the end of the first cycle there are the two original DNA strands plus two DNA strands whose 5′ end is the sequence of either the sense or the antisense primer but whose 3′ end is undetermined. The cycle is repeated, at the end of the second cycle, there are additionally DNA strands whose 5′ end is the sequence of one primer and whose 3′ end is complementary to the other primer. It is this targeted sequence whose number doubles with each cycle.

the exponential accumulation of the target sequence whose ends are defined by the primer sequences.

The use of PCR is limited for most applications only by the need to know at least some sequence of the target DNA in order to design the primers. In evolutionary studies, the purpose of amplification frequently is to sequence the target DNA. With the availability of 'universal primers' designed to take advantage of highly conserved sequences in various templates of phylogenetic interest, specific sequence information may not be limiting (Medlin *et al.*, 1988; Kocher *et al.*, 1989; White *et al.*, 1990). Another purpose relevant to evolutionary studies is to amplify DNA as a means to detect the expression of specific genes. For the latter application, knowledge of specific sequences is a limiting factor and is discussed below.

DNA sequencing

In the standard PCR reaction both strands of the target region of DNA are amplified; the PCR product then can be isolated and sequenced directly. Double-stranded PCR products are not the most easily sequenced templates. Since PCR products are linear, not circular, pieces of DNA, the two strands tend to reanneal to each other which interferes with hybridization of the sequencing primers to the template DNA. In our laboratory, we routinely sequence double-stranded PCR products ranging in size from 220 to 3000 bp (= lower and upper limits of templates attempted) using a modification of the detergent protocol developed by Bachmann (Bachmann *et al.*, 1990). Since Taq polymerase can introduce errors in nucleotide sequence as DNA strands are copied, the PCR product itself is the most reliable template to sequence. This is because the great majority of the nucleotides at a given site among the many molecules of target sequence generated during the PCR will be correct. In a sequencing reaction, when the PCR-generated fragment is the template, a population of template molecules is present. Those molecules containing a Taq-generated error at a specific nucleotide will be proportionately few and therefore undetectable.

It is easier to sequence circular double-stranded DNA or single-stranded DNA than linear double-stranded DNA. Sequencing kits, based on dideoxynucleotide chain-termination (Sanger *et al.*, 1977) are available from a variety of companies; two in common use are Sequenase (United States Biochemical) and T7 Sequencing (Pharmacia). Circular double-stranded DNA will supercoil; once the strands are separated, they will not reanneal to each other as readily as in linear DNA. In single-stranded DNA, there is only one strand and reannealing is not a problem; however, it is more difficult to obtain a single-stranded PCR product than one that is double-stranded. One way of obtaining either a circular double- or single-stranded template for sequencing is to subclone the PCR product into a vector. Commonly used vectors are circular pieces of DNA such as a plasmid or a bacteriophage like M13. The primers used for the amplification can be designed to contain a restriction enzyme site to facilitate subcloning of the DNA fragment amplified. In this case, the PCR product is isolated and subsequently digested with the appropriate restriction enzyme. This will leave a 5′ or 3′ overhang depending on the enzyme used. Such overhangs are sometimes called 'sticky ends' because the overhanging nucleotides will anneal to a matching overhang generated by the same enzyme. Thus, if one digests both a PCR product and a vector with the same enzyme one can subsequently ligate the PCR product into the vector. Once the PCR product is inserted into the vector, bacteria are used to replicate the vector containing the PCR DNA and thereby generate many copies for sequencing. Subcloning into a plasmid will yield a circular double-stranded template for sequencing. Subcloning into M13 will yield a single-stranded template for sequencing. In most instances, only a single clone is selected for sequencing. Herein lies the danger of obtaining incorrect sequence information, *i.e.* by selecting a single clone, only a single molecule of the PCR-generated DNA is represented. If PCR products must be subcloned, then sequencing should either be done on multiple individual clones or sequencing reactions can be performed prior to single clone selection. For example, there is a point in the subcloning process where multiple clones are grown in the same bacterial culture tube. If this mixture of clones is sequenced, then Taq polymerase infidelity is less likely to be a problem.

It is also possible to perform PCR reactions that yield predominantly single-stranded products. In asymmetric PCR, one of the primers is added in limiting quantities (Gyllensten & Erlich, 1988). Until the limiting primer in the reaction is used up, both strands will be copied; subsequently only the strand that is produced by elongation of the non-limiting primer is generated in abundance. In this case amplification of the single-stranded DNA is linear rather than exponential.

Other methods for generating single-stranded PCR templates are available. Most rely on modification of one of the primers used for PCR. For example, one primer can be biotinylated, the double-stranded PCR reaction performed and the product isolated. The DNA fragment is then exposed to a solid substrate on which avidin is bound. The strand to which the biotin is bound, in turn binds to the avidin but the other does not thereby allowing the strands to be separated from each other (Syvanen *et al.*, 1989). The strand bound to the solid substrate is sequenced.

Sequence templates

Phylogenetically informative templates include ribosomal RNA [(rRNA) or the nuclear and mitochondrial genes that code for ribosomal RNA], regulatory sequences, and protein encoding genes.

As mentioned above, since DNA templates tend to be easier to work with than RNA templates, it may be preferable to determine ribosomal RNA sequence information using the nuclear or mitochondrial genes that code for them as template. There are several nuclear-encoded ribosomal RNA subunits, 28S, 5S, and 18S (the latter is also variously referred to as 16S, 16S-like, and small subunit ribosomal RNA). There are also two mitochondrial encoded subunits, 12S and 16S. The mitochondrial rRNA genes do not code for the same ribosomal subunits as do nuclear rRNA genes.

For the Nemertina, or any other largely untried group, one cannot predict the taxonomic level at which a given template will be phylogenetically informative. Both nuclear and mitochondrial DNA (mtDNA) have conserved as well as highly variable regions (see Dowling *et al.*, 1990). In some organisms, notably vertebrates, single-copy nuclear DNA evolves more slowly than mitochondrial DNA (Brown *et al.*, 1979, 1982; Wilson *et al.*, 1985); in others, *e.g.* sea urchins and *Drosophila*, rates appear to be similar (Vawter & Brown, 1976; Caccone *et al.*, 1988).

Nuclear 28S and small subunit rRNA are useful for comparisons of distantly related organisms (Field *et al.*, 1988; Qu *et al.*, 1988; Lake, 1990) and mitochondrial rRNA for ordinal or family level comparisons (*e.g.*, Xiong & Kocher, 1991). Nuclear rRNA genes are arranged as repeating transcription units; each transcription unit comprises coding regions for the 18S and 28S ribosomal subunits separated by a transcribed spacer [internal transcribed spacer (ITS)]. Within the ITS is sequence for a 5.8S molecule. Between repeating transcription units are nontranscribed spacers or intergenic spacers (IGS). Nuclear ITS and IGS evolve even faster than mitochondrial rRNA genes and would be suitable for species level or population level comparisons. The 5S subunit is transcribed independently in most eucaryotes. The 5S subunit is not considered to be useful for evolutionary analysis due to too great a proportion of random deviations for the size of this small molecule (Hendricks *et al.*, 1986).

Mitochondrial DNA (mtDNA) sequences are commonly used in the study of evolutionary relationships at the individual, population, and species level (Avise *et al.*, 1987, 1989; Harrison, 1989; Wilson *et al.*, 1989). In addition to a high rate of point mutations, mtDNA is distinguished by a constant number of single-copy genes, predominantly maternal inheritance, and an apparent lack of recombination. The mitochondrial genome comprises a noncoding regulatory region or D-loop, coding regions for the two ribosomal RNA subunits mentioned above as well as 13 protein encoding genes and 22 transfer RNA genes. Both conserved and highly variable regions are present. In the D-loop, two peripheral domains have evolved in a species-specific manner while the central domain has been highly conserved (Saccone *et al.*, 1991). Regions of the cytochrome b gene are useful from the interspecific to the intergeneric level (Kocher *et al.*, 1989), whereas hypervariable sites can elucidate relationships among populations or among individuals (Wrischnik *et al.*, 1987; Avise *et al.*, 1987, 1989).

Sequence comparisons can also be made of specific protein-coding genes; however, their utility will depend upon the particular sequence chosen. Rates of evolution vary greatly among protein-coding genes. Additionally, rates will vary within a given gene depending upon the part of the gene sequenced *e.g.*, coding regions or various types of noncoding regions. Coding regions are more highly conserved than noncoding sequence. Among noncoding regions are introns, 5′ and 3′ untranslated regions, and 5′ and 3′ flanking regions. Introns are transcribed regions of genes that separate exons (the amino acid coding sequences). Introns are spliced out of the transcribed messenger RNA (mRNA) before it leaves the nucleus as mature mRNA. Flanking regions and untranslated regions are also noncoding but only the untranslated regions are transcribed. Unlike introns, untranslated regions are not removed from the mRNA and are present at the 5′ and 3′ ends, *i.e.*, immediately 5′ of the first exon and 3′ of the last exon. Flanking regions frequently contain the regulatory regions of genes *i.e.*, the parts of genes that control the expression of the gene. Flanking regions 'flank' the 5′ and 3′ untranslated regions of the gene. For a comparative discussion of nucleotide substitution rates see Li & Graur (1991).

Expression studies

Another approach to the study of evolution in organisms involves study of the processes of speciation. For example, the presence of a consistent environmental stress in combination with existing genetic variability leads to the selection of genetic variants. A subset of the above example would be a case in which populations of organisms found at the extremes of environmental clines *e.g.*, temperature or salinity, become isolated thus leading to speciation. Occasionally, these populations are found to have altered protein structure or enzyme kinetics. More commonly, mechanisms of adaptive evolution involve genetic change. While genetic change can be accomplished through mutation in the primary structure of proteins, the major mechanism is thought to be through change in regulatory genes (Wilson, 1976). Thus, by initiating the transcription of a gene into messenger RNA or by altering the rate at which it is transcribed the amount of an enzyme can be increased; this will increase the activity of the enzyme without change in the primary structure or the properties of the enzyme. If a gene has a regulatory region that is responsive to the stress, that gene will be adaptive. Various individuals in a species will most probably have the same genes, but if there is genetic change in gene regulation then some individuals may express a gene that is adaptive while others do not. A study of the transcription or expression of a gene is performed by determination of the specific mRNA of that gene.

An example of the way in which expression studies might be used in a group of organisms, such as nemertines, where virtually no molecular information is available might be to use a developmentally or an environmentally regulated gene as described below. First, however, it should be pointed out that although it is commonly recognized that ontogeny is related to phylogeny, it is not generally recognized that the physiology of an animal group reflects the evolutionary history of that group. As common ancestry is indicated by the presence of homologous structures, physiological homology indicates common ancestry (Prosser, 1990). An example of physiological homology would be members of a protein family, like globins, having the same or similar function among diverse animals. This is in contrast to physiological analogy which exemplifies convergent evolution, where a variety of different proteins, such as metalloproteins, are used for the same or similar function, *i.e.*, oxygen transport.

Little is known about the environmental responses of nemertines but we do know that those species examined are capable of both intra- and extracellular volume regulation (Ferraris & Schmidt-Nielsen, 1982; Ferraris, 1984, 1985). This is not surprising since nemertines are osmoconformers. That is, they do not osmoregulate or maintain their body fluids at a constant osmolality when the salinity of the surrounding seawater changes. Thus, when confronted with an increase in salinity their cells will shrink due to efflux of water. In order to regain cell volume, inorganic ions are first taken up by the cells to increase cell solute; water passively follows which helps to restore cell volume. Nemertines use osmolytes such as inorganic ions to regulate cell volume initially. Subsequently, the accumulated inorganic ions are replaced with organic osmolytes such as amino acids (Ferraris & Roderick, 1982). Replacement of inorganic ions with compatible organic osmolytes is a common strategy employed to avoid the perturbing effects of elevated inorganic ion concentrations (Yancey *et al.*, 1982). The accumulation of organic osmolytes like glycine betaine, taurine, and sorbitol is under the control of osmotically regulated genes (Garcia-Perez *et al.*, 1989; Cowley *et al.*, 1990; Robey *et al.*, 1991; Uchida *et al.*, 1991). Most information is known for the accumulation of sorbitol in mammalian kidney cells so I will use it as an example even though we do not yet know that it is accumulated in nemertines under hyperosmotic conditions. Glucose is converted to sorbitol through a reaction catalyzed by the enzyme aldose reductase. Transcription of the gene for aldose reductase, which we have now cloned (Ferraris *et al.*, 1990a, b), is osmotically regulated (Smardo *et al.*, 1990). Thus, when intracellular ionic strength increases, as it does when cells are shrunken, tran-

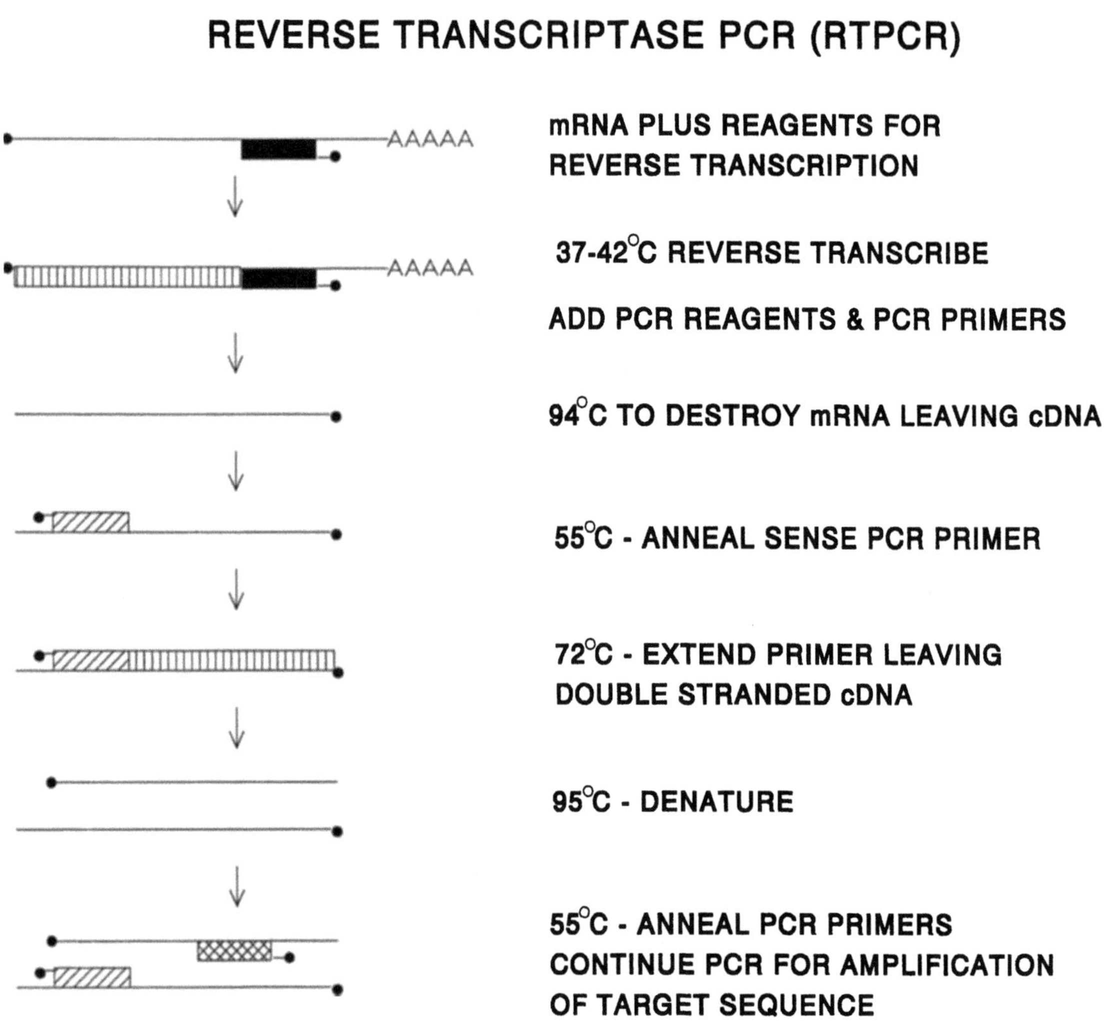

Fig. 2. Schematic representation of a reverse transcriptase polymerase chain reaction (RTPCR). In RTPCR, the initial template is single-stranded messenger RNA (mRNA) which has a poly(A)-tail. A primer is annealed to the mRNA. The primer may be an mRNA specific sequence, an oligo-dT sequence, or random sequences. The primer is extended by the action of reverse transcriptase forming DNA strands complementary (cDNA) to the mRNA. All of the cDNA strands so formed are antisense. The reaction is heated to destroy the RNA template. The sense PCR primer anneals to the antisense cDNA and is extended by Taq polymerase forming a sense DNA strand. At this point, double-stranded DNA becomes the template and the PCR reaction proceeds as described in Fig. 1.

scription of aldose reductase increases (Garcia-Perez *et al.*, 1989; Uchida *et al.*, 1989; Cowley *et al.*, 1990; Smardo *et al.*, 1990). This leads to the accumulation of sorbitol. The ability to accumulate a compatible organic osmolyte in the face of increased salinity confers an advantage. For example, bay and ocean populations of the oyster *Crassostrea virginica* differ in their salinity tolerance and in their ability to accumulate amino acids and betaine (Pierce *et al.*, 1990). These populations probably do not differ in the presence or absence of the genes that code for the accumulation of these osmolytes, however, it is reasonable to hypothesize that the populations do differ in their ability to express those genes in response to a salinity challenge.

There remains the question of how an investigator might make the connection between a gene that is osmotically regulated but whose nucleotide sequence is only known for distantly related organisms. The technique that I am currently using is reverse transcriptase PCR (RTPCR) using degenerate primers (Lee *et al.*, 1988; Belyavsky *et al.*, 1989). Unlike PCR as described above, in RTPCR the initial template is mRNA (Fig. 2). The first part of the reaction is to reverse transcribe all of the mRNA in a tissue sample using the enzyme reverse transcriptase. The primer in the reverse transcription reaction can be oligonucleotides that are dTTPs (which will hybridize to the series of dATPs found at the 3′ end of mRNAs) or hexamers of random dNTPs (these will hybridize wherever there are complementary sequences). The reverse transcription reaction results in cDNAs, *i.e.*, DNAs that are complementary to each of the mRNAs expressed in the cells. PCR is then performed on the cDNAs using degenerate primers. Degenerate primers are so named because they are a mix of primers representing all possible nucleotide combinations that would code for an amino acid sequence, *i.e.*, they are based upon the degeneracy of the genetic code where one amino acid may be coded for by as many as six nucleotide triplets. Thus, I compare the available amino acid sequences of a given osmotically regulated gene to first determine those amino acids that are highly conserved among organisms. Then sense and antisense primers are designed; if possible they are chosen to be on different exons. If they are on different exons, then they will be separated by an intron in the genomic DNA but not in the cDNA. The presence of an intron will make the genomic PCR product larger than the cDNA PCR product and therefore distinguishable. The cDNA PCR product is then sequenced for identification. In this manner I can determine if an osmotically regulated gene is expressed in response to increased salinity. Depending on the hypothesis, physiological homology can be examined across phyla or processes of speciation can be elucidated in the comparison of populations.

References

Avise, J. C., J. Arnold, R. M. Ball, E. Bermingham, T. Lamb, J. E. Neigel, C. A. Reeb & N. C. Saunders, 1987. Intraspecific phylogeography: the mitochondrial bridge between population genetics and systematics. Annu. Rev. Ecol. Syst. 18: 489–522.

Avise, J. C., B. W. Bowen & T. Lamb, 1989. DNA fingerprints from hypervariable mitochondrial genotypes. Mol. Biol. Evol. 6: 258–269.

Bachmann, B., W. Luke & G. Hunsmann, 1990. Improvement of PCR amplified DNA sequencing with the aid of detergents. Nucleic Acids Res. 18: 1309.

Belyavsky, A., T. Vinograova & K. Rajewsky, 1989. PCR-based cDNA library construction: general cDNA libraries at the level of a few cells. Nucleic Acids Res. 17: 2919–2932.

Britten, R. J., D. E. Graham & B. R. Neufeld, 1974. Analysis of repeating DNA sequences by reassociation. In L. Grossman & K. Moldave (eds), Methods in Enzymology. Academic Press, New York: 363–406.

Brown, W. M., M. George Jr. & A. C. Wilson, 1979. Rapid evolution of animal mitochondrial DNA. Proc. Natn. Acad. Sci. U.S.A. 76: 1967–1971.

Brown, W. M., E. M. Prager, A. Wang & A. C. Wilson, 1982. Mitochondrial DNA sequences of primates: tempo and mode of evolution. J. Mol. Evol. 18: 225–239.

Caccone, A., G. D. Amato & J. R. Powell, 1988. Rates and patterns of scnDNA and mtDNA divergence within the *Drosophila melanogaster* subgroup. Genetics 118: 671–683.

Cowley, B. D., Jr., J. D. Ferraris, D. Carper & M. B. Burg, 1990. *In vivo* osmoregulation of aldose reductase mRNA, protein, and sorbitol in renal medulla. Am. J. Physiol. 258: F154–F161.

Dowling, T. E., C. Moritz & J. D. Palmer, 1990. Nucleic acids

II: restriction site analysis. In D. M. Hillis & C. Moritz (eds.), Molecular Systematics. Sinauer, Sunderland, Mass.: 250–317.

Ferraris, J. D., 1984. Volume regulation in intertidal *Procephalothrix spiralis* (Nemertina) and *Clitellio arenarius* (Oligochaeta). J. Comp. Physiol. B 154: 125–137.

Ferraris, J. D., 1985. Effects of ablation of the cerebral organs and cerebral ganglia on volume regulation in an intertidal nemertine, *Paranemertes Peregrina*. Physiol. Zool. 58: 117–128.

Ferraris, J. D. & H. K. Roderick, 1982. Involvement of ninhydrin positive substances in regulatory volume decrease in intertidal Nemertina and Oligochaeta: independence of decerebration. Mt. Desert Island Biol. Lab. Bull. 22: 33–35.

Ferraris, J. D. & B. Schmidt-Nielsen, 1982. Volume regulation in an intertidal nemertine, *Procephalothrix spiralis* Coe. I. Short-term effects and independence of decerebration. J. exp. Zool. 224: 307–319.

Ferraris, J. D., B. D. Cowley, Jr., C. K. Williams, B. M. Martin, M. B. Burg & A. Garcia-Perez, 1990a. Rabbit aldose reductase, an osmoregulatory protein is encoded by a single gene. FASEB J. 4: A1823.

Ferraris, J. D., C. K. Williams, B. M. Martin, M. B. Burg & A. Garcia-Perez, 1990. Cloning and structure of the rabbet aldose reductase gene. J. am. Soc. Nephrol. 1: 739.

Field, K. G., G. J. Olsen, D. J. Lane, S. J. Giovannoni, M. T-. Ghiselin, E. C. Raff, N. R. Pace & R. A. Raff, 1988. Molecular phylogeny of the animal kingdom. Science 239: 748–753.

Garcia-Perez, A., B. Martin, H. R. Murphy, S. Uchida, H. Murer, B. D. Cowley, J. S. Handler & M. B. Burg, 1989. Molecular cloning of cDNA coding for kidney aldose reductase: Regulation of specific mRNA accumulation by NaCl-mediated osmotic stress. J. Biol. Chem. 264: 16815–16821.

Gyllensten, U. B. & H. A. Erlich, 1988. Generation of single-stranded DNA by the polymerase chain reaction and its application to direct sequencing of the HLA-DOA locus. Proc. natn. Acad. Sci. U.S.A. 85: 7652–7656.

Harrison, R. G., 1989. Mitochondrial DNA as a genetic marker in population and evolutionary biology. Trends Ecol Evol. 4: 6–11.

Hendriks, L., E. Huysmans, A. Vandenberghe & R. De Wachter, 1986. Primary structures of the 5S ribosomal RNAs of 11 arthropods and applicability of 5S RNA to the study of metazoan evolution. J. Mol. Evol. 24: 103–109.

Kocher, T. D., W. K. Thomas, A. Meyer, S. V. Edwards, S. Paabo, F. X. Villablanca & A. C. Wilson, 1989. Dynamics of mitochondrial DNA evolution in animals: amplification and sequencing with conserved primers. Proc. Natl. Acad. Sci. U.S.A. 86: 6196–6200.

Lake, J. A., 1990. Origin of the metazoa. Proc. natn. Acad. Sci. U.S.A. 87: 763–766.

Lane, D. J., B. Pace, G. J. Olsen, D. A. Stahl, M. L. Sogin & N. R. Pace, 1985. Rapid determination of 16S ribosomal RNA sequences for phylogenetic analyses. Proc. natn. Acad. Sci. U.S.A. 82: 6955–6959.

Lee, C. C., X. W. Wu, R. A. Gibbs, R. G. Cook, D. M-. Muzny & C. T. Caskey, 1988. Generation of cDNA probes directed by amino acid sequence: cloning of urate oxidase. Science 239: 1288–1291.

Li, W-H. & D. Graur, 1991. Fundamentals of molecular evolution. Sinauer, Sunderland, Mass. U.S.A.

Medlin, L., H. J. Elwood, S. Stickel & M. L. Sogin, 1988. The characterization of enzymatically amplified eukaryotic 16S-like rRNA-coding regions. Gene 71: 491–499.

Mullis, K. B., F. Faloona, S. J. Scharf, R. K. Saiki, G. T. H-orn & H. A. Erlich, 1986. Specific enzymatic amplification of DNA *in vitro*: the polymerase chain reaction. Cold Spring Harbor Symp. Quant. Biol. 51: 263–273.

Pierce, S. K., L. R. Faux & S. M. O'Brien, 1990. Salinity tolerance differences between Chesapeake Bay and Atlantic oysters: differences in amino acid and glycine betaine metabolism. Am. Physiol. Soc.: A–37.

Prosser, C. L., 1990. Comparative physiology: theory of adaptation. In C. L. Prosser (ed.) Environmental and metabolic animal physiology, comparative animal physiology, 4th edition. Wiley-Liss, New York: 1–11.

Qu, L. H., B. Michot & J. P. Bachellerie, 1983. Improved methods for structure probing in large RNAs: a rapid 'heterologous' sequencing approach is coupled to the direct mapping of nuclease accessible sites. Application to the 5′ terminal domain of eucaryotic 28S rRNA. Nucleic Acids Res. 11: 5903–5920.

Qu, L. H., M. Nicoloso & J. P. Bachellerie, 1986. Phylogenetic calibration of the 5′ terminal domain of large rRNA achieved by determining twenty eucaryotic sequences. J. Mol. Evol. 28: 113–124.

Robey, R., M. Kwon, J. Handler, A. Garcia-Perez & M. Burg, 1991. Induction of glycinebetaine uptake into *Xenopus* oocytes by injection of poly(A)$^+$RNA from renal cells exposed to high extracellular NaCl. J. Biol. Chem. 266: 10400–10405.

Saccone, D., G. Pesole & E. Sbisa, 1991. The main regulatory region of mammalian mitochondrial DNA: structure-function model and evolutionary patter. J. Mol. Evol. 33: 83–91.

Sanger, F., S. Nicklen & A. R. Coulson, 1977. DNA sequencing with chain-terminating inhibitors. Proc. natn. Acad. Sci. U.S.A. 74: 5463–5467.

Saiki, R. K., S. J. Scharf, F. A. Faloona, K. B. Mullis, G. T. Horn, H. A. Erlich & N. Arnheim, 1985. Enzymatic amplification of β-globin genomic sequences and restriction site analysis for diagnosis of sickle cell anemia. Science 230: 1350–1354.

Saiki, R. K., D. H. Gelfand, S. Stoffel, R. Higuchi, G. T. H-orn, K. B. Mullis & H. A. Erlich, 1988. Primer-directed enzymatic amplification of DNA with a thermostable DNA polymerase. Science 239: 487–491.

Smardo, F. M., Jr., M. B. Burg & A. Garcia-Perez, 1990. Osmotic regulation of kidney aldose reductase (AR) gene transcription. FASEB J. 4: A2096.

Springer, M. & C. Krajewski, 1989. DNA hybridization in animal taxonomy: a critique from first principles. Quart. Rev. Biol. 64: 291–318.

Syvanen, A. C., K. Aalto-Setala, K. Kontula & H. Soderlund, 1989. Direct sequencing of affinity-captured amplified human DNA application to the detection of apolipoprotein E polymorphism. FEBS Letters 258: 71–74.

Uchida, S., A. Garcia-Perez, H. Murphy & M. B. Burg, 1989. Signal for induction of aldose reductase in renal medullary cells by high external NaCl. Am. J. Physiol. 256: C614–C620.

Uchida, S., M. Kwon, A. Preston & J. Handler, 1991. Expression of MDCK cell Na^+ and Cl^- dependent taurine transporter in *Xenopus laevis* oocytes. J. Biol. Chem. 266: 9605–9609.

Upholt, W. B., 1977. Estimation of DNA sequence divergence from comparison of restriction endonuclease digests. Nucleic Acids Res. 4: 1257–1265.

Vawter, L. & W. M. Brown, 1986. Nuclear and mitochondrial DNA comparisons reveal extreme rate variation in the molecular clock. Science 234: 194–196.

Werman, S. D., M. S. Springer & R. J. Britten, 1990. Nucleic acids I: DNA-DNA hybridization. In D. M. Hillis & C. Moritz (eds), Molecular Systematics. Sinauer, Sunderland: 204–249.

White, T. J., T. Bruns, S. Lee & J. Taylor, 1990. Amplification and direct sequencing of fungal ribosomal RNA genes for phylogenetics. In M. A. Innis, D. H. Gelfand, J. J. Sninsky & T. J. White (eds), PCR protocols: a guide to methods and applications. Academic Press, N.Y.: 315–321.

Wilson, A. C., 1976. Gene regulation in evolution. In F. J. Ayala (ed.), Molecular Evolution. Sinauer, Sunderland Mass., U.S.A.: 225–234.

Wilson, A. C., R. L. Cann, S. M. Carr, M. George, U. B. Gyllensten, K. M. Helm-Bychowski, R. G. Higuchi, S. R. Palumbi, E. M. Prager, R. D. Sage & M. Stoneking, 1985. Mitochondrial DNA and two perspectives on evolutionary genetics. Biol. J. linn. Soc. 26: 375–400.

Wilson, A. C., E. A. Zimmer, E. M. Prager & T. D. Kocher, 1989. Restriction mapping in the molecular systematics of mammals: a retrospective salute. In B. Fernholm, K. Bremer & H. Jornvall (eds), The hierarchy of life. Elsevier, Amsterdam: 407–419.

Wrischnik, L. A., R. G. Higuchi, M. Stoneking, H. A. Erlich, N. Arnheim & A. C. Wilson, 1987. Length mutations in human mitochondrial DNA: direct sequencing of enzymatically amplified DNA. Nucleic Acids Res. 15: 529–542.

Xiong, B. & T. D. Kocher, 1991. Comparison of mitochondrial DNA sequences of seven morphospecies of black flies (Diptera: Simulidae). Genome 34: 306–311.

Hydrobiologia **266**: 267–271, 1993.
R. Gibson, J. Moore & P. Sundberg (eds), Advances in Nemertean Biology

The influence of light and sea water temperature on the reproductive cycle of *Lineus ruber* (Heteronemertea)

Guy Vernet[1] & Jacques Bierne[2]
[1]*Laboratoire de Zoologie et des Sciences de l'Environnement, UR4R, Faculté des Sciences, Université de Reims Champagne Ardenne, BP 347, 51062 Reims cedex, France;* [2]*Laboratoire de Biologie Cellulaire, UR4R, Faculté des Sciences, Université de Reims Champagne Ardenne, BP 347, 51062 Reims cedex, France*

Key words: nemerteans, sexual reproduction, internal control, external control, ecophysiology

Abstract

The cerebral ganglia of *Lineus ruber* secretes a neurohormone named gonad-inhibiting hormone (GIH) controlling reproduction. As *L. ruber* undergoes repeated annual cycles of reproduction during its life, the way the neuroendocrine activity of the worm is seasonally regulated is an important ecophysiological question. In other words, what environmental 'key factors' influences the beginning of spawning season in *L. ruber*?

In the present work we show that light is not a timer which controls the annual reproductive cycle progress of *L. ruber.* On the other hand our results suggest that thermoperiod is a good candidate for this function. Further work is still necessary to clarify the transduction mechanism of thermal stimuli in GIH discharge.

Introduction

The annual reproductive cycle of marine invertebrates is a result of the interplay of endogenous rhythms and exogenous timers (Giese & Kanatani, 1987). Where climatic conditions remain much the same throughout the year, as in the tropics and in the deep sea, one might expect breeding and spawning to occur throughout the year. Where climatic conditions vary considerably, breeding is likely to be seasonal. Most of the organisms from temperate waters spawn at a well-defined period year after year.

The reproductive cycle of *Lineus ruber* is annual. In the area of the French Marine Biological station of Roscoff the development of gonads takes place from October to December. Reproduction of *L. ruber* in the field is largely restricted to the period from late December to mid-March with a peak in January. From mid-march to October worms are in a resting period of reproduction. Experimental studies have shown that the cerebral ganglia of *L. ruber* secretes a neurohormone named gonad-inhibiting hormone (GIH) controlling reproduction (Bierne, 1964, 1966, 1970a, b, c, 1973, 1983; Rué & Bierne, 1978; Bierne & Rué, 1979). As male and female *L. ruber* worms undergoes repeated cycles of reproduction during their life, the method by which the neuroendocrine activity of the worm is seasonally regulated is an ecophysiological question that immediately springs to mind. Nothing is known about the ecological conditions which make the winter period of the year more appropriate for reproduction than others in *L. ruber*. To identify the environmental factors which determine the limits of the extended breeding season, the effects of environmental 'key-factors', such as light and sea water temperature on reproductive activity were studied in the laboratory.

From this point of view numerous scientific works, like those of Spears & Clarke, 1987; Giese & Kanatani, 1987; Beacham & Murray, 1988; Awaji & Hanyu, 1988; Aiken & Waddy, 1989; Chu & Levin, 1989; Paulet & Boucher, 1991, have proved that photo- and thermoperiod are the two main exogenous factors which control the sexual reproductive cycle in the animal kingdom. Nevertheless, as far as we know, few of them have been concerned with lucifugous species (Franke, 1986), in particular, none has been devoted to the nemerteans.

Material and methods

At the end of November 1989, mature specimens of male and female *L. ruber* were collected from the littoral of the English Channel, in the area of Roscoff. The worms were reared under standard laboratory conditions for nemerteans: constant temperature (12°), continuous darkness, and food once a week until reproduction time.

To try and delineate the influence of two of the main ecological factors, light and temperature on neuroendocrine activity of *L. ruber* we have selected a simple procedure.

From mid-March 1990, just after reproduction, to early May 1991, three sets of 10 males and 10 females worms were maintained in the dark (LD 0:24) in three separate isothermal rooms under the contrasting influences of a winter temperature (5° ± 0.5°), a summer temperature (18° ± 0.5°) and a medium temperature (12° ± 0.5°). At 12°, 2 other groups of 20 worms (10 ♂ and 10 ♀) were exposed to light, the former continually (LD 24:0), the latter for 12 hours a day (LD 12:12).

The sexual development was examined monthly. The worms were recorded as 1) sexually undifferentiated when gametogenesis was not identifiable, 2) sexually differentiated when gametogenesis was starting so that the sex (either oogenesis or spermatogenesis) could be known, 3) submature when the spermiogenesis or vitellogenesis was ended, 4) mature when testes were exclusively full of spermatozoa and ovaries endowed with eggs and 5) postspawning either when oocytes in females and sperm in males were released or when oosorption and sperm resorption was starting, if no egg and semen emission occurred.

The worms were regularly fed once a week for the 14-month duration of the experiment. The light sources used were fluorescent tubes day light of 40 W, giving 300 Lux to the worms.

Results

At the beginning of the mid March to October resting period, after egg-laying, we performed experiments according to the procedure described above. Results are summarized in Table 1.

Table 1. Influence of light and sea water temperature on sexual developmental in *Lineus ruber*.

Time (months)	Photoperiod				
	LD 0:24			LD 12:12	LD 24:0
Temperature	5°	12°	18°	12°	12°
8	–(20)	+ +(4)–(16)	–(20)	–(20)	–(20)
10	–(20)	+ +(8)–(12)	+ +(2)–(18)	+ +(4)–(16)	–(20)
12	+ + +(7)–(13)	+ + +(10)–(10)	+ + +(8)–(12)	+ + +(4)–(16)	–(20)
14	0(7)–(13)	0(10)–(10)	0(8)–(12)	0(4)–(16)	–(20)

– = sexually undifferentiated; + = sexually differentiated; + + = submature; + + + = mature; 0 = postspawning.

Worms remained sexually undifferentiated for several months, until October. Gonadogenesis subsequently took place in some worms. At the end of a year (Fig. 1) under 12° and in darkness 50 percent of the animals show reproductive development. At the same temperature under LD 12:12, only 20 percent worms develop gonads. Under 12° and continuous light 0 percent animals are sexualized. These results suggest that light might be a disruptive factor in the GIH production. Indeed continuous light can profoundly disturb the timing of the gonadogenesis as well as the annual cycle of reproduction by continually inhibiting the gonad development, i.e. by allowing the permanent release of GIH at a high level.

Moreover, one can see that at 18°, 'summer' sea water temperature, as at 5°, 'winter' sea water temperature at Roscoff, exposing the worms to continuous darkness is not sufficient to produce further sexual development than at 12° and continuous darkness. In fact 60 percent animals in the first case and 65 percent in the second had their maturation and spawning inhibited presumably by continuous GIH secretion.

Discussion

The environmental stimuli that impact the biological reactions of marine invertebrates are: daylength, temperature changes, food, phases of the moon and tidal rhythms. 'Daylength is perhaps the most invariant factor in the environment repeatable from year to year. It may serve to regulate breeding. Moreover since the sea is so vast and has a high heat capacity, temperature variations are muted and the temperature changes are rather similar year after year. Therefore temperature changes could serve as a cue for breeding' (Giese & Kanatani, 1987).

The present study tried to clarify the influence of the two environmental 'key factors': sea water

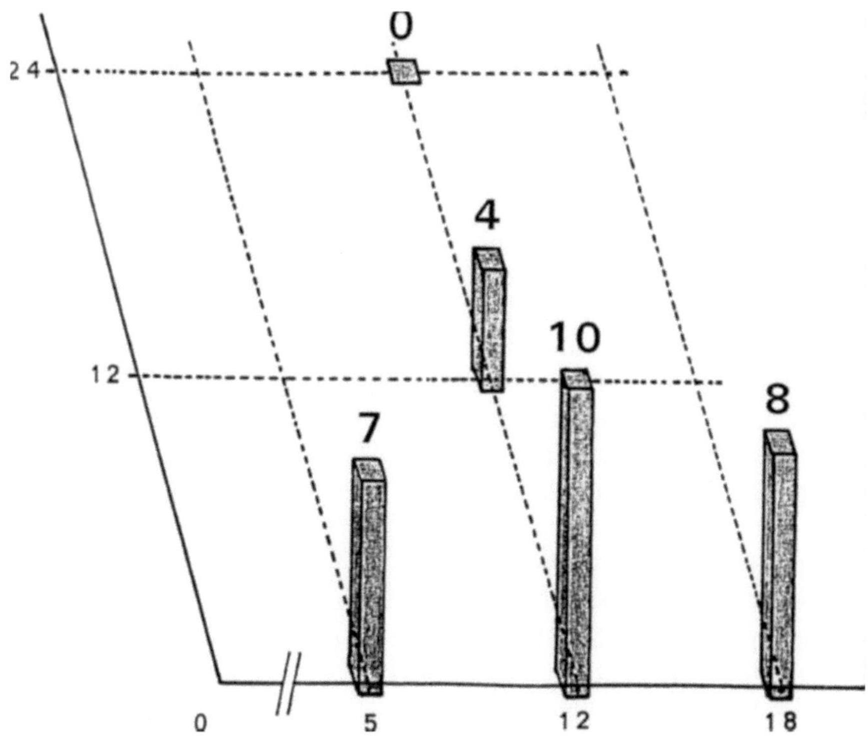

Fig. 1. Number of *Lineus ruber* which develop mature gonads at the end of 1 year after transfer to various daylength and temperature regimes.

temperature and photoperiod for the timing of the annual reproductive cycle of *L. ruber*.

The results show that only 50 percent of the worms developed mature ovaries or testes at the end of a year, at temperatures below 12 °C and continuous darkness. The laboratory experimental conditions appear to disturb the biological clock which controls the reproductive period more in *Lineus ruber*, a northern species, than in *L. lacteus*, a southern species, because under the same experimental conditions, 12° and LD 0:24, 100% *L. lacteus* develops mature gonads (Vernet & Bierne, 1988; Vernet, 1989; Vernet, 1990; Vernet & Anadon, 1992), *i.e.*, twice as much as *L. ruber*.

The presence of a photosensitive structure in the brain of *L. ruber* (Vernet, 1974) added to the results observed under 12° and continuous darkness as well as DL 12:12 or continuous light at first suggested that in a natural environment, light or more likely photoperiod could be the regulating factor of GIH production by the cerebral ganglia. However the effects of continuous light, as well as intermittent light, are always negative for reproduction. So positive effects have to be derived from other regulative parameters such as sea water temperature which also acts on GIH production, as shown by results obtained at 18° and 5° and continuous darkness.

Lineus ruber is a lucifugous species which lives permanently in darkness in gravel and/or beneath the littoral cobbles and boulders (though *L. ruber* lives in the dark, it is not insensitive to light). In such conditions, the zeitgeiber which controls the annual reproductive cycle progress of the worm cannot be photoperiod, but must be sea water temperature; the *L. ruber* reproduction neurohormonal cycle thus depend upon sea water thermoperiod.

Lineus ruber is reported to be a cosmopolitan species. Its annual breeding period is related to latitude; this species reproduces in March and April in the Gulf of Maine (Coe, 1889, Riser, 1974) when sea water temperature is about 9° (Riser, pers. comm.), in July and August in the Barents Sea (Schmidt & Jankovskaia, 1938) when sea water temperature is about 9° (Kulikova, pers. comm.) and in December and January in the English Channel (Oxner, 1911; Gontcharoff, 1951, Bierne, 1970a) also when the sea water temperature reaches 9°.

The strictly seasonal reproduction of *L. ruber* specimens in the field disappears under laboratory conditions of constant temperature and continuous darkness. Moreover, larvae breeding under laboratory conditions never permit the advent of a normal reproductive cycle.

Further work is still necessary to clarify the series of events that occur over the transduction of external stimuli, particularly of thermoperiod, in internal neurohormone release. As in the vertebrates (Collin *et al.*, 1988, pers. comm.) studies of transduction of thermal signals in GIH discharge by the brain of *L. ruber* would be a preliminary step to understanding the effects of external stimuli on release of neurohormones for reproduction.

References

Aiken, D. E. & S. L. Waddy, 1989. Interaction of temperature and photoperiod in the regulation of spawning by American lobster *Homarus americanus*. Can. J. Fish. aquat. Sci. 46: 145–148.

Awaji, M. & I. Hanyu, 1988. Effects of water temperature and photoperiod on the beginning of spawning season in the orange-red type Medaka. Zool. Sci. 5: 1059–1064.

Beacham, T. D. & C. B. Murray, 1988. Influence of photoperiod and temperature on timing of sexual maturity of pink salmon *Oncorhynchus gorbuscha*. Can. J. Zool. 66: 1729–1732.

Bierne, J., 1964. Maturation sexuelle anticipée par décapitation de la femelle chez l'Hétéronémerte *Lineus ruber* Müller. C. r. Acad. Sci., Paris 259: 4841–4843.

Bierne, J., 1966. Localisation dans les ganglions cérébroïdes du centre régulateur de la maturation sexuelle chez la femelle de *Lineus ruber* (Hétéronémertes) Müller. C. r. Acad. Sci., Paris 262: 1572–1575.

Bierne, J., 1970a. Recherches sur la différenciation sexuelle au cours de l'ontogenèse et de la régénération chez le Némertien *Lineus ruber* Müller. Ann. Sci. Nat. Zool. 12: 181–298.

Bierne, J., 1970b. Aspects expérimentaux de la différenciation sexuelle chez *Lineus ruber* (Hétéronémertes). Bull. Soc. zool. Fr. 95: 529–543.

Bierne, J., 1970c. Influence des facteurs hormonaux gonado-inhibiteur et androgène sur la différenciation sexuelle des parabiontes hétérosexués chez un Némertien. Ann. Biol. 11: 155–160.

Bierne, J., 1973. Contrôle neuroendocrinien de la puberté chez le mâle de *Lineus ruber* (Hétéronémertes) C. r. Acad. Sci., Paris 276: 363–366.

Bierne, J., 1983. Nemertina. In K. G. & R. G. Adiyodi (eds), Reproductive biology of Invertebrates, I: Oogenesis, oviposition, and oosorption. John Wiley, New York: 147–167.

Bierne, J. & G. Rué, 1979. Endocrine control of reproduction in two rhynchocoelan worms. Int. J. Invert. Reprod. 1: 109–120.

Chu, J. W. & L. A. Levin, 1989. Photoperiod and temperature regulation of growth and reproduction in *Streblospio benedicti* (Polychaeta: Spionidae). Invert. Reprod. Dev. 15: 131–142.

Coe, W. R., 1889. Notes on the times of breeding of some common New England Nemerteans. Science 9: 167–169.

Collin, J. P., J. Arendt & W. A. Gern, 1988. Le 'troisième oeil'. La Recherche 203: 1154–1165.

Franke, H. D., 1986. The role of light and endogenous factors in the timing of the reproductive cycle of *Typosyllis prolifera* and some other polychaetes. Am. Zool. 26: 433–446.

Giese, A. C. & H. Kanatani, 1987. Maturation and spawning. In A. C. Giese, J. S. Pearse & V. B. Pearse (eds), Reproduction of marine Invertebrates, Vol IX, Blackwell Scientific Publications, Palo Alto: 251–329.

Gontcharoff, M., 1951. Biologie de la régénération et de la reproduction chez quelques Lineïdae de France. Ann. Sci. Nat. Zool. 13: 149–235.

Oxner, M., 1911. Analyse biologique d'une série d'expériences concernant l'avènement de la maturité sexuelle, la régénération et l'inanition chez les Némertiens, *Lineus ruber* (Müller) et *Lineus lacteus* (Rathke). C. r. Acad. Sci., Paris 153: 1168–1171.

Paulet, Y. M. & J. Boucher, 1991. Is reproduction mainly regulated by temperature or photoperiod in *Pecten maximus*? Invert. Reprod. Dev. 19: 61–70.

Riser, N. W., 1974. Nemertinea. In A. C. Giese & J. S. Pearse (eds), Reproduction in marine Invertebrates, Vol. I, Academic Press, New York & London: 359–389.

Rué, G. & J. Bierne, 1978. Effet du facteur cérébral gonado-inhibiteur sur le métabolisme des cellules sexualisées germinales et somatiques du Némertien *Lineus ruber* Müller. C. r. Acad. Sci., Paris 286: 293–296.

Schmidt, G. A. & L. A. Jankovskaia, 1938. Biologie de la reproduction de *Lineus gesserensis-ruber* de Roscoff et du golfe de Kola. Arch. Zool. exp. Gén. 79: 487–513.

Spears, N. & J. R. Clarke, 1987. Effect of nutrition, temperature and photoperiod on the rate of sexual maturation of the field vole *Microtus agrestis*. J. Reprod. Fert. 80: 175–181.

Vernet, G., 1974. Etude ultrastructurale de cellules présumées photoréceptrices dans les ganglions cérébroïdes de Lineidae (Hétéronémertes). Ann. Sci. Nat. Zool. 16: 27–36.

Vernet, G., 1989. Influence de la photopériode et de la température sur le cycle de reproduction de *Lineus lacteus* (Hétéronémertes). Bull. Soc. zool. Fr. 114: 63–68.

Vernet, G., 1990. Le rôle de la lumière et de la température sur le déroulement du cycle annuel de reproduction de *Lineus lacteus* (Hétéronémertes, Lineidae). Dans 'Régulation des cycles saisonniers chez les Invertébrés'. Les colloques de l'INRA 52: 199–202.

Vernet, G. & J. Bierne, 1988. Neuroendocrine control of gonadogenesis in regenerating *Lineus lacteus* (Hétéronémertea), Hydrobiologia 156: 53–60.

Vernet, G. & N. Anadon, 1992. Effets de la température de l'eau de mer et de la photopériode sur le déroulement du cycle annuel de reproduction de *Lineus lacteus* (Hétéronémertes) de la ría de Vigo. (Espagne). Bull. Soc. Zool. Fr.. 117: 335–336.

Hydrobiologia **266**: 273–280, 1993.
R. Gibson, J. Moore & P. Sundberg (eds), Advances in Nemertean Biology

Patterns of trace metal accumulation in Swedish marine nemerteans

Eric G. McEvoy[1] & Per Sundberg[2]
[1] *School of Biological & Earth Sciences, Liverpool Polytechnic, Byrom Street, Liverpool L3 3AF, U.K.;*
[2] *University of Gothenburg, Department of Zoology, Medicinaregatan 18, S-413 90 Göteborg, Sweden*

Key words: Nemertea, heavy metals

Abstract

Total body burdens of Zn, Pb, Al and Cu were measured in Swedish marine nemerteans sampled from an area close to a major source of pollution. Accumulation patterns varied both seasonally and spatially. The results suggest that nemerteans may be useful bioindicator species in environmental monitoring programmes.

Introduction

The deposition of heavy metals in the world's ocean sediments is a consequence of fluvial discharge. Whitfield (1982) estimated that these elements have passed through the oceans six times during the earth's history and he assumed that the levels of heavy metals in the open oceans have more or less reached a steady state. These levels are largely independent of metals entering from rivers, or the atmosphere, but depend on the rate at which metals are removed from the sediments by geochemical and biochemical processes. The situation, however, differs in coastal waters where both fluvial and atmospheric input are often increased by anthropogenic activity. In such areas variability in trace metal levels may be due to river runoff and the pollution source. If this input is mainly of industrial origin, a relatively steady total load per time unit may be anticipated depending on the type of process and type of discharge. Periods of high runoff (e.g. spring flood) can increase the amount of organic and inorganic particulates in the water (Phillips, 1979) which may transport adsorbed and complexed trace metals a large distance from the source.

Bioindicator organisms are often used in environmental monitoring programmes to estimate metal burden in a particular area. Many pollution studies use filter feeders as the indicator organisms due to their ability to ingest large quantities of particulates. However, such organisms may not respond to trace metals which exist primarily in solution, like for example cadmium (Cd). Popham & D'Auria (1982) showed how the sabellid polychaete *Eudistylia vancouveri* accumulated significant amounts of titanium (Ti) and vanadium (V), whereas samples of *Mytilis* spp. from the same area displayed no comparable accumulation. The use of nemerteans (phylum Nemertea) as bioindicators of metal pollution has been described by McEvoy & Gibson (1982) and McEvoy (1988) and appear to fulfill the major criteria required of potential indicator organisms suggested by Butler *et al.* (1971). In addition, nemerteans secrete significant quantities of epidermal mucus which may serve as an initial barrier to the absorption of heavy metals. This adsorption of metals in the mucus matrix may protect against potentially toxic environmental levels; at the same time, this metal accumulation makes nemerteans useful as indicator species for particular elements even at low ambient concentrations. Protection of this nature may also confer a competitive advantage when pollution increases and nemerteans may be abundant in polluted areas where other organisms are scarce.

There are indications of high levels of trace metals in the most northern part of the Swedish west coast (Naes, 1983; Knutzen, 1984; Afzelius, pers. comm.). A possible source for this pollution is a number of industries along the river Glomma in Norway (Fig. 1). The output from this river affects the area in Sweden south of Norway, especially during spring flood when humus particules from the river colours the water dark brown some distance down the Koster fiord (east the Koster islands, Fig. 1). A sampling programme was organised in order to establish if the river Glomma is in fact a major source of some of the pollutants present in the area. Nemerteans were used as bioindicators, and we sampled in total 17 stations (only enough specimens were obtained from five of these stations; see below) in both the autumn and early summer. We predicted that (1) trace metal concentrations should be higher in the early summer due to the effect of the previous spring flood; and that (2) there should be a gradient of trace metal body burdens in nemerteans sampled further south as the distance increased from the suspected source.

The results show that there is a general increase in the heavy metal levels in the summertime, but we could not establish any gradient in respect to the pollution source. The metal levels from both summer and autumn samples are higher than control samples from an unpolluted area, thus confirming the general pollution of the area shown by Naes (1983) and Knutzen (1984).

Table 1. The species, and number of specimens, from the five sampling sites (Fig. 1) used in the analysis.

Site		Species	Number of specimens
7	Autumn	*Lineus viridis*	7
		L. ruber	4
	Summer	*L. viridis*	4
		Cephalothrix rufifrons	3
		Amphiporus lactifloreus	6
8	Autumn	*Lineus bilineatus*	3
		Micrura fasciolata	20
	Summer	*M. fasciolata*	14
9	Autumn	*Micrura fasciolata*	22
		M. purpurea	15
	Summer	*M. fasciolata*	6
10	Autumn	*Micrura fasciolata*	12
		Lineus bilineatus	4
	Summer	*Nipponnemertes pulcher*	10
		Tetrastemma sp.	23
		Micrura fasciolata	12
		Lineus bilineatus	3
14	Autumn	*Emplectonema gracile*	35
	Summer	*Cephalothrix rufifrons*	4
		Micrura fasciolata	7

Material and methods

Worm samples were taken from stations 1–17 (Fig. 1) in October/November 1989 and in June 1990. Enough specimens for the metal analysis were only obtained from stations 7, 8, 9, 10 and 14. It was planned to sample for the same species at each site during both seasons, but this was not possible. We have therefore used pooled samples and the species may vary between sites (Table 1). Nemerteans were collected from littoral samples of algae taken by dredging or SCUBA diving. The worms were placed in clean seawater for three days to allow for the expulsion of gut contents,

Table 2. Average metal levels ± S.D. in the sediment (5 samples) from station 7 (Fig. 1) expressed as μg/g dry weight.

Element	Autumn	Summer
Zn (zinc)	129.6 ± 5.6	130.0 ± 5.9
Pb (lead)	19.8 ± 2.7	17.4 ± 2.3
Al (aluminium)	11990 ± 31	10977 ± 36
Cu (copper)	23.4 ± 3.1	15.8 ± 3.2

Table 3. Total body burdens of control samples of *Lineus ruber*, *L. viridis*, and *Amphiporus lactifloreus* from an unpolluted are at the isle of Anglesey, expressed as $\mu g/g$ dry weight ± S.D.

Zn	Pb	Al	Cu
720 ± 130	85 ± 27	317 ± 94	22 ± 12

identified, fixed in absolute ethanol, and later digested in concentrated nitric acid (Williams, 1978). Digests were analysed using a Phillips PU 7450 induction coupled plasma arc spectrometer under standard operational settings. Sediment samples from station 7 (which was the only site with suitable substrate) were prepared using the method of Adrian (1973) and analysed for metal content (Table 2). Control samples were obtained from an unpolluted site on the isle of Anglesey, Wales, U.K., and were subjected to the same preparation and analytical protocols as the Swedish samples (Table 3). Total body burdens of the four trace metals zink (Zn), lead (Pb), aluminium (Al), and copper (Cu) were measured and expressed as $\mu g\ g^{-1}$ dry body weight.

We have not done any statistical tests of differences in metal concentrations for two reasons. First, the concentrations of the four metals at each station were measured on the same individuals, thus violating the assumption of independent samples. Second, we have been forced to pool samples and differences between sampling stations and seasons may, at least in part, be due to differences in species composition.

Results and discussion

Sufficient worms were obtained from stations 7, 8, 9, 10 and 14 (Table 1), we did not found any nemerteans at all from some of the localities (stations 1, 2, 4, 5, 6, 11, 12). The current in general low levels of littoral nemerteans may be an effect of an algal bloom in May/June 1988 when exceptional high and fatal concentrations of the toxin-producing flagellate *Chrysocromulina polylepis* changed the littoral ecosystems along large parts the Swedish west coast. The major effects of this bloom, and discussions of some of its causes, are summarized in Rosenberg *et al.* (1989). It is still not known what caused this particular algae to become poisonous but the toxins it produced changed the membrane permeability in cells and thus the ion equilibrium. There was an immediate decline in the number of the littoral nemertean *Oerstedia dorsalis*. At one locality, 43 specimens were found on May 15 among 81 of algae. Three days later, on May 18, no single specimen could be found in 30 l of seaweed from the same locality. Also other nemertean species were affected by this bloom, and we can still (five years later) see its effects on the ecosystem.

Concentrations in Zn, Pb and Cu are much higher in the nemerteans collected from site 7 in comparison to the substrate levels (Tables 2, 3, 4), whilst sediment Al levels are around ten times higher than total body burdens. This differential metal accumulation often occurs in a wide range of marine invertebrates and often reflects (particularly in the case of aluminium) the bioavailability

Table 4. Total body burdens of Zn, Pb, Al and Cu at sites 7, 8, 9, 10, and 14 (Fig. 1) in autumn (A) and summer (S). Expresed as average $\mu g\ g^{-1}$ dry body weight ± S.D.

Station	Zn		Pb		Al		Cu	
	A	S	A	S	A	S	A	S
7	2023 ± 655	2465 ± 207	628 ± 86	71 ± 14	932 ± 196	508 ± 129	251 ± 12	327 ± 52
8	2665 ± 534	3214 ± 445	370 ± 196	47 ± 15	1462 ± 838	380 ± 87	607 ± 41	309 ± 9
9	2623 ± 67	2803 ± 94	755 ± 111	106 ± 22	797 ± 41	852 ± 117	375 ± 84	485 ± 104
10	5285 ± 127	2233 ± 411	1145 ± 784	64 ± 26	7294 ± 2200	540 ± 124	2160 ± 1154	350 ± 90
14	2618 ± 324	950 ± 144	8.9 ± 2.0	250 ± 85	525 ± 163	722 ± 217	280 ± 64	620 ± 150

of the element and its chemical speciation. Aluminium is often the most abundant ambient metal in the geochemical cycle but its mobilisation depends largely on local environmental conditions.

There is no obvious correlation between metal average concentration at a site and its distance from suspected source of pollution for any of the four metals. However, there appears to be a general increase in total body burdens of all trace metals in the summer samples when compared with the winter ones, but not at all sites (Fig. 2, Table 4). This supports our prediction that metal

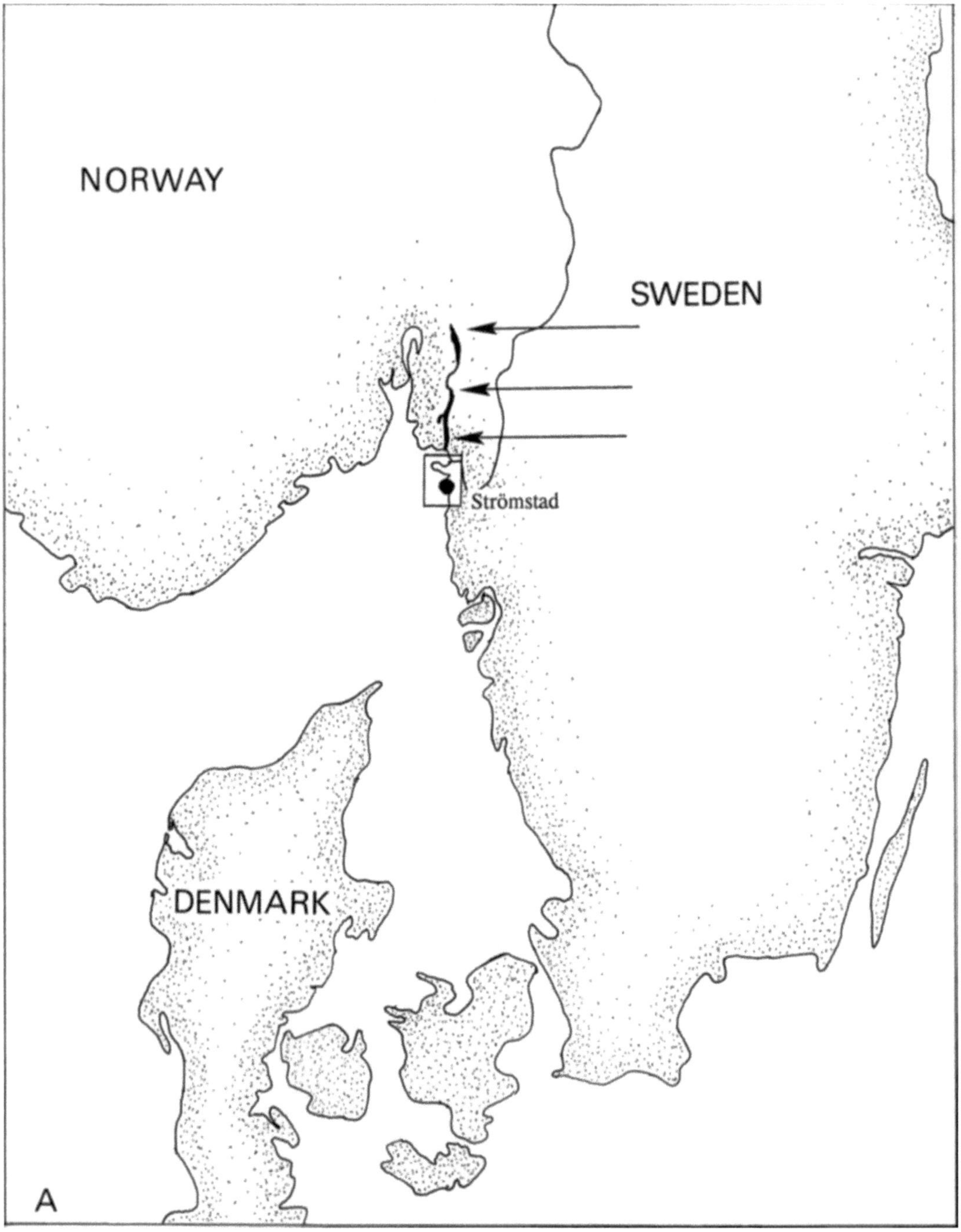

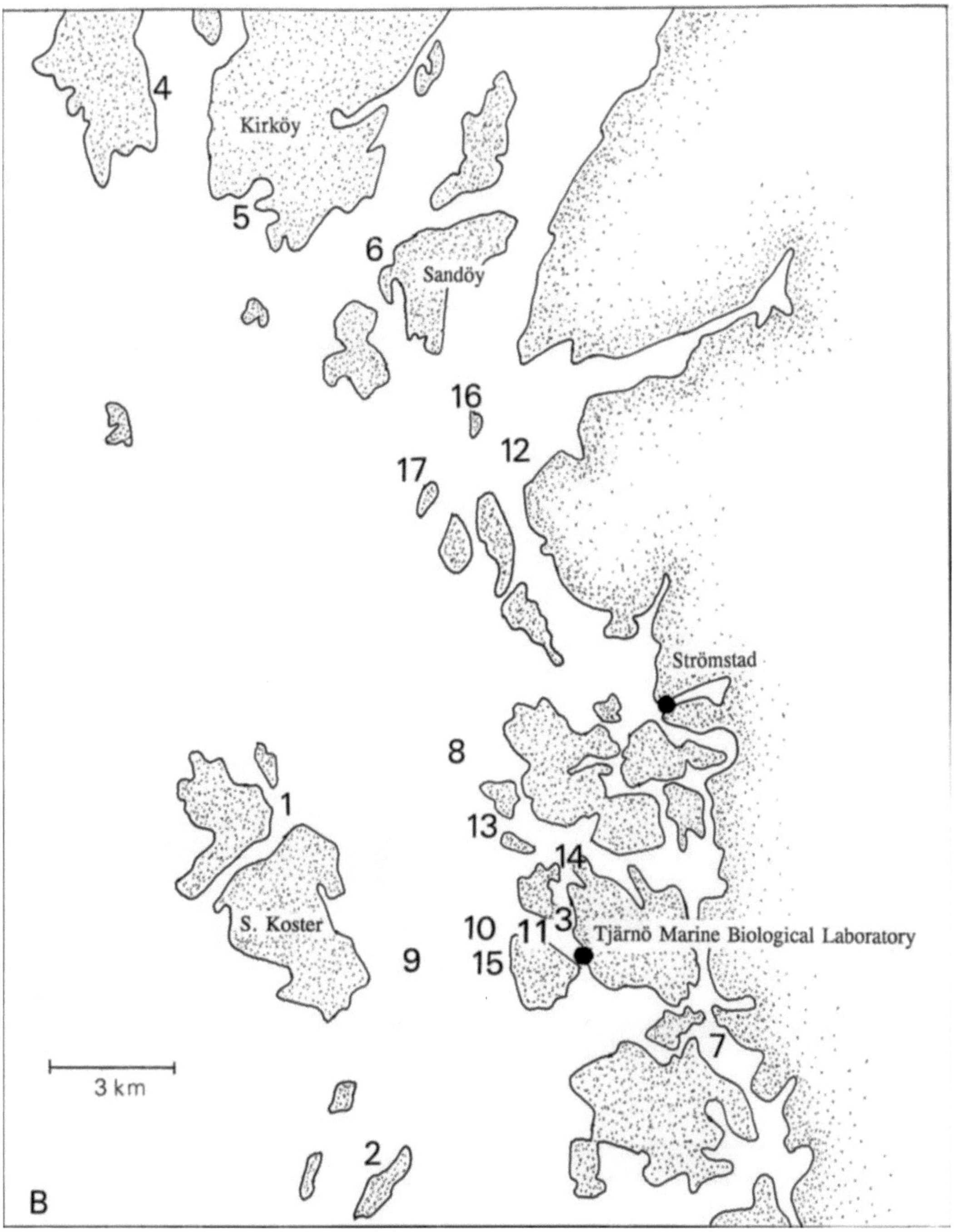

Fig. 1. A. A map showing the locality (square) of the sampling site, at the border between Sweden and Norway. Arrows pointing at the river Glomma discussed in the text. B. The sampling sites: 1. Saltholmen, the littoral; 2. Ramsö, the littoral in the harbour; 3. south Lindholmen, the littoral; 4. Asmalöy, 3–4 m; 5. Kirköy, 3–6 m; 6. Sandöy 2–3 m; 7. Styrsholmen, 1 m; 8. east Krugglö, 10 m; 9. Kostergrund, 10 m; 10. Between Inre Vattenholmen and Yttre Vattenhomen, 15–20 m; 11. North Saltö, 2 m; 12. Styrsö, 2–3 m; 13. Önbärsholmen, 4 m; 14. North Lindholmen, 3 m; 15. Yttre Vatennholmen, 6–10 m; 16. Tjurholmen, 10 m; 17. Nord-Hällsö, 4 m.

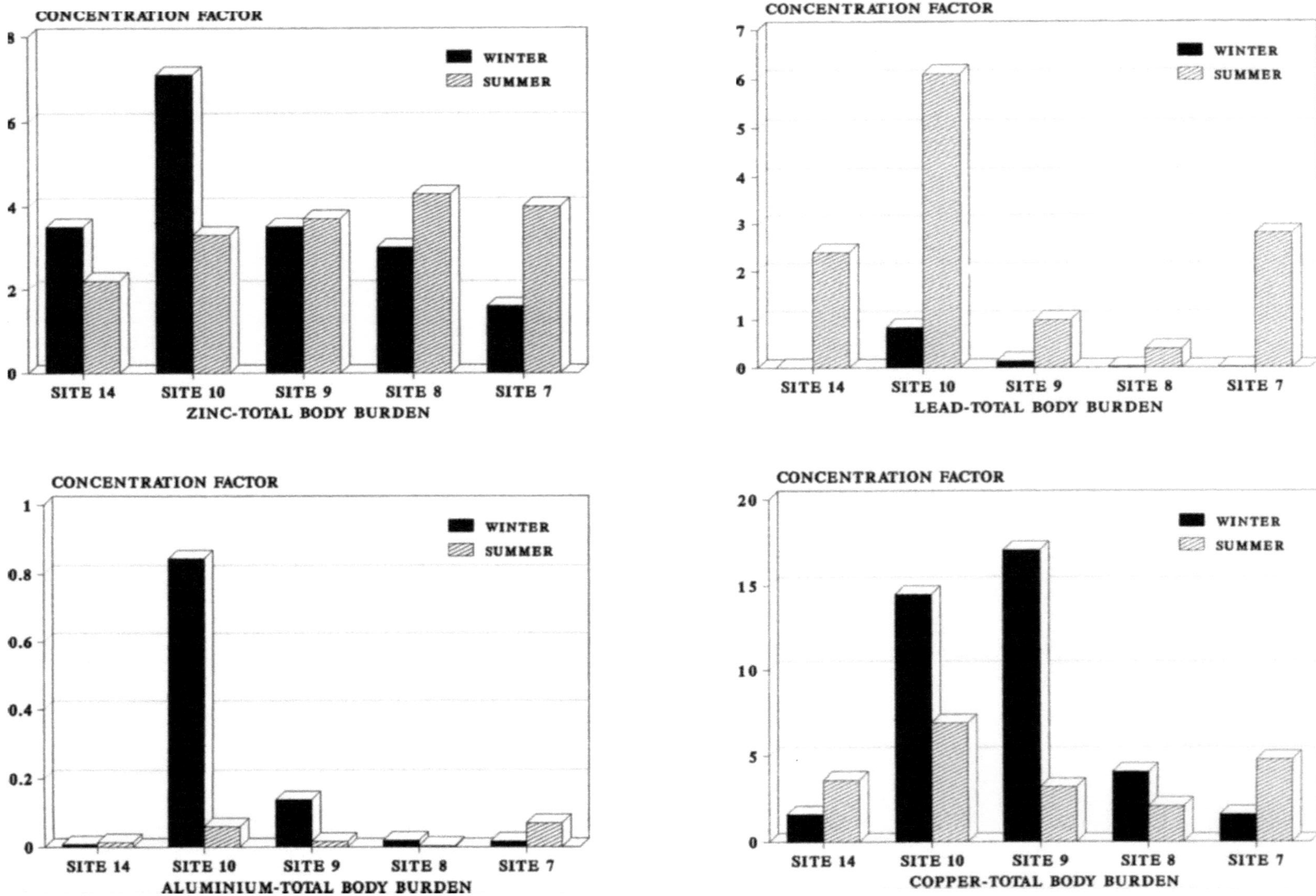

Fig. 2. Histograms showing the average metal concentrations from five sites (Fig. 1) from the autumn and summer samples. Levels are expressed relative to the substrate.

levels should be higher after the spring flood if the pollution source(s) is situated along a river. If the pollutants were washed out with the spring flood, and the main source is the Glomma river, we predicted that metal concentrations would decrease as we sampled further from the river. However, this pattern of pollutant accumulation could not be established. In fact station 8, which is the closest to the source, showed in general the lowest concentrations, and the samples from station 10 (Fig. 1), the highest.

There are several possible explanations for this. First, the main pollution source is not the factories discharging into the river Glomma. Second, McEvoy (1988) demonstrated that there is interspecific variation in nemerteans for metal accumulation, and by pooling worms from different species at each site we may hide spatial and temporal patterns in accumulation that do exist. A third explanation concerns fluvial sedimentation processes. We initially surmised that the highest concentrations of heavy metals would be nearer to the Glomma (assuming this is the main source) and that these elements would be diluted as the fresh water plume spreads over a much wider are. Naes (1983) found the highest levels of the elements iron (Fe), mangan (Mn), Titan (Ti), and Cu in the sediments taken near the mouth of the river Glomma. However, concentrations of Pb, Al, mercury (Hg) and Zn varied throughout the area with some unexpectedly high levels at distances up to 25 km from the river. Also Knutzen (1984) demonstrated variability in metal accumulation in *Mytilis edulis* sampled in this area. It is thus not certain that the highest levels will be found close to the source. One reason is that the outflow of fresh water will stay on the top of the saline sea water and a sharp boundary will form between the two water bodies. This boundary may stop the sedimentation from the fresh water and make the heavy metal sink through first when the water flow has reduced, *i.e.* at a further distance from the mouth.

In comparison with unpolluted areas and the control samples (Table 3), the concentrations of heavy metals are higher in the samples from Sweden. McEvoy (1988) reports Zn levels around 80 $\mu g\, g^{-1}$ dry weight and Pb levels around 0.1 $\mu g\, g^{-1}$ dry weight in nemerteans from the waters around the Scilly Isles; an area with no industrial effluent discharges. McEvoy & Gibson (1982) measured Zn and Pb levels in *Lineus ruber* collected close to station 14 to be around 170 $\mu g\, g^{-1}$ dry weight and 38 $\mu g\, g^{-1}$ dry weight respectively in 1981. The current values are much higher, but it is not clear whether this reflect differences in the species sampled (Table 1), or indicates an increase in industrial discharge during this period. The overall conclusion of this, and previous studies in the area, is that there is an increased level of heavy metals in these waters. Although we did not manage to establish whether the industries along Glomma is a source of this pollution, the levels indicate a serious pollution problem which must be taken care of.

Acknowledgements

The sampling was undertaken from the Tjärnö Marine Biology Laboratory, and we thank the director, L. Afzelius, and the staff for their generous support. We also thank Prof. N. Lepp for information on metal cycling and Ms D. Gibbons for sample analysis.

References

Adrian, W. J., 1973. A comparison of wet pressure digestion method with other commonly used wet and dry ashing methods. Analyst 98: 213–216.

Butler, P. A., L. Andren, G. J. Bonde, A. Jernelov & D. J. Reish, 1971. Monitoring organisms. In FAO Technical Conference on Marine Pollution and its Effects on Living Resources and Fishing. Rome, Italy: 101–112. FAO Fisheries Report No. 99, Suppl. 1, Rome.

Knutzen, K., 1984. Basisundersökelse i Hvalerområdet og Singlefjorde. Miljögifter i organismer. Overvakningsrapport 122/84. NIVA, Oslo, 85 pp.

Naes, K., 1983. Basisundersökelser i Hvalerområdet og Singlefjorden. Löste metaller, suspendert material og sedimenter. Rapport 70/83. NIVA, Oslo, 100 pp.

McEvoy, E. G. & R. Gibson, 1982. Heavy metal levels in six species of Swedish marine nemerteans. Sarsia 67: 61–62.

McEvoy, E. G., 1988. Heavy metals in marine nemerteans. Hydrobiologia 156: 135–143.

Phillips, D. J. H., 1980. Quantitative Aquatic Biological Indicators. Applied Sciences, London, 488 pp.

Popham, J. D. & J. M. D'Auria, 1982. A new sentinel organism for vanadium and titanium. Mar. Pollut. Bull. 13: 25–27.

Rosenberg, R., O. Lindahl & H. Blanck, 1988. Silent spring in the sea. Ambio 17: 289–290.

Whitfield, M., 1982. The salt sea – accident or design. New Scientist 9th May: 14–17.

Williams, E. V., 1978. New techniques for the digestion of biological materials – application to the determination of tin, iron and lead in canned foods. J. Fd Technol. 13: 367–384.

MIX
Papier aus verantwortungsvollen Quellen
Paper from responsible sources
FSC® C105338

If you have any concerns about our products,
you can contact us on
ProductSafety@springernature.com

In case Publisher is established outside the EU,
the EU authorized representative is:
Springer Nature Customer Service Center GmbH
Europaplatz 3, 69115 Heidelberg, Germany

Printed by Libri Plureos GmbH
in Hamburg, Germany

Advances in Nemertean Biology

Developments in Hydrobiology 89

Series editor

H. J. Dumont